Fish Defenses

Fish Defenses

Volume 2

Pathogens, Parasites and Predators

Editors

Giacomo Zaccone
Department of Animal Biology and Marine Ecology
Messina University
Italy

C. Perrière
Laboratoire de Biologie Animale. Insectes et Toxins
Facultè de Pharmacie
Chatenay-Malabry Cedex
France

A. Mathis
Department of Biology
Missouri State University
Springfield, Missouri
USA

B.G. Kapoor
Formerly Professor of Zoology
The University of Jodhpur
Jodhpur, India

CRC Press
Taylor & Francis Group
Boca Raton London New York

CRC Press is an imprint of the
Taylor & Francis Group, an **informa** business
A SCIENCE PUBLISHERS BOOK

First published 2009 by Science Publishers

Published 2018 by CRC Press
Taylor & Francis Group
6000 Broken Sound Parkway NW, Suite 300
Boca Raton, FL 33487-2742

First issued in paperback 2018

© 2009 reserved
CRC Press is an imprint of Taylor & Francis Group, an Informa business

No claim to original U.S. Government works

ISBN-13: 978-1-138-11353-4 (pbk)
ISBN-13: 978-1-57808-407-4 (hbk)

Visit the Taylor & Francis Web site at
http://www.taylorandfrancis.com

and the CRC Press Web site at
http://www.crcpress.com

Cover illustration: Reproduced from Chapter 8 of Jörgen I. Johnsson with kind permission of the authors.

Library of Congress Cataloging-in-Publication Data

Fish defenses/editors, Giacomo Zaccone ...[et al.].
    v. cm.
  Includes bibliographical references.
  Contents: v. 2. Pathogens, Parasites and Predators
  ISBN 978-1-57808-407-4 (hardcover)
  1. Fishes--Defenses. I. Zaccone, Giacomo.
  QL639.3.F578 2008
  571.9'617--dc22
                                                            2008016632

Dir blieb kein Wunsch, kein Hoffen, kein Verlangen,
Hier war das Ziel des innigsten Bestrebens,
Und in dem Anschaun dieses einzig Schönen,
Versiegte gleich der Quell sehnsüchtiger Tränen.

From: Elegie, W. GOETHE, 1823
In memory of W. A. MOZART

Preface

Over the past few decades, biologists have been forced to consider a dramatically altered view of the natural world. Many seasoned researchers were trained during a time when most populations were thought to be at equilibrium. Understanding the factors that shaped the current community structure and explained co-existence was a common theme for ecologists. Now, dramatic environmental changes, including habitat degradation and climate change, have led to a focus on understanding how individuals and populations respond to a shifting biotic and abiotic landscape. A critical step toward meeting this goal is a clear understanding of the capacity of individuals to defend themselves against threats.

For fishes and other aquatic species, changes in water quality (including temperature) can have both direct and indirect effects. Direct effects include toxic responses or physiological or behavioral responses to sub-optimal temperature regimes. Indirect effects can include changes in the community structure, including different arrays of prey, predators, parasites and pathogens that arise due to changes in habitat or because of accidental or deliberate acts of humans (e.g., species introductions). In this volume, we will focus on defensive responses of individuals to the biotic threats of pathogens, parasites, and predators.

Defensive responses can occur at many levels, from cellular to behavioral actions. Different levels of defenses often work together, making it somewhat difficult to categorize defensive mechanisms. Nonetheless, we generally consider defenses as either sub-organismal (occurring primarily at the molecular, cellular or tissue level) or behavioral (overt actions of individuals). Defenses against pathogens and parasites can occur at both levels, with sub-organism defenses primarily occurring after the fish has been attacked and behavioral defenses primarily leading to avoidance of attack. Defenses against predation can also occur at both levels, with sub-organism levels including productions of toxins or other secretions and behavioral levels functioning in avoidance or escape.

We first present two chapters that review certain broad categories of molecular defenses against pathogens and then two chapters that focus on specific cases. The first overview is provided by Dickerson who reviews the immune defenses that occur at the boundary between the individual and the environment: the mucosal layer. To infect an individual, pathogens must pass through mucosal barriers of the skin, gills or gut. Although these frontline defenses are of critical importance, surprisingly large gaps remain in our understanding of both the mechanisms and actions involved in these defenses. Patrzykat and Hancock focus on the incredible variety of peptide defenses against pathogens, many of which have been identified only since the advent of new molecular techniques. The most commonly identified functions are antimicrobial, but other functions include antiviral activity, wound-healing, and even an anticancer role. The authors have also noted some interesting similarities between the genetics and structure of fish and nonfish peptides.

Estepa, Tafalla and Coll concentrate on antiviral defenses, using the trout viral haemorrhagic septicemia virus VHSV, a common pathogen in aquaculture, as their primary case study. The authors provide a detailed description of both nonspecific and specific antiviral defenses; basic knowledge of defenses against viruses is critical because treatments for viral diseases are sorely lacking. Romalde and his colleagues also focus on a pathogen that is common in cultured fishes: streptococcal bacteria. In this chapter, background information about this re-emerging fish disease is presented along with several images of infected fishes. The focus of this overview is a discussion of the current state of knowledge concerning vaccination strategies for streptococcal diseases. Understanding the biology of fish defenses against aquaculture-related pathogens is essential, particularly as aquaculture plays an increasing role in human food supplies due to declining marine fisheries and increased human population sizes.

Defenses against pathogens and parasites can also occur at the behavioral level, although this mechanism has been much less studied than sub-organism defenses. Wisenden, Goater and James review both avoidance behaviors and behaviors that reduce parasite loads post infection. This chapter also considers whether constraints and trade-offs may have influenced the evolution of anti-parasite/pathogen behaviors.

Our coverage of defense against predators includes primarily toxicity and behavioral defensives. Two chapters provide details about different toxicity systems. Kalmanzon and Zlotkin provide a general overview of the anatomy of fish toxicity (secretory cells and venom glands) and the structure and function of toxic skin secretions, including neurotoxins ichthyocrinotoxins and surfactants. The chapter by Marin examines the defenses of opisthobranch slugs against fish predators. Although he primarily discusses toxic defenses, including aposematism, some mechanical defenses such as spicules and autotomy are also mentioned.

Behavioral defenses against predators are covered by two chapters in this volume. Johnsson takes a classic approach in his review by summarizing the antipredator behavioral defenses that occur at each stage of a predation event. These defenses include mechanisms to assist in encounter avoidance, detection avoidance, attack deterrence, flight, and escape following capture. Knouft provides a look at an often neglected area of consideration: the role of parental care behavior in defense of eggs and juveniles. Guarding against nest predators is a long-recognized phenomenon, but secretion of antimicrobial compounds that protect eggs and embryos is also becoming well documented.

The last two chapters concern behavioral responses to conspecific and heterospecific chemical alarm cues. Production of alarm chemicals in fishes of the Superorder Ostariophysi has been known since Karl von Frisch's description of 'Schrekstoff' over half a century ago; several papers have reviewed this alarm system over the years. Rather than provide an extensive review of species, Mirza categorizes the different types of cues (disturbance, damage-released, diet-based), and describes the role that the cues play in various aspects of the ecology of ostariophysan fishes. In contrast, production of alarm cues by nonostariophysan fishes is only beginning to be well understood. Mathis' chapter is a taxonomic overview of alarm cues in 13 families of nonostariophysan fishes. Her review includes a discussion of the possible sources of production of the chemicals and some cautionary notes for other researchers in this area.

The authors of this volume have attempted to provide an overview of the current state of knowledge of fish defenses with respect to pathogens, parasites, and predators, and to point out the existing gaps in need of further study. We hope that the chapters in this volume will stimulate further research in this important field.

Giacomo Zaccone
C. Perrière
A. Mathis
B.G. Kapoor

Contents

List of Contributors

Coll J.M.

INIA, Dept Biotecnología-Crt. Coruña Km 7–28040 Madrid, Spain.
E-mail: juliocoll@inia.es

Dickerson Harry W.

Associate Dean for Research and Graduate Affairs, UGA College of Veterinary Medicine, USA.
E-mail: hwd@uga.edu

Estepa Amparo

UMH, IBMC, Miguel Hernández University, 03202, Elche, Spain.
E-mail: aestepa@umh.es

Goater Cameron P.

Department of Biological Sciences, Lethbridge University, Lethbridge, AB, Canada.

Hancock E.W.

Department of Microbiology and Immunology, University of British Columbia, Centre for Microbial Diseases and Immunity Research, Lower Mall Research Station, UBC, Room 232 - 2259 Lower Mall, Vancouver, BC V6T 1Z4, Canada.
E-mail: bob@cmdr.ubc.ca

James Clayton T.

Department of Biological Sciences, Lethbridge University, Lethbridge, AB, Canada.

Johnsson Jörgen I.

Department of Zoology, University of Gothenburg, SE-405 30 Göteborg, Sweden.
E-mail: jorgen.johnsson@zool.gu.se

Kalmanzon Eliahu

Present address: Nitzana – Educational Community, Doar Na, Halutza, 84901, Nitzana, Israel, Jerusalem 91904, Israel.
E-mail: elizon@gmail.com

Knouft Jason H.

Department of Biology, Saint Louis University, 3507 Laclede Avenue, St. Louis, Missouri, 63103-2010, USA.
E-mail: jknouft@slu.edu

Magariños Beatriz

Departamento de Microbiología y Parasitología. C1BUS-Facultad de Biología. Universidad de Santiago de Compostela. 15782, Santiago de Compostela, Spain.

Marin Arnaldo

Departamento de Ecología e Hidrología, Facultad de Biología, Universidad de Murcia, 30100-Murcia, Spain.

Mathis Alicia

Department of Biology, Missouri State University, Springfield, Missouri, USA.
E-mail: aliciamathis@missouristate.edu

Mirza Reehan S.

Department of Biology, Nipissing University, North Bay, ON, Canada P1B 8L7.
E-mail: reehanm@nipissingu.ca

Patrzykat Aleksander

National Research Council Institute for Marine Biosciences, 1441 Oxford Street, Halifax, Nova Scotia, B3H3Z1, Canada.
E-mail: aleks.patrzykat@nrc-cnrc.gc.ca

Ravelo Carmen

Laboratorio de Ictiopatología, Estación de Investigaciones Hidrobiológicas de Guayana, Fundación La Salle de C.N. 8051, Ciudad Guayana, Venezuela.

Romalde Jesús L.

Departamento de Microbiología y Parasitología, C1BUS-Facultad de Biología. Universidad de Santiago de Compostela, 15782, Santiago de Compostela, Spain.
E-mail: jesus.romalde@usc.es

Tafalla Carolina

INIA, CISA Valdeolmos–28130 Madrid, Spain.
E-mail: tafalla@inia.es

Toranzo Alicia E.

Departamento de Microbiología y Parasitología. C1BUS-Facultad de Biología. Universidad de Santiago de Compostela. 15782, Santiago de Compostela, Spain.

Wisenden Brian D.

Department of Biosciences, Minnesota State University Moorhead, Moorhead, MN, USA.
E-mail: wisenden@mnstate.edu

Zlotkin Eliahu

Dept. of Cell and Animal Biology, Institute of Life Sciences, Hebrew University of Jerusalem, Jerusalem 91904, Israel.
E-mail: zlotkin@vms.hujl.ac.il

The Biology of Teleost Mucosal Immunity

Harry W. Dickerson

INTRODUCTION

The mucosal surfaces of teleosts (bony fishes) are the major interface between fishes and their immediate environment and serve as primary sites of entry for most pathogens. The mucosal surfaces of fishes include the epithelia and associated tissues of the gills, skin, gut, and the reproductive tract. In mammals, the mucosal system consists of an integrated network of tissues with associated immune cells referred to as the mucosa-associated lymphoid tissue (MALT). It is generally accepted that a comparable system exists in teleosts, although much less is known about its cellular and molecular components and the extent to which they function independently from the systemic immune response. Although a general understanding of the teleost immune system is emerging, fundamental questions still remain regarding primary lymphoid organ

Author's address: Office of Associate Dean for Research and Graduate Affairs, University of Georgia College of Veterinary Medicine, Athens, Georgia, USA.
E-mail: hwd@uga.edu

development, the induction, amplification and differentiation of local mucosal immune responses, the production of mucosal antibodies and effector lymphocytes, and immune memory. Answers to these questions will lead to a greater understanding of the evolution of basic immunological mechanisms as well as insights of immediate relevance to applied vaccines and the protection of farm-reared fish from microbial infections.

A number of laboratories have been or are currently engaged in research on mucosal immunity in various fishes, including, carp (*Cyprinus carpio*) (Rombout *et al.*, 1993), channel catfish (*Ictalurus punctatus*) (Lobb, 1987; Hebert *et al.*, 2002), rainbow trout (*Oncorhynchus mykiss*) (Bromage, 2004), Atlantic salmon (*Salmo salar*) (Lin *et al.*, 1998), sea bass (*Dicentrarchus labrax*) (Picchietti *et al.*, 1997), zebrafish (*Danio rerio*) (Danilova and Steiner, 2002) and others because teleosts are a diverse group of fishes, an understanding of the biology of their immune system requires a comparative approach. From the synthesis of research from various laboratories on multiple fish species, a general understanding of mucosal immunity exists and these concepts are presented in each of the chapter sections.

ORGANIZATION OF MUCOSAL TISSUES AND ASSOCIATED IMMUNE CELLS

Gastrointestinal Tract (Fig. 1.1)

The respiratory and digestive systems share the mouth and buccal cavity. The lining of the buccal cavity consists of a stratified mucoid epithelium on a thick basement membrane with a dermis that connects the epithelium to the underlying bone or muscle tissues (Roberts, 2001). The esophagus has an epithelial lining with large numbers of mucus cells. The stomach varies in size, depending on the species of the fish under study. The gastric mucosa is mucoid with numerous glands in the crypts of the mucosal folds (Roberts, 2001).

Although the intestinal morphology of teleosts varies depending on the species and diet, the intestinal tract has a common basic structure. The intestine is a single tube without the anatomically distinct colon found in mammals (Roberts, 2001). The rectum has a thicker muscle wall than the intestine and is very mucogenic (Roberts, 2001). The esophagus, stomach, and intestine have four basic layers that vary in composition

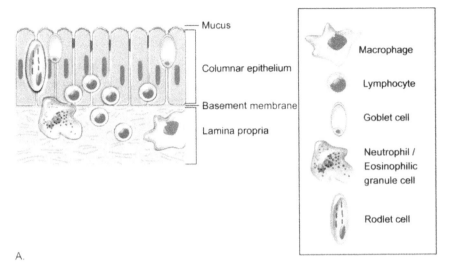

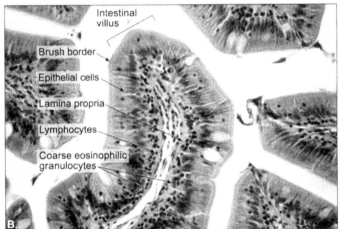

Fig. 1.1 Intestinal epithelium

A. Diagram of the basic anatomical structures of the intestinal epithelium and the identification and location of immune-related associated cells.

B. Photomicrograph of the intestinal villus of a channel catfish. Note the mucosal brush border, tall columnar epithelial cells (enterocytes), and supporting lamina propria containing migrating lymphocytes and coarse eosinophilic granulocytes (hematoxylin and eosin [H & E] stain).

among and within each of these organs. The innermost layer is the mucosa, which is composed of epithelium, a lamina propria of fibrous connective tissue, and sometimes a muscularis mucosae. The submucosa, comprised of fibrous connective tissue, lies between the mucosa and the muscularis, which is completely made of muscle. The outer layer of the serosa is composed of fibrous connective tissue covered with a simple squamous mesothelium (Grizzle and Rogers, 1976).

The intestinal mucosa is considered to be an immunologically important site in teleosts (Cain et al., 2000). In carp, the posterior segment of the gut, referred to as the second gut segment, plays a significant role in mucosal immunity (Rombout and van den Berg 1989; Rombout et al., 1989; Rombout et al., 1989) and comprises 20-25% the length of the gut (Rombout et al., 1993; Press and Evensen, 1999). The gut-associated lymphoid tissue of most teleosts, including rainbow trout (McMillan and Secombes, 1997), carp (Rombout et al., 1993), and sea bass (Picchietti et al., 1997) is comprised of cells with lymphoid morphology residing between the gut epithelial cells. These are predominantly intraepithelial T lymphocytes (Bernard et al., 2006; Huttenhuis et al., 2006), but Ig^+ lymphocytes are also found with the predominant number residing in the lamina propria (Rombout et al., 1993; Danilova and Steiner, 2002; Huttenhuis et al., 2006). Lymphoid aggregations that resemble the ileal or Peyer's patches in mammals are absent. The GALT of teleosts principally consists of lymphocytes of various sizes, plasma cells, macrophages as well as different types of granulocytes (Du Pasquier and Litman, 2000). Periodic acid Schiff (PAS) positive cells and eosinophilic granular cells are present, and may serve to modulate immune-hypersensitive responses that occur in the gut. In the intestinal epithelium and lamina propria, macrophages function as scavengers and antigen presenters. In carp, intestinal macrophages are different from the macrophages isolated from other lymphoid organs in the sense that they adhere poorly to glass and plastic, form clusters with lymphocytes, express antigenic determinants on their outer membranes and bind immunoglobulin (Ig) (Rombout et al., 1986, 1989 a, b, 1993).

The biliary system of the liver begins with intracellular bile canaliculi that anastamose extracellularly to form bile ducts. These fuse into the gall bladder, which directs bile into the intestine through the common bile duct. The gall bladder is lined with transitional epithelium. Hematopoietic tissue with melanomacrophage centers is associated with larger blood vessels of the liver (Roberts, 2001).

Skin (Fig. 1.2)

The skin of fishes provides protection against physical, chemical and biological damage. It consists of two anatomical layers, the epidermis and

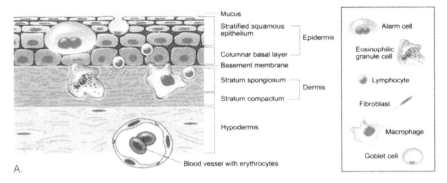

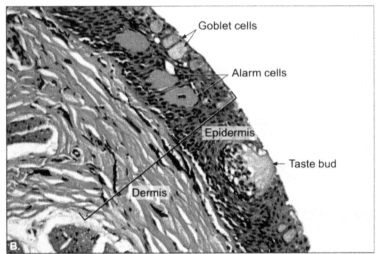

Fig. 1.2 Skin epithelium
A. Diagram of the basic anatomical structures of the skin and the identification and location of immune-related associated cells.
B. Photomicrograph of channel catfish skin (sensory barbel) (H & E stain).

dermis. The thickness of the stratified epithelium of the epidermis varies, depending on the area of the body, age, sex, maturation and environmental stresses (Grizzle and Rogers, 1976; Yasutake and Wales, 1983). On average, it has a thickness of 10-12 cells. Cells in the basal columnar layer of the epidermis, referred to as the stratum germinativum, replicate and move toward the surface of the fish. This basal layer lies immediately above a basement membrane. At least six types of cells have been described in the epidermis of teleosts, including filament-containing malpighian cells (keratinocytes), mucus cells, chemosensory cells, club cells (alarm substance cells), granule cells and chloride cells (Grizzle and

Rogers, 1976; Yasutake and Wales, 1983; see for review zaccone *et al.* 2001). The malpighian cells are the most abundant in the epithelium. These cells are rounded in shape with bundles of fibers and mitochondria around a generally ovoid nucleus (Roberts, 2001). At the epithelial surface, keratinocytes become more flattened and their cytoplasm consisting predominantly of oblong vesicles, degenerating mitochondria and denser bundles of fibers. The outermost layer of cells is not keratinised. The surfaces of the outermost cells have convoluted microridges of an unknown function that possibly assist in holding mucous secretions to the skin. Mucus cells begin differentiating in the stratum germinativum and migrate to the surface of the skin where they release their contents. Packets of mucus are bound by membranes and progressively fill the cell as they move toward the surface. At the surface of the epithelium, the mucus cell (a holocrine gland) moves between the keratinocytes and discharges its contents. The epidermis is covered by a glycocalyx or cuticle, consisting of a thin (1.0 μm) mucopolysaccharide layer. It is a complex mixture of molecules derived primarily from the contents of sloughed surface epithelial cells and mucus secreted from goblet cells (Roberts, 2001).

The deeper layers of the epidermis contain alarm substance cells and melanophores, which do not reach the surface. The contents of alarm substance cells are only released when the epidermis is physically damaged (Grizzle and Rogers, 1976). Capillaries extend into the epidermis from dermal papilli, and come within 10 cell layers of the surface (Lobb, 1987).

The dermis is composed of two layers. The upper layer, referred to as the stratum spongiosum, consists of a loose network of collagen and reticulum fibers and is contiguous with the epidermal basement membrane that lies just above it. It contains chromatophores, mast cells and the cells of the scale beds. The lower layer, the stratum compactum, is a dense matrix of collagen that provides the structural strength to the skin. The hypodermis, lying beneath the dermis, is composed of loose connective tissue. It is more vascular than the overlying dermis. Melanophores occur in the hypodermis, dermis and sometimes in the epidermis.

No organised lymphoid germinal centers have been found in the skin (Flajnik, 1998), although cells with the morphology of lymphocytes can be detected by light microscopy in stained tissue sections of channel catfish skin (Lobb, 1987). These cells occur throughout the epidermis and are located predominantly near the stratum germinativum at the junction of the epidermis and dermis (Lobb, 1987). Antigen-specific and total antibody secreting cells (ASC) have been isolated from the skin of channel catfish and detected by ELISPOT (Zhao *et al.*, 2007). B cells isolated from the skin of channel catfish can be stimulated with LPS to replicate and secrete antibody *in vitro*, a response that, in turn, is abrogated by the addition of hydroxyurea to the culture medium (Zhao *et al.*, 2007). Macrophages are also present in the skin (Roberts, 2001).

Gills (Fig. 1.3)

The gills consist of gill arches, gill filaments (primary lamellae), and gill lamellae (secondary lamellae). Two rows of filaments are present on each arch and the secondary lamellae branch out perpendicularly from the filaments (Grizzle and Rogers, 1976; Yasutake and Wales, 1983). The gill arches and filaments are supported by a branching system of cartilaginous rods. A stratified squamous epithelium covers both the gill filament and the gill lamellae. The lamellae provide the actual respiratory surfaces. Each lamella comprises a network of interconnected spaces that are separated and supported by pillar cells. Blood enters the lamellae from the afferent arterioles of the filaments and exits into the efferent arteriole. The lamellar intercellular spaces through which blood flows are lined with endothelial cells. A basement membrane lies over the endothelial cells and pillar cells, which form supportive 'flanges' around the intra-lamellar spaces (Grizzle and Rogers, 1976). The stratified epithelium itself is only one to two cells thick in order to allow gas exchange, a degree of thinness that makes the tissue vulnerable to invasion by pathogens.

Different cell types are associated with the gill epithelium. Chloride cells function in the transport of Cl⁻ and other ions across the epithelium. These cells are more spherical than those that surround them in the epithelium; they project somewhat above the surface (Yasutake and Wales, 1983), and their cytoplasm is more eosinophilic (in hematoxylin and eosin stained sections) than is the case with other epithelial cells. Chloride cells are abundant in the gill filament epithelium between lamellae (Grizzle and Rogers, 1976). Mucus cells are abundant in the lamellar epithelium, and appear under light microscopy as mucus-filled domes or vacuolated cells (Yasutake and Wales, 1983). Goblet cells are most abundant on the margins near arterioles. Alarm substance cells are absent in gill epithelia (Grizzle and Rogers, 1976). Although the surface of the gill lamellar epithelium is irregular, it does not have the distinctive

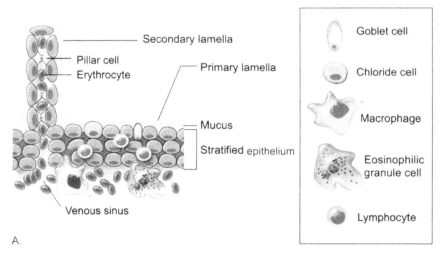

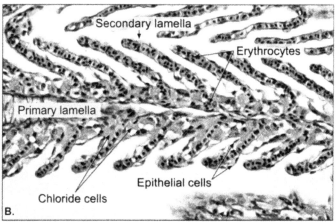

Fig. 1.3 Gill epithelium

A. Diagram of the basic anatomical structures of the gill epithelium and the identification and location of associated immune-related cells.

B. Photomicrograph of fish gill. Note the capillaries with erythrocytes in secondary lamellae and chloride cells concentrated in lamellar troughs (H & E stain).

microridges seen on the surface of the skin epidermis (Roberts, 2001). Nevertheless, these irregularities are sufficient to aid in attachment of mucus, which in addition to its role in reducing invasion of microorganisms, also serves to regulate the transfusion of gases, ions, and water across the epithelial membrane (Roberts, 2001).

Similar to the situation that exists in skin, there is no evidence to indicate the existence of organised aggregations of lymphoid tissue in the gills. Nevertheless, there have been a number of studies to show the functional immunological activity in gills as well as gill-associated leucocytes and lymphocytes (Goldes *et al.*, 1986; Powell *et al.*, 1990; Lumsden *et al.*, 1995; Davidson *et al.*, 1997; Lin *et al.*, 1998; Rombout *et al.*, 1998; Dos Santos *et al.*, 2001 a,b). Considerable numbers of lymphocytes, ASC, and macrophages were found to reside in the gill tissue of Atlantic salmon and dab (Lin *et al.*, 1998). In leucocyte suspensions from carp gill (as in skin), Rombout *et al.* (1998) found an abundant population of intraepithelial lymphocytes (IEL) that reacted with a monoclonal antibody (mAb WCL38), which is specific for IEL T cells in the carp intestine. In gill IEL leucocyte suspensions, WCL38$^+$ cells comprised the major population of lymphoid cells. Lymphocytes with surface immunoglobulin (i.e., B cells) were a minor component of these cell populations. In cryosections, many of the WCL38$^+$ cells were detected at the base of the gill lamellae. Immunogold labeling showed that the WCL38$^+$ cells had the ultrastructure of lymphoid cells, although two morphologically different cell types were found: small lymphocytes with a high nucleus/cytoplasm ration, and larger granular lymphocytes with a lower nucleus cytoplasm ration and a variable amount of electron-dense, lysosome-like material.

Ontogeny of Mucosal Lymphocytes and Immune-associated Cells

Lymphocytes and other cells (such as macrophages) that function in acquired immune responses of teleosts are present in gut-associated immune tissues and other mucosal tissues and most likely evolved in these sites early in the development of the vertebrate adaptive immune response (Matsunaga, 1998; Matsunaga and Rahman, 1998; Cheroutre, 2004). In present-day teleosts, however, the ontogeny of mucosal lymphocytes has not been resolved and the extent to which they develop and remain resident in mucosal tissues or migrate to and from primary and secondary lymphoid organs, such as the head kidney and spleen remains to be determined.

The mammalian gut can function as a primary lymphoid organ and intraepithelial lymphocytes (IEL) develop at this site (Lefrancois and Puddington, 1995) and, as indicated above, it is likely that the early adaptive immune system of vertebrates also evolved in gut epithelium (and possibly skin and gill epithelia as well) (Matsunaga and Rahman, 1998; Cheroutre, 2004). With the evolving adaptive immune system, however, the thymus acquired the mechanisms of lymphocyte maturation and selection and subsumed this function from mucosal sites. Thus, gut mucosal tissues were eventually relieved by the thymus of the responsibility to educate the developing IEL regarding self and non-self (Cheroutre, 2004).

Immune mechanisms of mucosal surfaces have been extensively studied in higher vertebrates and the roles of specific T and B cells located in epithelia are elucidated (Cheroutre, 2004). For example, in mice, it has been shown that αβ T cells localised in epithelia migrate from the GALT and peripheral lymphoid tissues following antigen stimulation (Kim et al., 1998). In this process, specialised cells in the follicle-associated epithelium of the gut, referred to as M cells, sample the lumen of the gut and transport antigens to the subepithelial tissues and GALT (Neutra et al., 1996). Local dendritic cells then process these antigens and further distribute them to peripheral lymphoid tissues in which resident naïve αβ T cells become activated and proliferate. These antigen-specific, differentiated T cells then migrate to the gut where they seed the epithelium as effector and memory cells.

There also are specialised IEL in the mammalian gut that develop via an extrathymic pathway (Lefrancois and Puddington, 1995). These IELs mostly consist of γδ T cells with a oligoclonal TCR repertoire (Regnault et al., 1994; Cheroutre, 2004; Bernard et al., 2006). The mechanisms responsible for the limited repertoire is unknown, but is believed to be the result of selection during lymphocyte development in the gut (Takimoto et al., 1992), a process involving resident microflora (Helgeland et al., 2004). It has been suggested that extrathymic development of T cells occurs in teleosts as well, at least in carp, a species in which the first studies of mucosal lymphocyte ontogeny have been systematically carried out (Huttenhuis et al., 2006). These studies showed that IELs develop in embryonic gut epithelia before the development of the thymus. In addition, the expression of rag-1 in intestinal tissues was seen to occur concurrently with the early appearance of these intestinal IELs. However, there may be species-specific differences among teleosts regarding IEL ontogeny. A recent immunoscope-based analysis (Pannetier et al., 1995) of the VβJβ spectratypes of IEL and systemic T cell receptors (TCR) in trout showed that intraepithelial T lymphocytes isolated from the gut of naïve fish have similar TCR repertoires to T cells found in the blood and spleen (Bernard et al., 2006). While this finding does not preclude an extrathymic development pathway for IEL or a subpopulation of IEL not surveyed in this study, it suggests (at least in trout) that αβ IEL correspond to random samples of systemic αβ T cells (Bernard *et al.*, 2006).

The predominant population of IEL in the mammalian gut consists of γδ T cells, which are suggested to have evolved before αβ T cells in the development of adaptive immunity. Although the genes encoding the γδ TCR have been identified in the Japanese flounder (*Paralichthys olivaceus*) (Nam *et al.*, 2003), the extent to which teleost lymphocytes equivalent to the γδ T cells exist in populations of IEL is still not known. The development of reagents such as monoclonal antibodies to identify characteristic cell surface receptors and ancillary proteins in teleosts will be necessary to answer this question (Miller *et al.*, 1998).

B cells occur in mucosal tissues, but current evidence suggests that they develop in the primary lymphoid tissues of the head kidney. In zebrafish, B cells are first found to appear in the embryonic pancreas, and then the head kidney (Danilova and Steiner, 2002). In carp, B cells first appear in the head kidney and spleen of embryos, and later in mucosal organs and the thymus, but Ig[+] lymphocytes are never abundant in the thymus and intestine (Huttenhuis *et al.*, 2006).

MUCOSAL INNATE IMMUNITY

The mucosal surfaces of the skin, gills and intestine are constantly exposed to environmental pathogens; yet, under normal circumstances they remain free from infection and life-threatening lesions. Epithelia also heal quickly following mechanical or chemical injury. Resistance to infection and recovery from traumatic insult is facilitated by innate non-specific immunity that consists of a plethora of constitutively expressed elements as well as induced components of the inflammatory response. The physical factors of innate immunity consist of the membrane-anchored surface mucus barrier (glycocalyx) and the contiguous underlying epithelial cells with their tight junctions. The components of innate immunity can be generally classified as either cellular or humoral effectors.

Cellular Components of Mucosal Innate Immunity

Teleosts have interacting leukocyte subpopulations that mediate both innate and adaptive immune responses (Miller *et al.*, 1998). Cell populations involved in the innate immune response include phagocytic cells (macrophages and neutrophils), non-phagocytic cells (natural killer (NK) cells and non-specific cytotoxic (NCC) cells), and other cells (mast cells/eosinophilic granule cells and rodlet cells). Mast cells/eosinophilic granular cells have structural and functional properties similar to those of mammalian mast cells (Reite, 1997), and store a number of inflammatory and anti-microbial compounds, including phospholipids, alkaline phosphatase, peroxidase and lysozyme (Silphaduang et al., 2006). Rodlet cells occur in blood and epithelia of a large number of teleost species (Reite, 1997, 2005) and have a characteristic morphology with cytoplasmic club-like crystalline inclusions that are released at epithelial, mesothelial and endothelial surfaces. Although there is still some question as to the origin and function of these cells, most recent studies interpret rodlet cells to be elements of the host defense system, appearing in association with insult from various stressors including parasites, neoplasia, viral infections, and general tissue damage (Reite, 1997, 2005; Manera and Dezfuli, 2004; Bielek, 2005; Reite and Evensen, 2006; Silphaduang et al., 2006).

The innate cell inflammatory response of teleosts is usually biphasic, beginning with an influx of neutrophils followed by the arrival of monocytes and macrophages (Sharp et al., 1991; Neuman et al., 2001). Neutrophils originate from the head kidney, while macrophages originate from blood-derived monocytes that migrate to the relevant tissues following inflammatory insult. Monocytes develop from hematopoietic stem cells in the head kidney and/or the spleen. In addition to the phagocytic cells that extravaginate and migrate to tissues during inflammation, mucosal tissues also have resident macrophages that are involved in the ingestion of antigens and antigen presentation and are postulated to play an important role in both innate and acquired immune responses (Huttenhuis et al., 2006).

Gastrointestinal Tract: The gastrointestinal tract of teleosts contains intraepithelial macrophages as well as neutrophils and mast cells/eosinophilic granular cells (MC/ECG) located in the lamina propria (Georgopoulou and Vernier, 1986; Vallejo et al., 1989; Rombout et al., 1989, 1993b; Davidson et al., 1991; Powell et al., 1991; Dorin et al., 1993; Sveinbjornsson et al., 1996; Hebert et al., 2002; Leknes, 2002; Grove et al., 2006). In experiments carried out in platy (*Xiphophorus maculatus*), horse-spleen ferritin injected into the coelomic cavity was taken up by macrophages located primarily in the lamina propria of the gut (Leknes, 2002). A MC/ECG submucosal layer is well developed in salmonids.

MCs/ECGs can move from the submucosal layers of the intestine into the villi or mucosa, as also in certain allergic and bacterial infections marked degranulation of these cells occurs. Experimental intracoelomic injection of extracellular products from culture supernatants of the bacterium *Aeromonas salmonicida* elicited vasodilatation of blood vessels in the lamina propria of the gut with concomitant dissemination and degranulation of the MC/ECG cells (Ellis *et al.*, 1981). Rodlet cells often occur associated with the presence of adult parasitic trematodes or cestodes in the intestine and with encysted helminth larvae in the intestine or its adjacent tissues (Reite, 1997; Dezfuli *et al.*, 1998; Bielek, 2005).

A significant number of neutrophils (> 64% of leukocyte cells isolated from collagenase-digested intestine) appear to reside in the gut of healthy juvenile channel catfish, suggesting that innate immunity plays an important role in host defense in this species (Hebert *et al.*, 2002). Likewise, in gilthead seabream (*Sparus aurata*), acidophilic granulocytes (considered equivalent to neutrophils in this species) occur principally dispersed in the lamina propria of the mucosa in the posterior intestine (Mulero *et al.*, 2007). It is hypothesised that these cells play an important role in innate immunity and immune surveillance and studies have shown that the administration of probiotics to gilthead seabream elicits an increased number of these cells in the gut (Picchietti *et al.*, 2007).

Ig⁻ lymphoid cells are diffusely distributed within the epithelia of the gut. Although this population consists mainly of intra-epithelial lymphocytes (IEL, primarily putative T cells), NK cells are postulated to occur here as well (Rombout *et al.*, 1993). Isolation of cytotoxic IELs from the intestine of rainbow trout have been isolated and functionally characterised with regard to non-specific killing of target cells. These cells did not contain cytotoxic granules analogous to those seen in mammalian NK cells, suggesting an alternative mechanism for cell killing (McMillan and Secombes, 1997).

Skin: Macrophages, neutrophils and other granulocytes such as MC/ECG appear in the deeper layers of the epidermis, particularly in response to inflammatory events such as parasitic infection (Cross and Matthews, 1993; Buchmann, 1999; Reite and Evensen, 2006). In rainbow trout and channel catfish, migratory macrophages and lymphocytes are present in the skin (Lobb, 1987; Peleteiro and Richards, 1990). Activation of fish leucocytes *in vitro* elicits the production of leukotriene B4 which, in turn, induces the migration of neutrophils (Hunt and Rowley, 1986). Teleost macrophages and neutrophils secrete interleukin 1, which affects other macrophages (Secombes and Fletcher, 1992). These signaling molecules are likely to play a role in the induction and activation of the cellular innate immune response in the skin.

Langerhans cells are dendritic antigen-trapping cells found in the human skin and have the ability to process and present antigen to lymphocytes (Koch et al., 2006). These cells have a typical granular cytoplasm and defined cell surface determinants. Reports of resident antigen-trapping phagocytic cells in teleost skin are rare (Peleteiro and Richards, 1990), with only one reference to epidermal cells with membrane folding that resembles the Birbeck's granules typical of human Langerhans cells (Mittal et al., 1980). Although phagocytic cells with the typical morphology of Langerhans cells apparently do not occur in the epidermis of fish, this does not preclude the possibility that dermal macrophages, which migrate across the basement membrane into the epithelium, do trap and process antigen. Indeed, phagocytic cells that share cell surface determinants with Langerhans cells (referred to as indeterminate or agranular dendritic cells) exist in human epithelia (Rowden et al., 1979), and it is postulated that these are monocyte-derived dermal macrophages that migrate into the epidermis and develop into Langerhans cells (expression of surface determinants and formation of Birbeck's granules) under the influence of chemokine gradients and a particular epithelial micro-environment (Koch et al., 2006). It has been hypothesised that the migration of macrophages into the epidermis of fish could be the equivalent of these non-differentiated Langerhans precursor cells seen in human skin (Peleteiro and Richards, 1990).

NK or NCC cells have not been reported in the skin, but it is possible that activated cells recruited from the head kidney into the peripheral blood could end up in this peripheral site (Graves et al., 1985).

Gills: In addition to epithelial cells, mucus-secreting goblet cells and chloride cells described earlier, various types of leukocytes have been isolated from the gills of teleosts. Macrophages, eosinophilic granular cells (EGC) and neutrophils have been isolated and characterised in perfused gill tissue from Atlantic salmon and dab (*Limanda limanda*) (Lin et al., 1998). In experiments carried out in platy (*Xiphophorus maculates*), horse-spleen ferritin injected intracoelomically was taken up by macrophages located within the gill filament, but not the gill lamellae (Leknes, 2002).

Thus, while the main functions of gill phagocytes are presumably to capture foreign substances and kill infectious agents that gain entry from the water, these cells also apparently participate in the clearance of foreign substances from the blood (Leknes, 2002). Although resident dendritic cells have not been described in gill epithelia, gill macrophages most likely process and present antigenic material to lymphocytes to initiate a specific, acquired immune response (Davidson *et al.*, 1997; Lin *et al.*, 1998).

Humoral Components of Mucosal Innate Immunity

The mucus coating of fish skin, gills and gut epithelia is a complex mixture comprising molecules secreted by goblet cells and cellular contents released from effete surface epithelial cells. The major component of mucus is mucin, which is composed mainly of glycoproteins. Also present are lysozyme, proteolytic enzymes, and C-reactive proteins (Ingram, 1980; Fletcher, 1981). Mucus acts as both a physical and chemical barrier to microbial invasion and environmental insult.

Non-immunoglobulin humoral defense factors in fish have been classified into four general categories based on their effects on invading pathogens: (1) microbial growth inhibitory substances, (2) enzyme inhibitors, (3) lytic agents (lysins), and (4) agglutinins/precipitins (Alexander and Ingram, 1992). Various antimicrobial compounds in these categories including trypsin, lysozyme, lectins, complement, and other lytic factors are present in mucus and mucosal tissues where they serve to prevent adherence and colonisation of pathogenic microorganisms (Alexander and Ingram, 1992; Dalmo *et al.*, 1997). These factors are described below with specific indications of their roles in mucosal innate immunity, if known to occur.

Microbial Growth Inhibitory Substances: The microbial growth inhibitors—transferrin, caeruloplasmin, metallothionein, and interferon—are all present in fish tissues (Alexander and Ingram, 1992). Transferrin is an acute phase protein that is elicited during inflammation to remove iron from damaged tissues, and activate macrophages (Magnadottir, 2006). It is expressed constitutively in liver cells. Lactoferrin, a protein related to transferrin, occurs in mucus secretions of mammals, but has not been reported in fish mucus or epithelial cells (Alexander and Ingram, 1992). Interferons (IFN) are secreted proteins that activate cells to an anti-virus state, inducing the expression of Mx and other antiviral proteins (Leong *et al.*, 1998; Robertson, 2006). Type I IFN α and β and type II IFN γ have been detected or deduced in a number of different fish species (Graham and Secombes, 1990; Alexander and Ingram, 1992; Robertson, 2006). IFN γ produced in NK cells modulates innate immune responses; but as indicated earlier, there have been no studies to indicate whether or not NK cells are found in mucosal tissues.

Enzyme Inhibitors: The basic function of enzyme inhibitors is to maintain homeostasis of blood and other body fluids through the regulation of enzyme activities including those involved in the functions of complement activation and coagulation (Alexander and Ingram, 1992). Following invasion by pathogens, destructive enzymes are actively secreted into tissues by parasites and passively released from damaged host cells including neutrophils and macrophages that have migrated to the site of infection. These released proteases require inactivation to prevent and reduce secondary tissue destruction. A plethora of proteinase inhibitors (serine-, cysteine-, and metalloproteinases) have been isolated and characterised in mammals, but few have been described in fishes. The most widely studied in fishes is α2 macroglobulin, which has broad inhibitory effect through encapsulation of protease molecules (Armstrong and Quigley, 1999; Magnadottir, 2006). The extent to which enzyme inhibitors function at the mucosal surfaces is currently unknown.

Lytic Agents: The lytic components of humoral innate immunity are enzymes that exist as either single molecular entities, such as lysozyme, or a cascade of component enzymes as occurs in the complement system.

Lysozyme has been found in tissues and secretions of fish including the gut, cutaneous mucus and gills (Alexander and Ingram, 1992; Magnadottir, 2006), where it is produced by macrophages, neutrophils, and eosinophilic granule cells (Murray and Fletcher, 1976). Lysozyme attacks structures containing β 1-4 linked N-acetylmuramamine and N-acetylglucosamine, (the peptidoglycan components of bacterial cell walls), as well as chiton, a component of fungal cells and is, thus, both antibacterial and antifungal. It also functions as an opsonin with subsequent activation of complement and phagocytes (Magnadottir, 2006). The amount of enzyme varies among tissues and species of fish (Alexander and Ingram, 1992). Lysozyme has been described in the cutaneous mucus of a number of fish species, including carp and channel catfish.

The teleost complement system consists of more than 35 soluble plasma proteins that play roles in both innate and acquired immunity (Boshra et al., 2006). Complement activation products initiate or are involved in the innate immune functions of phagocytosis and cytolysis of pathogens, solubilisation of immune complexes, and inflammation (Boshra et al., 2006). There are only a few experimental studies that address the extent to which the components and functions of complement occur in mucosal tissues and secretions. A study showing that the parasitic monogenetic trematode Gyrodactylus salaris was killed following incubation in cutaneous mucus of Atlantic salmon suggests that components of the complement system are involved in the innate immune responses of the skin. In this study, mucus activity was approximately one twentieth of that found in serum. Activity (in serum) was not dependent on the immune status of the fish and opsonisation of parasites with antibodies did not enhance killing, suggesting that complement was activated by the alternative pathway (Harris et al., 1998) or by the lectin pathway (Buchmann, 1998, 1999). Transcripts of complement factors C3 (rainbow trout) and C7, P (FP), Bf/C2A, C4, and D (FD) (carp) were detected in the skin following infection with the ciliated protozoan parasite Ichthyophthirius multifliis (Sigh et al., 2004; Gonzalez et al., 2007 a, b). These studies also suggest that parasite infection elicits expression of a subset of extrahepatic complement genes in the skin. It is postulated that the proteins are produced in macrophages (Buchmann, 1999).

Cutaneous mucus of Japanese eels (Anguilla japonica) contains a locally produced hemolysin that could have a non-specific protective role, although this has not been determined (Alexander and Ingram, 1992). Trypsin has been found in mucus and mucus-secreting cell layers of the skin, gill lamellae, and anterior intestine of Atlantic salmon and rainbow trout, where it is hypothesised to play a role in non-specific immunity against microbial invasion at these surfaces (Hjelmeland et al., 1983; Braun et al., 1990). It should be noted that the presence of active trypsin at these surfaces suggests that enzyme inhibitors are not present.

Agglutinins: Agglutinins are agglutinating factors (non-immunoglobulin) produced in the absence of defined antigenic stimuli (Ingram, 1980). These carbohydrate-binding proteins elicit opsonisation, phagocytosis and activation of the complement system (Buchmann, 1999). Mucosal agglutinins and precipitins consist primarily of lectins such as C-type lectins and pentraxins. In the presence of Ca^+, C-type lectins bind mannose, N-acetylglucosamine and fucose leading to opsonisation, phagocytosis and activation of the complement system (Magnadottir, 2006). Pentraxins, which include C-reactive proteins, are commonly associated with the acute phase inflammatory response and take part in innate immunity by activating complement pathways. A hemagglutinin is found in the cutaneous mucus of Japanese eels but the extent to which it is involved in innate immunity is not known (Magnadottir, 2006). Lectins found in cutaneous mucus appear to play a role in the innate immune response against parasites of the skin, such as the ciliate *I. multifiliis*, and the trematode *Gyrodactylus* (Yano, 1996; Buchmann, 1999; Buchmann *et al.*, 2001; Xu *et al.*, 2001). Nevertheless, the roles of mucus lectins remain unresolved in many cases and it is possible that they could work independently or in cooperation with other biologically active molecules (Alexander and Ingram, 1992).

Natural Antibodies: Although antibodies (immunoglobulins) are generally considered to be the primary effector mechanism of the humoral acquired immune response, natural antibodies are also considered to be components of the innate immune system. There are different sources of natural antibodies including: adoptive transfer, environmental antigen exposure, and production by gene rearrangement without specific antigen stimulation (Sinyakov *et al.*, 2002; Magnadottir, 2006). Natural antibodies have increasingly been shown to play a role in mammalian immunity and their occurrence and function in immunity in fishes also has been well documented (Sinyakov *et al.*, 2002; Magnadottir, 2006). The fact that specific antibodies are produced locally in mucosal tissues would suggest that natural antibodies also could occur in these sites, although no systematic studies have been done to determine this. In vaccine studies with channel catfish, however, a relatively small but consistent number of antibody secreting cells (plasma cells) that produce antibody against the major surface antigen of *I. multifiliis* have been detected in skin epithelia of naïve fish (Dickerson, unpubl. data). These could be natural antibodies. Given the importance of the surface mucosa as a first line of defense against pathogens, it seems logical to expect that natural antibodies would occur in these sites. More research is necessary in this area.

Antimicrobial Peptides: Low molecular weight antibacterial peptides in vertebrates are usually associated with peripheral blood leucocytes or mucosal surfaces (Bevins, 1994; Cole *et al.*, 1997; Smith *et al.*, 2000; Silphaduang *et al.*, 2006). They have a number of useful characteristics for innate immune responses, namely, broad spectra of activity against microorganisms, low toxicity for host cells, ease of synthesis, and rapid diffusion rates (Smith et al., 2000). Antimicrobial peptides have been described in the skin from a number of different fish species, including rainbow trout, where mucus extracts were shown to have muramidase and non-muramidase lytic activity against selected bacteria (Smith et al., 2000; Ellis, 2001). The peptide piscidin has recently been found in a wide range of teleost species and is produced in gill, skin, stomach and intestinal epithelia. Piscidin is produced in MC/eosinophilic cells and rodlet cells (Cole et al., 1997; Silphaduang et al., 2006). The presence of piscidins in eosinophilic cells, which occur in epithelial tissues, suggests that they play an important function in innate defenses in these tissues (Silphaduang et al., 2006).

MUCOSAL ADAPTIVE IMMUNITY

The adaptive mucosal immune response of teleosts, which is postulated to have appeared early in the evolution of acquired immunity, plays an important role in protection against infection. Fishes are the earliest vertebrates to have both innate and adaptive immunity, and acquired immunity is postulated to have evolved earliest in the gut of jawed fishes (Matsunaga, 1998; Matsunaga and Rahman, 1998; Cheroutre, 2004). However, relatively few immunologists have focused their efforts on the study of mucosal immunity of fishes, and consequently, there is much less basic knowledge when compared to that known about the mammalian system. Also, necessarily, the experimental data generated from fish are in many cases more descriptive than mechanistic due to a paucity of immunological reagents available for quantitative studies (e.g., antibodies against cell surface antigens and signaling molecules, and knock-out, isogenic experimental animals) (Rombout et al., 1993; Lin et al., 1998; Huttenhuis et al., 2006). For instance, although it is known that antigen is absorbed preferentially in the posterior intestine (Rombout et al. 1985; Georgopoulou and Vernier 1986; Otake et al., 1995), the precise sites where antigen is processed and presented by phagocytic cells, and where B and T lymphocytes interact, proliferate and differentiate remain unknown. Relatively few cell-signaling molecules such as cytokines and chemokines have been identified. Lymphocytes and antibody-secreting plasma cells have been described in the intestinal epithelia and lamina propria (Rombout et al., 1993; Hebert et al., 2002), but the extent to which phagocytes and lymphocytes traffic between peripheral (mucosal) and central (pronephros and spleen) tissues is largely undetermined. Questions as basic as how antibodies produced at mucosal sites are translocated across intact epithelial cell layers also remain unanswered. It is clear that compared to the substantial amount of experimental data that have contributed to the elucidation of the basic mechanisms of mucosal immunity in mammals, there is much less data available for fish. Most of the experimental work on basic immunity in fishes has focused on the systemic immune response, and what is known on mucosal immunity has been gleaned primarily from studies of the fish intestine, with less information available on the gills and skin.

As the elements of teleost mucosal immunity are presented in each section below, the mucosal immune response in mammals is briefly reviewed as necessary in order to point out the notable anatomical and functional differences (or similarities) that exist between the two groups. It should be emphasised, however, that contemporary fish have a mucosal adaptive immune system that is as effective in preventing infections as that of mammals. Comparative immunological studies are intended to shed light on evolutionary adaptations as well as provide insights into shared and unique mechanisms that exist among these different groups of animals.

Induction and Initiation of Mucosal Adaptive Immunity (Fig. 1.4)

There is experimental evidence to suggest that the induction of mucosal immunity occurs by mechanisms similar to those that exist in higher vertebrates, namely, antigen processing and presentation by phagocytic cells, followed by priming of B cells and T cells, induction of B cell proliferation and differentiation with T cell help, and production of antibody by fully differentiated plasma cells (Miller et al., 1998). The precise sites of antigen induction, however, and the degree to which the mucosal and systemic immune response interact, are still unknown. The sections below present current knowledge and hypotheses regarding the induction of mucosal immunity in the various mucosal tissues of teleosts.

Gastrointestinal Tract: The initiation of the mucosal immune response begins with uptake of antigen. The distal intestine of teleost fishes (referred to as the second intestinal segment) is the primary site of antigen uptake, and enterocytes in this region are postulated to function similarly to the specialised membranous epithelial cells (M cells) found in the gut

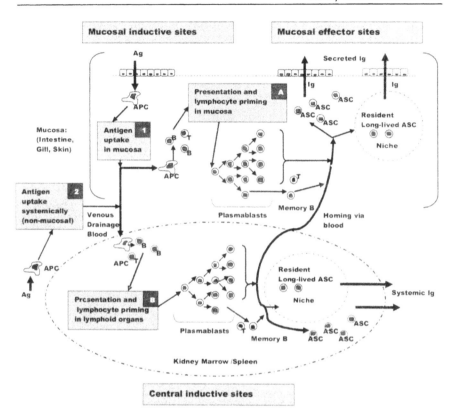

Fig. 1.4 Conceptualised elements of adaptive mucosal humoral immunity in teleosts.

In this model, which is derived from various studies in different fish species, both **mucosal (1)** and **systemic (2)** antigen (**Ag**) exposure are postulated to elicit a mucosal antibody (**Ab**) response. Mucosal exposure to antigen can elicit the production of systemic antibody as well. After entry through the mucosal epithelium or systemically (e.g., inoculation) antigen is phagocytosed by antigen-presenting cells (**APC**), processed, and presented in hypothetical mucosal inductive sites (**A**) and/or the central inductive sites of the pronephros kidney pulp and spleen (**B**). Plasmablasts generated with T cell help in the kidney pulp and spleen traffic through the blood to peripheral mucosal sites. It is postulated that plasmablasts generated in mucosal inductive sites can traffic to central lymphoid organs as well. Following surface antigen exposure, mucosal antibody responses can be elicited without production of any systemic antibody. Memory B cells, long-lived antibody secreting cells (**ASC**), humoral memory and long-lived ASC niches are discussed in the text.

of mammals (Davina *et al.*, 1982; Egberts *et al.*, 1985; Rombout and van den Berg, 1989; Rombout *et al.*, 1989). M cells, which are modified gut epithelial cells, serve as sites of antigen uptake (Egberts *et al.*, 1985; McLean and Donaldson, 1990), and have apical membranes with microvilli that are shorter and broader than those on surrounding enterocytes (McLean and Donaldson, 1990). Epithelial cells with similar morphology have not been described in fishes, but the functional aspects of the posterior segment of the fish intestine suggest analogous roles for intestinal cells in this region, namely, the ability to absorb intact proteins and the close association of lymphoid cells (Rombout et al., 1985).

Macrophages take up antigen from the posterior region of the gut, suggesting that this is a site of induction and initiation of the mucosal immune response (Rombout et al., 1985; Doggett and Harris, 1991). Lymphocytes (referred to as intra-epithelial lymphocytes or IEL) are diffusely disseminated within the columnar epithelium (Rombout et al., 1993; McMillan and Secombes, 1997; Picchietti et al., 1997). These are primarily T cells, expressing the $\alpha\beta$ T cell receptor (TCR), but a few antibody-secreting plasma cells are present as well (Scapigliati et al., 2000; Bernard et al., 2006). Macrophages and lymphocytes also are distributed diffusely in the underlying lamina propria. Organised germinal centers functionally and morphologically comparable to the ileal and Peyer's patches and regional lymph nodes of mammals are not present. Resident macrophages in the intestinal epithelium have been shown to take up antigen and display antigenic determinants on their outer membranes, suggesting an antigen-presenting function (Rombout and van den Berg, 1989). The differentiation and proliferation of resident or circulating antigen-specific B and T cells could occur locally following antigen presentation by resident macrophages, although this has not been shown experimentally. The population of $\alpha\beta$ T cells found in IEL populations were found to share functional and phenotypic similarity with $\alpha\beta$ T cells found in the peripheral circulation (Bernard et al., 2006), which allows the possibility that IEL circulate in the blood. It is also possible that following antigen uptake and processing (in the gut or elsewhere), antigen-presenting cells migrate to the central lymphoid organs of the pronephros (also referred to as the head kidney) and the spleen, where they subsequently present antigen to initiate the immune response (Rombout and Van den Berg, 1989). This latter possibility would predict that differentiated T cells, plasmablasts or plasma cells that originate and develop in the central lymphoid organs traffic via blood to the peripheral epithelia. Again, there is no direct experimental evidence to resolve where the sites of induction occur. Studies indicate, however, that anal administration of particulate bacterial antigen elicits mucosal as well as serum antibody responses (Rombout et al., 1989).

Skin and Gills: The skin is the site where the immune system encounters most environmental pathogens (Kupper, 2000), and in mammals it has been postulated to serve as an immune organ (Puri *et al.*, 2000). Mammalian skin has phagocytic dendritic cells (Langerhans cells) that extend pseudopodial processes between epithelial cells to reach close to the surface. These cells survey the epidermal barrier for the presence of foreign antigen intrusion. Once an antigen is encountered, internalised and processed, Langerhans cells migrate to regional lymph nodes to continue their development, which involves the production of additional co-stimulatory molecules (involved in T-cell activation) and the cessation of antigen processing (Kupper, 2000). The mature Langerhans cell no longer processes antigen to ensure that only the initial antigen encountered in the skin is displayed to initiate the immune response. Antigen is then presented to resident T cells, which when activated, home back to the skin in order to eliminate or prevent further antigen intrusion (Kupper, 2000). In mice, the epidermis also contains small numbers of specialised $\delta\gamma$ T cells, which are referred to as dendritic T cells. These cells have a restricted pattern of TCR usage and appear to play a unique role in cutaneous immune responses. Analogous cells are not found in humans (Bogen, 2004).

Fish have phagocytic cells and leucocytes that are associated with the epithelia of the skin and gills, either within the epithelium or immediately below it (Lobb, 1987; Iger and Wendelaar Bonga, 1994; Davidson *et al.*, 1997; Lin *et al.*, 1998; Moore *et al.*, 1998), and these cells are postulated to be involved in the initiation of the mucosal immune response. Cells with the morphology of mammalian dendritic cells have not been described in fishes, but analogous antigen-presenting and processing cells are postulated to occur based on evidence such as the relatively high expression levels of MHC II β chain mRNA in gills of Atlantic salmon (Koppang *et al.*, 1998). However, the precise sites of induction of mucosal immunity are unknown. Studies in sea bass have shown that immersion vaccination elicits large numbers of antibody-secreting cells in the gills without a concomitant response in the gut or systemic organs (Dos Santos *et al.*, 2001b). Similarly, it was shown in channel catfish that immersion vaccination in a soluble antigen elicited a mucosal antibody response without stimulating a serum antibody response (Lobb, 1987). These studies suggest that at some level, the development of the mucosal and systemic immune responses are partitioned, although it has been postulated that induction of an immune response at a particular mucosal site elicits stimulation in other remote mucosal tissues as well (Kawai *et al.*, 1981, Rombout *et al.*, 1989; Davidson *et al.*, 1993). Indeed, based on a number of studies in different fish species (St. Louis-Cormier *et al.*, 1984; Rombout *et al.*, 1989; Cain *et al.*, 2000; Maki and Dickerson, 2003), there clearly appears to be cellular communication between mucosal and systemic induction sites following immunisation at either place. For example, antibody containing lymphocytes were increased in the skin of rainbow trout following the intracoelomic injection (i.c.) of sheep erythrocytes (St. Louis-Cormier *et al.*, 1984). Similarly, i.c. injection of the major surface antigen of the parasite *I. multifiliis* in channel catfish elicits both serum and cutaneous antibodies (Maki and Dickerson, 2003). Pathways of migration of antigen-presenting cells and lymphocytes within epithelia and among these tissues and the pronephros and spleen are postulated to occur as described above for the intestinal MALT. Research is necessary to elucidate more precisely the sites and kinetics of induction following exposure to antigen at different sites. The various possible sites of antigen presentation are shown diagrammatically in Fig. 9.4.

Effector Mechanisms of Mucosal Adaptive Immunity

The effectors of adaptive immunity are antigen-specific antibodies and cytotoxic T lymphocytes, both of which exist in teleosts. While there is considerable experimental data regarding the molecular characterisation of antibodies and the kinetics of antibody expression, there is considerably less information available on antigen-specific cytotoxic T cell subsets (Nakanishi *et al.*, 2002a). Most experimental work on T cells has focused on lymphocytes isolated from peripheral blood, head kidney (pronephros), or spleen. Thus, the information presented below on the effector mechanisms of mucosal adaptive immunity is focused on mucosal antibodies, B cells and antibody-secreting plasma cells.

Mucosal Antibodies: The mucosal antibodies of mammals, which are predominantly dimeric molecules of the IgA isotype, are transported across epithelial layers to the mucosal surface by the polyclonal Ig receptor (pIgR) that binds the joining chain (J chain) of IgA and IgM molecules. A part of the pIg referred to as the secretory component is released together with the Ig into the mucosal secretions (Bogen, 2004).

In teleosts, the predominant antibody found in both mucus and blood is an IgM tetramer with a molecular mass ranging from 600-900 kDa, with each monomeric subunit consisting of two light chains (each light chain polypeptide ~ 25 kDa in size) and two heavy chains (each heavy chain polypeptide ~ 70 kDa in size) (Wilson and Warr, 2002). Although usually tetrameric in form under physiologic conditions, fish Ig has a degree of structural heterogeneity derived from non-uniform disulfide polymerisation of the monomeric or halfmeric (one light chain and one heavy chain) subunits (Kaattari et al., 1998; Bromage et al., 2004). This diversity is not related to isotypic differences (Bromage, 2005). Fish IgM is comparable to the pentameric mammalian IgM molecule with regard to heavy chain size, antigen affinity and avidity (Bromage, 2005).

J Chains and pIg receptors have not been reported in teleost fishes, except in one early study in a marine fish, Archosargus probatocephalus, in which a 95-kDa molecule was described covalently bound to the heavy chain of a dimeric Ig isolated from the cutaneous mucus (Lobb and Clem, 1981). Recent studies in puffer fish (Takifugu rubripes) (Hamuro et al., 2007) and carp (Rombout et al., 2008), however, suggest the expression of pIg receptors in skin and other mucosal tissues of teleosts, and a function in secretion of Ig.

Although tetrameric IgM is the most common antibody produced in vivo among different fish species and the only isotype shown to be an effector of protective immunity, new isotypes recently have been discovered. These include two transcribed μ genes in salmon, δ genes encoding IgD antibodies in salmon, channel catfish, cod and Japanese flounder, and ω and τ genes encoding IgZ and IgT in zebrafish and trout, respectively (Bromage, 2005). It has not yet been determined whether the functions of these isotypes occur in mucosal secretions.

Twenty years ago, a fundamental question that remained unanswered in fish was whether or not mucosal antibodies are produced locally in the mucosa or remotely in the head kidney and spleen (Lobb, 1987). Today, experimental evidence indicates that they are produced locally (Lobb and Clem, 1981a, b; Lobb, 1987; Rombout et al., 1993; Lin et al., 1996; Cain et al., 2000; Maki and Dickerson, 2003). For example, a localised cutaneous antibody response is generated against I. multifiliis, a protozoan parasite that infects the epithelial tissues of the skin and gills (Clark et al., 1992). Passive immunisation experiments with naïve channel catfish showed that mouse monoclonal antibodies (mAbs) against i-antigens confer protection against a lethal parasite challenge (Lin et al., 1996), but antibodies must be present at the site of infection. Antibody availability and function depended on the molecular size of the antibody, as mouse IgG, but not IgM, antibodies protected. Similarly, serum antibodies from actively immune fish, which are tetrameric IgM-like molecules of approximately 750,000 daltons (Wilson and Warr, 1992), also failed to protect following passive transfer into naïve animals, despite the fact that such antibodies strongly immobilise the parasite *in vitro* (Lin *et al.*, 1996). The ability to immobilise *in vitro* corresponds to protection *in vivo* (Clark *et al.*, 1995). These results indicate that antibodies must be present in the skin and presumably the gills where the parasite infects in order to afford protection.

In further studies using the *I. multifiliis* infection system, a two- to three-fold increase in IgM mRNA expression was demonstrated in skin at days 4 and 6 after *I. multifiliis* invasion, signifying an upregulation of Ig transcription in response to infection (Sigh *et al.*, 2004). These results suggest that antibodies are produced in the skin by resident antibody secreting cells (ASC). Additional experiments have shown directly that antibodies against *I. multifiliis* are produced in the skin (Xu and Klesius, 2003). Skin explants removed from immune fish, and placed into sterile tissue culture media, produced *I. multifiliis*-specific antibodies, which persisted for four days, suggesting cells in the skin actively produced that specific antibody. Cultures from skin explants of immune—but not control—fish contained antibodies that immobilised *I. multifiliis* and reacted with the predominant surface antigen on Western blots. In addition, similar experiments have shown that cutaneous antibodies against *F. columnare* are detected in cultures of skin explants from infected channel catfish, suggesting that antibodies also are involved in protective immunity against this bacterial pathogen (Shoemaker *et al.*, 2005).

While experimental evidence indicates that mucosal antibodies are produced locally, the extent to which they differ in structure and function to serum antibodies remains unclear. Research has shown that cutaneous mucosal antibodies are physically and immunologically identical or share similar molecular epitopes to those isolated from blood (Lobb and Clem, 1981, 1982; St. Louis-Cormier *et al.*, 1984; Itami *et al.*, 1988; Rombout *et al.*, 1993). Studies in carp using mAbs against purified Ig from mucus or serum, however, have revealed antigenic differences between cutaneous mucosal antibodies and serum antibodies (Rombout *et al.*, 1993). It has been suggested that alternate forms of Ig could be generated at mucosal surfaces that cannot be detected using current methods (Cain *et al.*, 2000; Bromage, 2006).

B Cell Differentiation in Mucosal Tissues: In mammals, differentiation of B cells is initiated by antigen presentation in secondary lymphoid tissues such as lymph nodes, mucosa-associated lymphoid tissue (MALT), and spleen. These lymphoid organs are organised to recruit naïve B and T lymphocytes from the blood and to promote their interaction with cognate antigen by activated antigen-presenting cells migrating to those sites from surrounding tissues. Once the lymphocytes have been activated and clonally expanded in centralised lymphoid organs, the resulting effector cells migrate to and localise in the infected or inflamed tissues. For example, B cells responding to respiratory pathogens first are detected in local lymph nodes draining the respiratory tract, and later are found in the lung (Moyron-Quiroz *et al.*, 2004).

Mucosal surfaces are particularly vulnerable to infection, as these epithelial surfaces are thin and permeable barriers to the interior of the body, and the vast majority of infectious agents invade through these routes. In mammals, mucosal-associated lymphoid tissue is organised to respond to pathogens invading through mucosal surfaces. In fish, the skin, gills and intestine comprise the major surface areas of the animal exposed directly to the environment and are, consequently, the site of entry of many pathogens. It is possible that lymphocytes and ASC directly underlying these surfaces serve as a primary site for antigen presentation to B cells, and consequently a site in which memory B cells differentiate and proliferate, facilitating rapid response to reinfection.

Affinity maturation of antibodies is a cornerstone of the acquired immune response in mammals, and an increase in affinity of IgG antibodies by several orders of magnitude results from clonal selection of B cells (Gourley *et al.*, 2004). In teleosts, only IgM is produced and class switching does not occur. Whether affinity maturation and somatic hypermutation (SHM) of IgM occurs in fishes has been debated, but recent reports clearly demonstrate that modest increases in antibody affinity occur for trout IgM and shark IgNAR following immunisation with model antigens (Cain *et al.*, 2002, Kaattari, 2002; Dooley, 2006). Sequence analysis of channel catfish heavy chain cDNAs has demonstrated SHM of both V_H and J_H encoded regions (Yang *et al.*, 2006). In mammals, activation-induced cytidine deaminase (AID) is an essential mediator of somatic hypermutation, class switch recombination, and gene conversion, all of which occur during the process of B cell differentiation and affinity maturation. AID is expressed exclusively in germinal centers and appears to be the only B-cell specific component required for these processes. It has been shown that AID is expressed in the skin of channel catfish suggesting that B cells may mature locally in the skin of this species

(Saunders and Magor, 2004). Undifferentiated B cells responsive to LPS stimulation have been isolated directly from the skin of channel catfish (Zhao *et al.*, 2008).

Immunological Memory and Mucosal Immunity (Fig. 1.4)

Activated B cells differentiate into populations of memory B cells and antibody secreting cells (ASC), which include plasmablasts, short-lived plasma cells and long-lived plasma cells. In mammals, long-lived plasma cells reside in the bone marrow, where they produce the majority of circulating serum antibodies (Manz *et al.*, 2002). Long-lived ASC may also occur in mucosal tissues (Etchart *et al.*, 2006). Long-lived plasma cells and memory B cells provide humoral immunological memory (Bernasconi *et al.*, 2002; Gourley *et al.*, 2004). Recent studies have provided evidence that subpopulations of antibody secreting lymphocytes, similar to those found in mammals, also occur in fishes (Bromage *et al.*, 2004). This work showed for the first time that long-lived ASCs reside in the head kidney of trout, and that these cells accumulate in this tissue and secrete antibody for as long as 35 weeks after immunisation. These cells are a source of serum antibodies. Such long-lived ASCs were not found in spleen or in the peripheral blood (PBL) population. Short-lived (i.e., weeks) plasma cells were found in both the spleen and head kidney.

As stated earlier, tissues comparable to mammalian lymph nodes do not exist in fishes, and other than the spleen and pronephros, the anatomical sites where B cells encounter foreign antigen have not been well defined (Bromage *et al.*, 2004). It has recently been shown that B cells in fish also have potent phagocytic and micobicidal activities, not observed in mammalian B cells (Jun *et al.*, 2006), suggesting that they play an even more central role in the initiation of immune responses than previously suspected. These findings raise questions as to where primary adaptive immune responses occur following infection, and where memory B cells and long-lived plasma cells are generated and ultimately reside. It is possible, although not yet tested, that the skin, gills and intestinal epithelia with their associated lymphoid tissues are the primary sites of antigen presentation to B cells for epithelial pathogens. They may not be the exclusive sites, however, as infection of the skin with *I. multifiliis* (as an example) leads to the production of antibodies in both the skin and serum, demonstrating that ASC localise to both skin and head kidney following infection (Maki and Dickerson, 2003). Nevertheless, it is possible that tissues directly underlying epithelial surfaces serve as sites for antigen presentation to B cells, and are consequently reservoirs for memory B and T cells, facilitating rapid response to re-infection, although as stated above, this remains to be tested.

Whether or not long-term humoral immunity in mucosa is provided by long-lived plasma cells remains an open question in mammals (Etchart *et al.*, 2006; Heipe and Radbruch, 2006). ASC residing in nasal mucosa contribute to both serum antibody as well as secretory mucosal IgA. Their longevity suggests that survival niches for plasma cells exist in mucosal tissue and that these ASC constitute a second set of long-lived plasma cells (not residing in the bone marrow) that contribute to humoral immunity at mucosal surfaces. It is possible that an analogous situation exists in fishes as well. For example, channel catfish immunised against *I. multifiliis* remain immune to surface infection for more than a year, suggesting that protective cutaneous antibodies are continually produced by resident, long-lived ASC (Burkart *et al.*, 1990; Zhao *et al.*, 2008).

MUCOSAL IMMUNITY AND VACCINES

A recent survey of the fish farming community indicates that commercially available vaccines against 15 bacterial diseases are used worldwide in aquaculture (Hastein *et al.*, 2005). The two main methods of vaccination are immersion and injection. Oral vaccination is less effective compared to the other methods, although an experimental method has recently been developed using a plant expression system that may increase the efficacy of this route (Companjen *et al.*, 2005). Immersion vaccination with inactivated bacteria or subunit antigens is used against the following bacterial diseases (Hastein *et al.*, 2005; Navot *et al.*, 2005): classical vibriosis (*Listonella anguillarum* or *Vibrio ordalii*) in sea bass, salmonids, catfish, ayu, and turbot; furunculosis (*Aeromonas salmonicida*) in salmonids, spotted sea wolf and goldfish; yersiniosis (*Yersinia ruckeri*) in salmonids, cyprinids, eels, sole and sturgeon; pasteurellosis (*Photobacterium damselae*) in sea bass and seabream; warm-water vibriosis (*Vibrio alginolyticus, V. parahaemolyticus, V. vulnificus*) in barramundi, grouper, sea bass, seabream, and snapper; edwardsiellosis (*Edwardsiella ictaluri*) in channel catfish; flavobacteriosis (*Flavobacterium columnare*) in salmonids; flexibacteriosis (*F. maritimus*) in salmonids and turbot; and streptococcosis (*Streptococcus iniae*) in rainbow trout, tilapia, turbot and yellowtail.

Viral vaccines licensed for aquaculture are all based on inactivated antigens in oil emulsions. Because the viruses or subunit components are non-replicating and non-infective, these vaccines are administered by injection (Biering et al., 2005). Antibodies are the primary response elicited following immunisation with these vaccines, which may not provide the most efficacious protection. Live attenuated viral vaccines comprise naturally occurring low-virulence isolates or virus that has been attenuated by other means. The advantage of these types of vaccines is that they can infect by natural routes and replicate in the host. Thus, they can be administered either by immersion or orally. The primary disadvantage is the risk of reversion by mutation to virulent forms (Biering et al., 2005). Currently, however, no viral vaccines for fishes are administered by immersion (Navot et al., 2005).

Vaccination by immersion has been used effectively to protect fishes against bacterial pathogens for many years, although the precise mechanisms of antigen uptake and protection remain unknown in many instances. It has been experimentally determined in some cases, however, that antigen passes through skin and gill epithelia directly or after hyperosmotic and/or ultrasound treatment to reach the blood and lymphoid tissues (Alexander et al., 1982; Ototake, 1996; Ototake et al., 1996; Moore et al., 1998; Navot et al., 2004). Ultrasound irradiation causes microscopic injuries to the skin (Navot et al., 2004, 2005), and it has been suggested that this treatment is comparable to intradermal immunisation, which in mammals is one of the most effective means of vaccination (Navot et al., 2005). Uptake also is enhanced following mild, controlled puncture or abrasion of the skin (Nakanishi et al., 2002b).

Immersion vaccines against pathogens that gain entry through the gill or skin epithelia have been effective when antibodies against surface antigens are elicited that block pathogen entry and colonisation. For example, immersion vaccines against *Photobacterium damselae* subspecies *piscicida* (formally *Pasteurella piscicida*) comprised of the over-expressed 97-kDa and 52-kDa bacterial proteins are effective with relative percentage of survival (RPS) rates of 50% when compared to controls (Barnes et al., 2005). Following immersion immunisation of sea bass, the gills were shown to be the primary sites for ASC, indicating that protective antibodies are produced (and perhaps stimulated) locally (Dos Santos et al., 2001b; Barnes et al., 2005). A non-commercial, experimental subunit vaccine comprised of the major surface antigen of *I. multifiliis* elicits a cutaneous antibody response and protective immunity against challenge (Wang and Dickerson, 2002; Wang et al., 2002).

Adjuvants and Delivery Methods That Enhance Mucosal Immunity: Adjuvants are compounds that aid immunity through accelerated, prolonged or enhanced responses to vaccine antigens. Although many different adjuvants have been tested in fish (mainly through trial and error), water-in-oil immersions in either mineral or non-mineral oils have proved to be the most successful in commercial aquaculture (Schijns and Tangeras, 2005). There is little information available in the literature on adjuvants and mucosal immunity, however. Approaches used in the human vaccinology field include the use of toll-like receptor agonists (e.g., CpG motifs and gylcans), as well as immunostimulants (e.g., cytokines and co-stimulatory molecules such as interleukin) (Toka *et al.*, 2004). ADP-ribosylating toxins have been used as effective mucosal adjuvants in higher vertebrates, but have not yet been tested or established as mucosal adjuvants in fishes.

A number of treatments (hypo- and hyper-osmotic baths, scarification of skin surfaces, ultrasound irradiation and combinations of hyperosmotic dips and ultrasound irradiation) have been used in combination with antigen immersion to enhance mucosal immune responses. These have been referenced in the preceding section.

SUMMARY

The epithelia of the intestine, skin and gills are critical barriers to infection by pathogens. These tissues comprise dynamic organs that provide protection through physical, chemical and physiological mechanisms. Although substantial progress has been made over the last 20 years in elucidating the innate and acquired mechanisms of mucosal immunity, much more research remains to be done in order to fully understand the processes of protection at mucosal surfaces. For instance, fundamental questions remain unanswered regarding the mechanisms of acquired mucosal immunity, such as absorption of antigen and location of sites where antigen is processed and presented by phagocytic cells. Likewise, it is still unknown where mucosal B and T lymphocytes interact, proliferate and differentiate. Relatively few cell-signaling molecules such as cytokines and chemokines have been identified. Lymphocytes and antibody secreting plasma cells have been described in epithelia, but the extent to which phagocytes and lymphocytes traffic between peripheral (mucosal) and central (pronephros and spleen) tissues is largely undetermined. Questions remain regarding the mechanisms of immune and humoral memory in teleosts, the diaspora of differentiated memory T and B cells, and how immune memory functions in relation to mucosal immunity. Finally, basic questions such as how antibodies produced at mucosal sites are translocated across intact epithelial cell layers still remain unanswered.

Research on the above questions and others regarding the biology of teleost mucosal immunity will provide basic answers of fundamental importance on the evolution of the vertebrate immune system. In addition, continued research on mucosal immunity is important in the development of new vaccines and delivery technologies that are critically needed for burgeoning worldwide aquaculture industries.

Acknowledgements

The author thanks Dr Al Camus (Department of Veterinary Pathology, College of Veterinary Medicine, University of Georgia) for providing the photomicrographs in Figures 1–3, and Mr Kip Carter (Educational Resources, College of Veterinary Medicine, University of Georgia) for rendering the artistic diagrams in Figures 1–3.

References

Alexander, J.B. and G.A. Ingram. 1992. Noncellular nonspecific defense mechanisms of fish. *Annual Review of Fish Diseases* 2: 249-279.

Alexander, J.B., A. Bowers, G.A. Ingram and S. M. Shamshoom. 1982. The portal of entry of bacteria into fish during hyperosmotic infiltration and the fate of antigens. *Developmental and Comparative Immunology* 2(Supplement): 41-46.

Armstrong, P. B. and J. P. Quigley. 1999. α2 macroglobulin: An evolutionarily conserved arm of the innate immune system. *Developmental and Comparative Immunology* 23: 375-390.

Barnes, A.C., N.M.S. dos Santos and A.E. Ellis. 2005. Update on bacterial vaccines: *Photobacterium damselae* supsp. *piscicida*. In: *Developments in Biologicals — Progress in Fish Vaccinology*, P.J. Midtlyng (ed.). S. Karger, Basel, Vol. 121, pp. 75-84.

Bernard, D., A. Six, L. Rigottier-Gois, S. Messiaen, S. Chilmonczk, E. Quillet, P. Boudinot and A. Benmansour. 2006. Phenotypic and functional similarity of gut intraepithelial and systemic T cells in a teleost fish. *Journal of Immunology* C176: 3942-3949.

Bernasconi, N.L., E. Traggiai and A. Lanzavecchia. 2002. Maintenance of serological memory by polyclonal activation of human memory B cells. *Science* 298: 2199-2202.

Bevins, C.L. 1994. *Antimicrobial Peptides as Agents of Mucosal Immunity, Antibacterial Peptides*. Jonn Wiley & Sons, Chichester, Vol. 186, pp. 260-261.

Bielek, E. 2005. Development of the endoplasmic reticulum in the rodlet cell of two teleost species. *Anatomical Record* A283: 239-249.

Biering, E., S. Villoing, I. Sommerset and K. E. Christie. 2005. Update on viral vaccines for fish. In: *Developments in Biologicals — Progress in Fish Vaccinology*, P.J. Midtlyng (ed.). S. Karger, Basel, Vol. 121, pp. 97-113.

Bogen, S.A. 2004. Organs and tissues of the immune system. In: *Immunology, Infection and Immunity*, G.B. Pier, J.B. Lyczak and L.M. Wetzler (eds.). ASM Press, Washington, D.C., pp. 67-84.

Boshra, H., J. Li and J.O. Sunyer. 2006. Recent advances on the complement system of teleost fish. *Fish and Shellfish Immunology* 20: 239-262.

Braun, R., J.A. Arnesen, A. Rinne and K. Hjelmeland. 1990. Immunohistological localization of trypsin in mucus-secreting cell layers of Atlantic salmon, *Salmo salar* L. *Journal of Fish Diseases* 13: 233-238.

Bromage, E.R., I.M. Kaattari, P. Zwollo and S.L. Kaattari. 2004. Plasmablast and plasma cell production and distribution in trout immune tissues. *Journal of Immunology* 173: 7317-7323.

Bromage, E.S. 2006. Antibody structural variation in rainbow trout fluids. *Comparative Biochemistry and Physiology* 143: 61-69.

Bromage, E.S., J. Ye, L. Owens, I.M. Kaattari and S. Kaattari. 2004. Use of staphylococcal protein A in the analysis of teleost immunoglobulin structural diversity. *Developmental and Comparative Immunology* 28: 803-814.

Buchmann, K. 1998. Binding and lethal effect of complement from *Oncorhynchus mykiss* on *Gyrodactylus derjavini* (Platyhelminthes Monogenea). *Diseases of Aquatic Organisms* 32: 195.

Buchmann, K. 1999. Immune mechanisms in fish skin against monogeneans — A model. *Folia Parasitologica* 46: 1-9.

Buchmann, K., J. Sigh, C.V. Nielsen and M. Dalgaard. 2001. Host responses against the fish parasitizing ciliate *Ichthyophthirius multifiliis*. *Veterinary Parasitology* 100: 105-116.

Burkart, M.A., T.G. Clark and H.W. Dickerson. 1990. Immunization of channel catfish *Ictalurus punctatus* Rafinesque, against *Ichthyophthirius multifiliis* (Fouquet): killed versus live vaccines. *Journal of Fish Diseases* 13: 401.

Cain, K.D., D.R. Jones and R.L. Raison. 2000. Characterization of mucosal and systemic immune responses in rainbow trout (*Oncorhynchus mykiss*) using surface plasmon resonance. *Fish and Shellfish Immunology* 10: 651-666.

Cain, K.D., D.R. Jones and R.L. Raison. 2002. Antibody-antigen kinetics following immunization of rainbow trout (*Oncorhynchus mykiss*) with a T-cell dependent antigen. *Developmental and Comparative Immunology* 26: 181-190.

Cheroutre, H. 2004. Starting at the beginning: New perspectives on the biology of mucosal T cells. *Annual Review of Immunology* 22: 217-246.

Clark, T.G., R.A. McGraw and H.W. Dickerson. 1992. Developmental expression of surface antigen genes in the parasitic ciliate *Ichthyophthirius multifiliis*. *Proceedings of the National Academies of Sciences of the United States of America* 89: 6363-6367.

Clark, T.G., T. Lin and H.W. Dickerson. 1995. Surface immobilization antigens of *Ichthyophthirius multifiliis*: Their role in protective immunity. *Annual Review of Fish Diseases* 5: 113.

Cole, A.M., P. Weis and G. Diamond. 1997. Isolation and characterization of pleurocidin, an antimicrobial peptide in the skin secretions of winter flounder. *Journal of Biological Chemistry* 272: 12008-12013.

Companjen, A.R., D.E.A. Florack, J.H.M.W. Bastiaans, C.I. Matos, D. Bosch and J.W. Rombout. 2005. Development of a cost-effective oral vaccination method against viral disease in fish. In: *Developments in Biologicals — Progress in Fish Vaccinology*, P.J. Midtlyng (ed.). S. Karger, Basel, Vol. 121, pp. 143-150.

Cross, M.L. and R.A. Matthews. 1993. Localized leukocyte response to *Ichthyophthirius multifiliis* establishment in immune carp *Cyprinus carpio* L. *Veterinary Immunology and Immunopathology* 38: 341.

Dalmo, R.A., K. Ingebrightsen and J. Bogwald. 1997. Non-specific defense mechanisms in fish, with particular reference to the reticuloendothelial system (RES). *Journal of Fish Diseases* 20: 241-273.

Danilova, N. and L.A. Steiner. 2002. B cells develop in the zebrafish pancreas. *Proceedings of the National Academies of Science of the United States of America* 99: 13711-13716.

Davidson, G.A., A.E. Ellis and C.J. Secombes. 1991. Cellular responses of leucocytes isolated from the gut of rainbow trout, *Oncorhynchus mykiss* (Walbaum). *Journal of Fish Diseases* 14: 651-659.

Davidson, G.A., A.E. Ellis and C.J. Secombes. 1993. Route of immunization influences the generation of antibody secreting cells in the gut of rainbow trout, *Oncorhynchus mykiss* (Walbaum, 1792). *Developmental and Comparative Immunology* 17: 373-376.

Davidson, G.A., S.-H. Lin, C.J. Secombes and A.E. Ellis. 1997. Detection of specific and 'constitutive' antibody secreting cells in the gills, head kidney and peripheral blood leucocytes of the dab (*Limanda limanda*). *Veterinary Immunology and Immunopathology* 58: 363-374.

Davina, J.H.M., H.K. Parmentier and L.P.M. Timermans. 1982. Effect of oral administration of Vibrio bacterin on the intestine of cyprinid fish. *Developmental and Comparative Immunology* 2: 157-166.

Dezfuli, B.S., S. Capuano and M. Manera. 1998. A description of rodlet cells from the alimentary canal of *Anguilla anguilla* and their relationship with parasitic helminths. *Journal of Fish Biology* 53: 1084-1095.

Doggett, T.A. and J.E. Harris. 1991. Morphology of the gut associated lymphoid tissue of *Oreochromis mossambicus* and its role in antigen absorption. *Fish and Shellfish Immunology* 1: 213-227.

Dooley, H. 2006. First molecular and biochemical analysis of in vivo affinity maturation in an ectothermic vertebrate. *Proceedings of the National Academies of Sciences of the United States of America* 103: 1846-1851.

Dorin, D., P. Martin, M.F. Sire, J. Smal and J.M. Vernier. 1993. Protein uptake by intestinal macrophages and eosinophilic granulocytes in trout: an in vivo study. *Biology of the Cell* 79: 37-44.

Dos Santos, N.M.S., J. J. Taverne-Thiele, A.C. Barnes, A.E. Ellis and J.H.W. M. Rombout. 2001a. Kinetics of juvenile seas bass (*Dicentrarchus labrax* L.) systemic and mucosal antibody secreting cell response to different antigens (*Photobacterium damselae* spp. piscicida, *Vibrio anguillarum* and DNP. *Fish and Shellfish Immunology* 11: 317-331.

Dos Santos, N.M., J.J. Taverne-Thiele, A.C. Barnes, W.B. Van Muiswinkel, A.E. Ellis and J.H.W.M. Rombout. 2001b. The gill is a major organ for antibody secreting cell production following direct immersion of sea bass (*Dicentrarchus labrax* L.) in a *Photobacterium damselae* ssp. *piscicida* bacterin: An ontogenetic study. *Fish and Shellfish Immunology* 11: 65-74.

Du Pasquier, L. and G.W. Litman (ed.). 2000. *Origin and Evolution of the Vertebrate Immune System*. Springer-Verlag, Berlin, Vol. 248: 93.

Egberts, H.J.A., M.G.M. Brinkhoff, J.M.V.M. Mouven, J.E. van Dijk and J.F.J.G. Koninkz. 1985. Biology and pathology of the intestinal M cell. A review. *Veterinary Quarterly* 7: 333-336.

Ellis, A.E. 2001. Innate host defense mechanisms of fish against viruses and bacteria. *Developmental and Comparative Immunology* 25: 827-839.

Ellis, A.E., T.S. Hastings and A.L.S. Munro. 1981. The role of *Aeromonas salmonicida* extracellular products in the pathology of furunculosis. *Journal of Fish Diseases* 4: 41-52.

Etchart, N., B. Beaten, S.R. Andersen, L. Hyland, S.Y. Wang and S. Hou. 2006. Intranasal immunization with inactivated RSV and bacterial adjuvants induces mucosal protection and abrogates eosinophilia upon challenge. *European Journal of Immunology* 36: 1136-1144.

Flajnik, M.F. 1998. Churchill and the immune system of ectothermic vertebrates. *Immunological Reviews* 166: 5-14.

Fletcher, T.C. 1981. The identification of nonspecific humoral factors in the plaice (*Pleuronectes platessa* L). *Development of Biological Standards* 49: 321-327.

Georgopoulou, U. and J.M. Vernier. 1986. Local immunological response in the posterior intestinal segment of the rainbow trout after oral administration of macromolecules. *Developmental and Comparative Immunology* 10: 529-537.

Goldes, S.A., H.W. Ferguson, P.Y. Daoust and R.D. Moccia. 1986. Phagocytosis of the inert suspended clay kaolin by the gills of rainbow trout, *Salmo gairdneri* Richardson. *Journal of Fish Diseases* 9: 147-152.

Gonzalez, S.F., K. Buchmann and M.E. Nielsen. 2007a. Complement expression in common carp (*Cyprinus carpio* L.) during infection with *Ichthyophthirius multifiliis*. *Developmental and Comparative Immunology* 31: 576-586.

Gonzalez, S.F., N. Chatziandreou, M.E. Nielsen, W. Li, J. Rogers, R. Taylor, Y. Santos and A. Cossins. 2007b. Cutaneous immune responses in the common carp detected using transcript analysis. *Molecular Immunology* 44: 1675-1690.

Gourley, T.S., E.J. Wherry, D. Masopust and R. Ahmed. 2004. Generation and maintenance of immunological memory. *Seminars in Immunology* 16: 323-333.

Graham, S. and C.J. Secombes. 1990. Do fish lymphocytes secrete interferon γ? *Journal of Fish Biology* 36: 563-573.

Graves, S.S., D.L. Evans and D.L. Dawe. 1985. Mobilization and activation of nonspecific cytotoxic cells (NCC) in the channel catfish (*Ictalurus punctatus*) infected with *Ichthyophthirius multifiliis*. *Comparative Immunology Microbiology and Infectious Diseases* 8: 43-51.

Grizzle, J.M. and W.A. Rogers. 1976. *Anatomy and Histology of the Channel Catfish*, Auburn University, Agricultural Experiment Station, Auburn, AL, USA.

Grove, S., R. Johansen, L.J. Reitan and C.M. Press. 2006. Immune- and enzyme histochemical characterization of leukocyte populations within lymphoid and mucosal tissues of Atlantic halibut (*Hippoglossus hippoglossus*). *Fish and Shellfish Immunology* 20: 693-708.

Harris, P.D., A. Soleng and T.A. Bakke. 1998. Killing of *Gyrodactylus salaris* (Platyhelminthes, Monogenea) mediated by host complement. *Parasitology* 117: 137-143.

Hastein, T., R. Gudding and O. Evensen. 2005. Bacterial vaccines for fish — an update of the current situation worldwide. In: *Developments in Biologicals Progress in Fish Vaccinology*, P.J. Midtlyng (ed.). S. Karger, Basel, Vol. 121, pp. 55-74.

Hebert, P., A.J. Ainsworth and B. Boyd. 2002. Histological enzyme and flow cytometric analysis of channel catfish intestinal tract immune cells. *Developmental and Comparative Immunology* 26: 53-62.

Heipe, F. and A. Radbruch. 2006. Is long-term humoral immunity in the mucosa provided by long-lived plasma cells? A question still open. *European Journal of Immunology* 36: 1136-1144.

Helgeland, L., E. Dissen, K.Z. Dai, T. Midtvedt, P. Brandtzaeg and J.T. Vaage. 2004. Microbial colonization induces oligoclonal expansions of intraepithelial CDA T cells in the gut. *European Journal of Immunology* 34: 3389-3400.

Hjelmeland, K., M. Christie and J. Raa. 1983. Skin mucus protease from rainbow trout, *Salmo gairdneri* Richardson, and its biological significance. *Journal of Fish Biology* 23: 13-22.

Hamuro, K., H. Suetake, N.R. Saha, K. Kikuchi and Y. Suzuki. 2007. A teleost polymeric Ig receptor exhibiting two Ig-like domains transports tetrameric IgM into the skin. *Journal of Immunology* 178: 5682-5686.

Hunt, T.C. and A.F. Rowley. 1986. Leukotriene B4 induces enhanced migration of fish leukocytes in vitro. *Immunology* 59: 563-568.

Huttenhuis, H.B.T., N. Romano, C.N. Van Oosterhoud, A.J. Taverne-Thiele, L. Mastrolia, W.B. Van Muiswinkel and J.H.W.M. Rombout. 2006. The ontogeny of mucosal immune cells in common carp (*Cyprinus carpio* L.). *Anatomy and Embryology* 211: 19-29.

Iger, Y. and A.F. Wendelaar Bonga. 1994. Cellular aspects of he skin of carp exposed to acidified water. *Cell and Tissue Research* 275: 481-492.

Ingram, G.A. 1980. Substances involved in the natural resistance of fish to infection — A review. *Journal of Fish Biology* 16: 23-60.

Itami, T., Y. Takahashi, T. Okamoto and K. Kubono. 1988. Purification and characterization of immunoglobulin in skin mucus and serum of ayu. *Nippon Suisan Gakkaishi* 54: 1611-1617.

Jun, L., D.R. Barreda, Y.A. Zhang, H. Boshra, A.E. Gelman, S. LaPatra, L. Tort and J.O. Sunyer. 2006. B lymphocytes from early vertebrates have potent phagocytic and microbicidal abilities. *Nature Immunology* 7: 1116-1124.

Kaattari, S., D. Evans and J. Klemer. 1998. Varied redox forms of teleost IgM: An alternative to isotype diversity? *Immunological Reviews* 166: 133-142.

Kaattari, S.L. 2002. Affinity maturation in trout: Clonal dominance of high affinity antibodies late in the immune response. *Developmental and Comparative Immunology* 26: 191-200.

Kawai, K., R. Kusudu and T. Itami. 1981. Mechanisms of protection in ayu orally vaccinated for vibriosis. *Fish Pathology* 15: 257-262.

Kim, S.K., D.S. Reed, S. Olson, M.J. Schnell, J.K. Rose. 1998. Generation of mucosal cytotoxic T cells against soluble protein by tissue-specific environmental co-stimulatory signals. *Proceedings of the National Academies of Sciences of the United States of America* 95: 10814-10819.

Koch, S., K. Kohl, E. Klein, D. von Bubnoff and T. Bieber. 2006. Skin homing of Langerhans cell precursors: Adhesion, chemotaxis, and migration. *Journal of Allergy and Clinical Immunology* 177: 163-168.

Koppang, E.O., M. Lundin, C.M. Press, K. Ronningen and O. Lie. 1998. Differing levels of MHC class II β chain expression in a range of tissues from vaccinated and non-vaccinated Atlantic salmon (*Salmo salar* L.). *Fish and Shellfish Immunology* 8: 183-196.

Kupper, T.S. 2000. T cells, immunosurveillance, and cutaneous immunity. *Journal of Dermatological Science* 24: S41-S45.

Lefrancois, L. and L. Puddington. 1995. Extrathymic intestinal T-cell development: Virtual reality? *Immunology Today* 16: 16-21.

Leknes, I.L. 2002. Uptake of foreign ferritin in platy *Xiphophorus maculatus* (Poeciliidae: Teleostei). *Diseases of Aquatic Organisms* 51: 233-237.

Leong, J.A.C., G.D. Tobridge, K.C.H.Y., M. Johnson and B. Simon. 1998. Interferon-inducible Mx proteins in fish. *Immunological Reviews* 166: 349-363.

Lin, S.H., G.A. Davidson, C.J. Secombes and A.E. Ellis. 1998. A morphological study of cells isolated from the perfused gill of dab and Atlantic salmon. *Journal of Fish Biology* 53: 560-568.

Lin, T.L., T.G. Clark and H. Dickerson. 1996. Passive immunization of channel catfish (*Ictalurus punctatus*) against the ciliated protozoan parasite *Ichthyophthirius multifiliis* by use of murine monoclonal antibodies. *Infection and Immunity* 64: 4085-4090.

Lobb, C.J. 1987. Secretory immunity induced in channel catfish, *Ictalurus punctatus*, following bath immunization. *Developmental and Comparative Immunology* 11: 727-738.

Lobb, C.J. and L.W. Clem. 1981a. The metabolic relationship of the immunoglobulins in fish serum, cutaneous mucus, and bile. *Journal of Immunology* 127: 1525-1529.

Lobb, C.J. and L.W. Clem. 1981b. Phylogeny of immunoglobulin structure and function. XII. Secretory immunoglobulins in bile of the marine teleost, *Archosargus probatocephalus*. *Molecular Immunology* 18: 615-619.

Lobb, C.J. and L.W. Clem. 1982. Fish lymphocytes differ in the expression of surface immunoglobulin. *Developmental and Comparative Immunology* 6: 473-479.

Lumsden, J.S., V.E. Ostland, D.D. MacPhee and H.W. Ferguson. 1995. Production of gill-associated and serum antibody by rainbow trout (*Oncorhynchus mykiss*) following immersion immunization with acetone killed *Flavobacterium branchophilium* and the relationship to protection from experimental challenge. *Fish and Shellfish Immunology* 5: 151-165.

Magnadottir, B. 2006. Innate immunity of fish (Overview). *Fish and Shellfish Immunology* 20: 137-151.

Maki, J.L. and H.W. Dickerson. 2003. Systemic and cutaneous mucus antibody responses of channel catfish immunized against the protozoan parasite *Ichthyophthirius multifiliis*. *Clinical and Diagnostic Laboratory Immunology* 10: 876-881.

Manera, M. and B.S. Dezfuli. 2004. Rodlet cells in teleosts: A new insight into their nature and function. *Journal of Fish Biology* 65: 597-619.

Manz, R.A., A. Sergio, C. Giuliana, A.E. Hauser, F. Hiepe and A. Radbruch. 2002. Humoral immunity and long-lived plasma cells. *Current Opinion in Immunology* 14: 517-521.

Matsunaga, T. 1998. Did the first adaptive immunity evolve in the gut of ancient jawed fish? *Cytogenetics and Cell Genetics* 80: 138-141.

Matsunaga, T. and A. Rahman. 1998. What brought the adaptive immune response to vertebrates? The jaw hypothesis and the seahorse. *Immunological Reviews* 166: 177-186.

McLean, E. and E.M. Donaldson. 1990. Absorption of bioactive proteins by the gastrointestinal tract of fish: A review. *Journal of Aquatic Animal Health* 2: 1-11.

McMillan, D.N. and C.J. Secombes. 1997. Isolation of rainbow trout (*Oncorhynchus mykiss*) intestinal intraepithelial lymphocytes (IEL) and measurement of their cytotoxic activity. *Fish and Shellfish Immunology* 7: 527-541.

Miller, N., M. Wilson, E. Bengten, T. Stuge, G. Warr and W. Clem. 1998. Functional and molecular characterization of teleost leukocytes. *Immunological Reviews* 166: 187-197.

Mittal, A.K., M. Whitear and S.K. Agarwal. 1980. Fine structure and histochemistry of the epidermis of the fish *Monopterus cuchia. Journal of Zoology* 191: 107-125.

Moore, J.D., M. Ototake and T. Makanishi. 1998. Particulate antigen uptake during immersion immunization of fish: The effectiveness of prolonged exposure and the roles of skin and gill. *Fish and Shellfish Immunology* 8: 393-407.

Moyron-Quiroz, J.E., J. Rangel-Moreno, K. Kusser, L. Hartson, F. Sprague, S. Goodrich, D.L. Woodland, F.E. Lund and T.D. Randell. 2004. Role of inducible bronchus associated lymphoid tissue (iBALT) in respiratory immunity. *Nature Medicine* 10: 927-934.

Mulero, I., E. Chaves-Pozo, A. Garcia-Alcazar, J. Meseguer, V. Mulero and A.G. Ayala. 2007. Distribution of the professional phagocytic granulocytes of the bony fish gilthead bream (*Sparus aurata* L.) during the ontogeny of lymphomyeloid organs and pathogen entry sites. *Developmental and Comparative Immunology* 31: 1024-1033.

Murray, C.K. and T.C. Fletcher. 1976. The immunohistochemical localization of lysozyme in plaice (*Pleuronectes Platessa* L.) tissues. *Journal of Fish Biology* 9: 329-334.

Nakanishi, T. and M. Ototake. 1996. Antigen uptake and immune responses after immersion vaccination. Developments in Biological Standardization — Progress in Fish. *Vaccinology* 90: 59-68.

Nakanishi, T., I. Kiryu and M. Ototake. 2002a. Development of a new vaccine delivery method for fish: Percutaneous administration by immersion with application of a multiple puncture instrument. *Vaccine* 20: 3764-3769.

Nakanishi, T., U. Fischer, J.M. Dijkstra, S. Hasegawa, T. Somamoto, N. Okamoto and M. Ototake. 2002b. Cytotoxic T cell function in fish. *Developmental and Comparative Immunology* 26: 131-139.

Nam, B.H., I. Hlono and T. Aoki. 2003. The four TCR genes of teleost fish: The cDNA and genomic DNA analysis of Japanese flounder (*Paralichthys olivaceus*) TCR alpha-beta-, gamma-, and delta-chains. *Journal of Immunology* 170: 3081-3090.

Navot, N., E. Kimmel and R.R. Avtalion. 2004. Enhancement of antigen uptake and antibody production in goldfish (*Carassius auratus*) following bath immunization and ultrasound treatment. *Vaccine* 22: 2660-2666.

Navot, N., E. Kimmel and R.R. Avtalion. 2005. Immunization of fish by bath immersion using ultrasound. In: *Developments in Biologicals — Progress in Fish Vaccinology*, P.J. Midtlyng (ed.). S. Karger, Basel, Vol. 121, pp. 135-142.

McMillan, D.N. and C.J. Secombes. 1997. Isolation of rainbow trout (*Oncorhynchus mykiss*) intestinal intraepithelial lymphocytes (IEL) and measurement of their cytotoxic activity. *Fish and Shellfish Immunology* 7: 527-541.

Neuman, N.F., J.L. Stafford, D. Barreda, A.J. Ainsworth and M. Belosevic. 2001. Antimicrobial mechanisms of fish phagocytes and their role in host defense. *Developmental and Comparative Immunology* 25: 807-825.

Neutra, M.R., E. Pringault and J.P. Kraehenbuhl. 1996. Antigen sampling across epithelial barriers and induction of mucosal immune responses. *Annual Review of Immunology* 14: 275-300.

Otake, T., J. Hirokawa, H. Fujimoto and K. Imaizumi. 1995. Fine structure and function of the gut epithelium of pike eel larvae. *Journal of Fish Biology* 47: 126-142.

Ototake, M., G. Iwama and T. Nakanishi. 1996. The uptake of bovine serum albumin by the skin of bath immunized rainbow trout *Oncorhynchus mykiss*. *Fish and Shellfish Immunology* 6: 321-333.

Pannetier, C., J. Even and P. Kourilsky. 1995. T-cell repertoire diversity and clonal expansions in normal and clinical samples. *Immunology Today* 16: 176-181.

Peleteiro, M.C. and R.H. Richards. 1990. Phagocytic cells in the epidermis of rainbow trout, *Salmo gairdneri* Richardson. *Journal of Fish Diseases* 13: 225-232.

Picchietti, S., L. Terribili, G. Mastrolia and L. Abelli. 1997. Expression of lymphocyte antigenic determinants in developing gut-associated lymphoid tissue of the sea bass *Dicentrarchus labrax* (L.). *Anatomy and Embryology* 196: 177-186.

Picchietti, S., M. Mazzini, A.R. Taddei, R. Renna, A.M. Fausto, V. Mulero, O. Carnevali, A. Cresci and L. Abelli. 2007. Effects of administration of probiotic strains on GALT of larval gilthead seabream: Immunohistochemical and ultrastructural studies. *Fish and Shellfish Immunology* 22: 57-67.

Powell, M.D., G.M. Wright and J.F. Burka. 1990. Eosinophilic granule cells in the gills of rainbow trout, *Oncorhynchus mykiss*: Evidence for migration? *Journal of Fish Biology* 37: 495-497.

Powell, M.D., G.M. Wright and J.F. Burka. 1991. Degranulation of eosinophilic granule cells induced by capsaicin and substance P in the intestine of the rainbow trout (*Oncorhynchus mykiss* Walbaum). *Cell and Tissue Research* 266: 469-474.

Press, C.M. and O. Evensen. 1999. The morphology of the immune system in teleost fishes. *Fish and Shellfish Immunology* 9: 309-318.

Puri, N., E.H. Weyand, S. Abdel-Rhaman and P.J. Sinko. 2000. An investigation of the intradermal route as an effective means of immunization for microparticulate vaccine delivery system. *Vaccine* 18: 2600-2612.

Regnault, A., A. Cumano, P. Vassalli, D. Guy-Grand and P. Kourilsky. 1994. Oligoclonal repertoire of the CD8 $\alpha\alpha$ and the CD8 $\alpha\beta$ TCR-$\alpha\beta$ murine intestinal intraepithelial T lymphocytes: Evidence for the random emergence of T cells. *Journal of Experimental Medicine* 180: 1345-1358.

Reite, O.B. 1997. Mast cells/eosinophilic granule cells of salmonids: staining properties and responses to noxious agents. *Fish and Shellfish Immunology* 7: 567-584.

Reite, O.B. 2005. The rodlet cells of teleostan fish: their potential role in host defense in relation to the role of mast cells/eosinophilic granule cells. *Fish and Shellfish Immunology* 19: 253-267.

Reite, O.B. and O. Evensen. 2006. Inflammatory cells of teleostean fish: A review focusing on mast cells/eosinophilic granule cells and rodlet cells. *Fish and Shellfish Immunology* 20: 192-208.

Roberts, R.J. 2001. *Fish Pathology*. W. B. Saunders, London. 3rd Edition

Robertson, B. 2006. The interferon system of teleost fish. *Fish and Shellfish Immunology* 20: 172-191.

Rombout, J.H.W.M. and A.A. van den Berg. 1989. Immunological importance of the second gut segment of carp. I. Uptake and processing of antigens by epithelial cells and macrophages. *Journal of Fish Biology* 35: 13-22.

Rombout, J.W., C.H. Lamers, M.H. Helfrich, A. Dekker and J.J. Taverne-Thiele. 1985a. Uptake and transport of intact macromolecules in the intestinal epithelium of carp (*Cyprinus carpio* L.) and the possible immunological implications. *Cell and Tissue Research* 239: 519-530.

Rombout, J.W., C.H.J. Lamers, M.H. Helfrich, A. Dekker and J. J. Taverne-Thiele. 1985b. Uptake and transport of intact macromolecules in the intestinal epithelium of carp (*Cyprinus carpio* L.) and the possible immunological implications. *Cell and Tissue Research* 239: 519-530.

Rombout, J.W., L.J. Block, C.H. Lamers and E. Egberts. 1986. Immunization of carp (*Cyprinus carpio*) with *Vibrio anguillarum* bacteria: Indications for a common mucosal immune system. *Developmental and Comparative Immunology* 10: 341-351.

Rombout, J.H.W.M., H.E. Bot and J. Taverne-Thiele. 1989. Immunological importance of the second gut segment of carp. II. Characterization of mucosal lymphocytes. *Journal of Fish Biology* 35: 167-178.

Rombout, J.H.W.M., A.A. van den Berg, C.T. G.A. van den Berg, P. Witte and E. Egberts. 1989. Immunological importance of the second gut segment of carp III. Systemic and/or mucosal immune responses after immunization with soluble or particulate antigen. *Journal of Fish Biology* 35: 179-189.

Rombout, H.W.M., N. Taverne, M. van de Kamp and A.J. Taverne-Theile. 1993a. Differences in mucus and serum immunoglobulin of carp (*Cyprinus carpio* L.). *Developmental and Comparative Immunology* 17: 309-317.

Rombout, J.H.W.M., A.J. Taverne-Thiele and M. Villena. 1993b. The gut-associated lymphoid tissue (GALT) of carp (*Cyprinus carpio* L.): An immunocytochemical analysis. *Developmental and Comparative Immunology* 17: 55-66.

Rombout, J.H.W.M., P.H.M. Joosten, M.Y. Engelsma, A.P. Vos, N. Taverne and J.J. Taverne-Thiele. 1998. Indications for a distinct putative T cell population in mucosal tissue of carp (*Cyprinus carpio* L.). *Developmental and Comparative Immunology* 22: 63-77.

Rombout, J.H.W.M., S.J.L. van der Tuin, G. Yang, N. Schopman, A. Mroczek, T. Hermsen and J.J. Taverne-Thiele. 2008. Expression of the polymeric Immunoglobulin Receptor (pIgR) in mucosal tissues of common carp (*Cyprinus carpio* L.). *Developmental and Comparative Immunology* 24: 620-628.

Rowden, G., T.M. Phillips and M.G. Lewis. 1979. Ia antigens on indeterminate cells of the epidermis: Immunoelectron microscopic studies of surface antigens. *British Journal of Dermatology* 100: 531-542.

Saunders, H.L. and B.G. Magor. 2004. Cloning and expression of the AID gene in the channel catfish. *Developmental and Comparative Immunology* 28: 657-663.

Scapigliati, G., N. Romano, L. Abelli, S. Meloni, A.G. Ficca, F. Buonocore, S. Bird and C.J. Secombes. 2000. Immunopurification of T-cells from sea bass *Dicentrarchus labrax* (L.). *Fish and Shellfish Immunology* 10: 329-341.

Schijns, V.E.J.C. and A. Tangeras. 2005. Vaccine adjuvant technology: From theoretical mechanisms to practical approaches. In: *Developments in Biologicals*, P.J. Midtlyng (ed.). Karger, Basel, Vol. 121, pp. 127-134.

Secombes, C.J. and T.C. Fletcher. 1992. The role of phagocytes in the protective mechanisms of fish. *Annual Review of Fish Diseases* 2: 53-71.

Sharp, G.J.E., A.W. Pike and C.J. Secombes. 1991. Leucocyte migration in rainbow trout (*Oncorhynchus mykiss* [Walbaum]): Optimization of migration conditions and responses to host and pathogen (*Diphyllobothrium dendriticum* [Nitzsch]) derived chemoattractants. *Developmental and Comparative Immunology* 15: 295-305.

Shoemaker, C.A., D.H. Xu, R.A. Shelby and P.H. Klesius. 2005. Detection of cutaneous antibodies against *Flavobacterium columnare* in channel catfish, *Ictalurus punctatus* (Rafinesque). *Aquaculture Research* 36: 813-818.

Sigh, J., T. Lindenstrom and K. Buchmann. 2004. The parasitic ciliate *Ichthyophthirius multifiliis* induces expression of immune relevant genes in rainbow trout, *Oncorhynchus mykiss* (Walbaum). *Journal of Fish Diseases* 27: 409-417.

Silphaduang, U., A. Colorni and E.J. Noga. 2006. Evidence for widespread distribution of piscidin antimicrobial peptides in teleost fish. *Diseases of Aquatic Organisms* 72: 241-252.

Sinyakov, M.S., M. Dror, H.M. Zhevelev, S. Margel and R.R. Avtalion. 2002. Natural antibodies and their significance in active immunization and protection against a defined pathogen of fish. *Vaccine* 20: 3668-3674.

Smith, V.J., J.M. Fernandes, S.J. Jones, G.D. Kemp and M.F. Tatner. 2000. Antibacterial proteins in rainbow trout, *Oncorhynchus mykiss*. *Fish and Shellfish Immunology* 10: 243-260.

St. Louis-Cormier, E.A., C.K. Osterland and P.D. Anderson. 1984. Evidence for a cutaneous secretory immune system in rainbow trout (*Salmo gairdneri*). *Developmental and Comparative Immunology* 8: 71-80.

Sveinbjornsson, B., R. Olsen and S. Paulsen. 1996. Immunocytochemical localization of lysozyme in eosinophilic granular cells (EGC) of Atlantic salmon, *Salmo salar* L. *Journal of Fish Diseases* 19: 349-355.

Takimoto, H., T. Nakamura, M. Takeuchi, Y. Sumi, T. Tanaka, K. Nomoto and Y. Yoshikai. 1992. Age-associated increase in number of CD4$^+$CD8$^+$ intestinal intraepithelial lymphocytes in rats. *European Journal of Immunology* 22: 159-164.

Toka, F.N., C.D. Pack and B.T. Rouse. 2004. Molecular adjuvants for mucosal immunity. *Immunological Reviews* 199: 100-112.

Vallejo, A.N., Jr. and A.E. Ellis. 1989. Ultrastructural study of the response of eosinophilic granule cells to *Aeromonas salmonicida* extracellular products and histamine liberators in rainbow trout *Salmo gairdneri* Richardson. *Developmental and Comparative Immunology* 13: 133-148.

Wang, X. and H.W. Dickerson. 2002. Surface immobilization antigen of the parasitic ciliate *Ichthyophthirius multifiliis* elicits protective immunity in channel catfish (*Ictalurus punctatus*). *Clinical and Diagnostic Laboratory Immunology* 9: 176-181.

Wang, X., T.G. Clark, J. Noe and H.W. Dickerson. 2002. Immunization of channel catfish, *Ictalurus punctatus*, with *Ichthyophthirius multifiliis* immobilization antigens elicits serotype-specific protection. *Fish and Shellfish Immunology* 13: 337-350.

Wilson, M. and G. Warr. 1992. Fish immunoglobulins and the genes that encode them. *Annual Review of Fish Diseases* 2: 201-221.

Xu, D.H. and P.H. Klesius. 2003. Protective effect of cutaneous antibody produced by channel catfish, *Ictalurus punctatus* (Rafinesque), immune to *Ichthyophthirius multifiliis* Fouquet on cohabited non-immune catfish. *Journal of Fish Diseases* 26: 287-291.

Xu, D.H., P.H. Klesius and C.A. Shoemaker. 2001. Effect of lectins on the invasion of *Ichthyophthirius* theront to channel catfish tissue. *Diseases of Aquatic Organisms* 45: 115-120.

Yang, F., G.C. Waldbieser and C.J. Lobb. 2006. The nucleotide targets of somatic mutation and the role of selection in immunoglobulin heavy chains of a teleost fish. *Journal of Immunology* 176: 1655-1667.

Yano, T. 1996. The nonspecific immune system: Humoral defense. In: *The Fish Immune System*, G.K. Iwama and T. Nakanishi (eds.). Academic Press, San Diego, pp. 106-159.

Yasutake, W.T. and J.H. Wales. 1983. *Microscopic Anatomy of Salmonids: An Atlas*. United States Department of the Interior Washington, D.C., Vol. 150.

Zaccone, G., B.G. Kapoor, S. Fasulo and L. Ainis. 2001. Structural, Histochemical and Functional aspects of the Epidermis of Fishes. In: *Advances in Marine Biology*, A.JH. Southward, P.A. Tyler, C.M. Young and L.A. Fuiman (eds.). Academic Press, London, Vol. 40, pp. 255-347.

Zhao, X., R.C. Findly and H.W. Dickerson. 2008. Cutaneous antibody-secreting cells and B cells in teleost fish. *Developmental and Comparative Immunology* 32: 500-508.

Host Defense Peptides in Fish: From the Peculiar to the Mainstream

Aleksander Patrzykat[1,*] and Robert E.W. Hancock[2]

INTRODUCTION

A lead science story published online by the Canadian Broadcasting Corporation (*Thousands of bacterial species discovered in oceans*; www.cbc.ca /story/science/national/2006/07/31/ocean-microbes.html, July 31, 2006) began with the following statement:

"The oceans could be teeming with 10 to 100 times more types of bacteria than thought.... . The international team of researchers concluded that there are more than 20,000 different microbial species in one litre of seawater taken from deep in the Atlantic and Pacific oceans".

Authors' addresses: [1]National Research Council Institute for Marine Biosciences, 1441 Oxford Street, Halifax, Nova Scotia, B3H3Z1, Canada.
[2]Department of Microbiology and Immunology, University of British Columbia, Centre for Microbial Diseases and Immunity Research, Lower Mall Research Station, UBC, Room 232 - 2259 Lower Mall, Vancouver, BC V6T 1Z4, Canada. E-mail: bob@cmdr.ubc.ca
*Corresponding author: E-mail: aleks.patrzykat@nrc-cnrc.gc.ca

The story refers to a study (Sogin *et al.*, 2006) which indeed suggests that microbial biodiversity of the oceans is likely to be underestimated, even with the mind-boggling numbers that scientists quote today. If we assume that even a portion of these microbes pose a threat to marine fish, and have thus been responsible for the co-evolution of natural antimicrobial defenses, we can anticipate a corresponding diversity of such defenses in fish.

Indeed, a technological leap in discovery techniques over the past decade has confirmed that we are only beginning to discover the wealth of immune defenses, including host defense peptides, in fish. Biochemical means of discovering new peptides, which were almost exclusively used 10 years ago, had severely limited our ability to discover novel compounds from fish. This is because such approaches concentrated on looking for known characteristics such as predetermined physical characteristics (charge, size, amphipathicity) or activity of the peptides, and given the great diversity (including some unusual characteristics) of fish peptides, these constraints became limiting. In addition, logistical problems were created by limited access to samples of fish biomass and the minimal options for handling those samples without degradation. We will also describe how new techniques based on gene discovery, which have been used more recently, are leading to substantially higher rates of discovery of novel peptides in fish.

We are now able to find genetic, structural and functional relatives of non-fish host defense peptides in fish, and we will devote a portion of our discussion to each of these topics. Table 2.1 provides a comparative timeframe of research milestones for non-fish peptides and fish peptides. Even more importantly, experiences to date suggest that the marine fish peptides discovered today represent only the tip of the iceberg. We shall conclude this chapter by venturing into the realm of speculation about what we anticipate will be discovered in the field of fish host defense peptides in the future.

THE FIRST FEW: ODD BALLS AND ILLEGITIMATE ANTIMICROBIAL PEPTIDES

Long before the first fish peptides were discovered, scores of peptides had been identified based on their *in vitro* antimicrobial activity, including a substantial number of well-described antimicrobial peptides such as, for example, cecropins and magainins. It is now recognised that such peptides

Table 2.1   Chronology of selected research milestones for fish and non-fish peptides. (Discussion and references can be found within the text.)

Research milestone	First report	First report in fish
Identification of cathelicidins	1993	2002
Identification of hepcidins	2001	2002
Immunomodulatory activities of peptides	1995	2005
Antiviral activities of peptides	1985	2004

are likely ubiquitous in nature and an antimicrobial peptide database (http://www.bbcm.univ.trieste.it/~tossi/pag1.htm) has more than 800 entries. These peptides can be defined as having the following characteristics: they are short (12 to approximately 50 amino acids in length); cationic (net charge of +2 to +9 due to excess basic lysine and arginine residues); and containing around 50% hydrophobic amino acids. They fold, often in the presence of membrane bilayers, into amphiphilic structures having patches of polar (including positively charged) and hydrophobic amino acids; these structures most usually are either β-sheets of 2-3 β strands or α-helices, or less frequently extended structures with over-representation of certain amino acids (e.g., W, P, R), or loops created by a single disulphide bridge. The most obvious biological activity is often their ability to kill a broad spectrum of microbes, and a single peptide such as horseshoe crab polyphemusin or cattle indolicidin has *in vitro* antimicrobial activity against Gram-negative and Gram-positive bacteria, fungi, and certain enveloped viruses (Jenssen *et al.*, 2006). However, more recently, it is becoming obvious that in mammals, certain peptides possess rather weak direct antimicrobial activities that are strongly blocked by physiological concentrations of monovalent and divalent cations at available concentrations and that the activities of these peptides in overcoming infections are more likely due to their profound and diverse immunomodulatory activities (Bowdish *et al.*, 2005). For example, it has been shown that peptides are able to increase the expression of hundreds of genes in mammalian monocytes, macrophages and epithelial cells to attract cells to the site of infections through a combination of direct chemokine activity and induction in immune cells of chemokine production, to counter potentially harmful endotoxin-induced production of pro-inflammatory cytokines, and to stimulate wound healing and angiogenesis. Many of these functions have been demonstrated in animal models.

Arguably the most convincing evidence for the importance of host defense peptides came from experiments in mice. Welling *et al.* (1998) demonstrated that tiny amounts (0.4 to 4 ng per animal) of human neutrophil peptide-1 could protect mice against *Klebsiella* and *Staphylococcus* infections, but in a neutrophil-dependent fashion (in that there was no protection in leukocytopenic mice). In another study, mice with a disruption of the gene encoding a cathelicidin peptide, CRAMP, were more susceptible to infection by Group A *Streptococcus* relative to mice with the intact CRAMP gene (Nizet *et al.*, 2001). In a third study, a group of transgenic mice expressing human defensin 5 was compared to a group lacking this peptide (Salzman *et al.*, 2003). Upon exposure to *Salmonella typhimurium*, mice lacking the defensin died within 2 days, while mice expressing the peptide survived. Depending on the peptide and experimental design, the direct activity of peptides on bacteria and/or the modulation of the host immune system have been credited with the host-protective function. Interestingly, protection experiments have also been performed in fish and delivery of an insect hybrid peptide by osmotic pump led to protection of fish against vibriosis (Jia *et al.*, 2001).

In 2001, we wrote a book chapter on antimicrobial peptides for fish disease control (Patrzykat and Hancock, 2002) in which we reviewed known fish antimicrobial peptides and their potential for preventing infections in aquaculture. The list of peptides from fish known at the time was very short and included the highly modified hagfish peptides (Shinnar *et al.*, 1996), sole-fish pardaxin (Shai *et al.*, 1988), the loach misgurin (Park *et al.*, 1997), and the winter flounder pleurocidin (Cole *et al.*, 1997), as well as several histone-derived peptides and a small assortment of non-cationic peptides from fish. The principal activities of those peptides were their ability to kill bacteria and inhibit bacterial growth and the prime avenue for fish peptide development was seen to be for applications in aquaculture. In the context of non-fish antimicrobial peptides, the number, variety and commercial potential of fish derived peptides were rather modest.

Based on the non-fish peptides known at the time when the first fish peptides were discovered, certain assumptions were made as to what antimicrobial peptides should look like. The first host defense peptides from fish did not fit these patterns and were largely ignored. Indeed, in the context of non-fish peptides, their fish relatives might have been considered oddballs. Pardaxin had only one positive charge and a structure that did not qualify it for grouping with classical microbial peptides, and misgurin as well as the hagfish peptides looked even more unlike the classic antimicrobial peptides. The hagfish peptides contained a post-translationally modified amino acid, bromo-tryptaphan, and were thus ignored. As will be discussed in the following paragraphs, this lack of attention was unjustified because the hagfish peptides have now been shown to belong to the cathelicidin family, the most important non-defensin family of host defense peptides, which also contains the well-studied human peptide LL-37 and mouse peptide CRAMP. Given the evolutionary relationship among vertebrates, this indicates that the ancestors of mammalian cathelicidins can be found among fish.

THE TIDAL WAVE OF RECOGNIZABLE HOST DEFENSE PEPTIDES

A great leap in understanding host defense peptides in fish came when genetic techniques replaced the previously-utilised tedious and expensive purification techniques, which relied on obtaining large quantity of the biomass sample, homogenising it, and extracting and purifying peptides in a series of multidimensional protein purification steps alongside activity assays.

Revolution in Discovery Techniques

This revolution in host defense peptide discovery techniques occurred in 2000-2001, as extensively reviewed in 2003 (Patrzykat and Douglas, 2003). The breakthrough involved replacing traditional screening methods, which involved resource- and time-intensive sampling, purification, and testing protocols with genomic screening. This leveled the playing field and brought about a flurry of new discoveries of fish peptides. The molecular biology techniques used to isolate peptides and coding sequences from marine organisms are variations on a routine that starts from the construction of a cDNA or genomic library. Those libraries are then screened using oligonucleotide probes based on the sequences of known host defense peptides. This technique was used to discover new pleruocidins from winter flounder (Douglas *et al.*, 2001) and hepcidin from bass (Shike *et al.*, 2002).

In addition, we now know that peptides can be encoded by families of multiple genes which share common characteristics in their flanking regions, such as the prepro region. By using primers based on these conserved sequences, entirely novel peptides can be identified. This approach was successful for pleurocidins (Patrzykat *et al.*, 2003: Figure 2.1), hepcidins (Douglas *et al.*, 2003) and cathelicidins (Chang *et al.*, 2005, 2006).

In the case of pleurocidins, a conserved sequence flanking the mature peptide was discovered. By designing primers based on this flanking sequence and screening both a DNA and a cDNA library of winter flounder, more than 20 additional peptide-encoding genes were discovered. Even more surprisingly, when the same primers were used to screen a library of cDNAs from related species of flounder, additional peptides were found (Patrzykat *et al.*, 2003). These related genes encode a great diversity of novel mature peptides with a variety of activities. None of these peptides could have been discovered without genetic techniques.

This use of a genetic template from a known peptide to identify novel related peptides has now been used repetitively and is described in greater detail below for the pleurocidin, hepcidin and cathelicidin families of peptides.

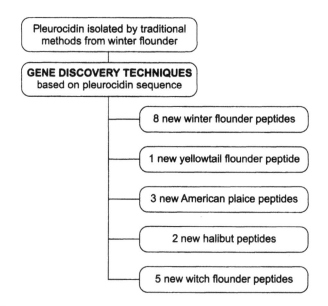

Fig. 2.1   An illustration of the role of gene discovery techniques in identifying new peptides. The diagram summarizes a process described in Patrzykat *et al.* (2003), in which 19 novel antimicrobial peptides from 5 different flatfish species have been discovered, synthesized and tested.

Relationship to Non-fish Peptides

The genetic mode of discovery has also brought to light similarities between the genes encoding fish and non-fish peptides. As described below, cathelicidins not only share common pre-pro sequences but also gene exon/intron structure.

In addition to looking at genetic relationships, similarities can exist in the amino acid sequence and in the structure of the peptide itself. For example, the winter flounder pleurocidin shows sequence and structural homology to dermaseptin B from frogs and cecropin from insects. While these peptides may be genetically unrelated, their structural characteristics and antimicrobial activities seem to coincide. This is remarkable given the evolutionary distance between the 3 groups of organisms, and indicates the possibility of functional 'convergence'.

In fact, aside from the genetic and structural relationships discussed above, a functional approach can also be used to identify relationships between fish and non-fish peptides. Clearly, there are both fish and non-fish peptides with antimicrobial activities. This is not surprising given that antibacterial activities have traditionally been a selection criterion in purifying the peptides. However, a remarkable range of other activities have been recently described in non-fish peptides. Those include antiviral activities, a broad range of immunomodulatory activities, wound-healing activities, angiogenesis-related activities, antiviral activities, anticancer activities and others. In the following paragraphs, the reader will find examples of those non-fish peptide activities which have also been identified among fish antimicrobial peptides.

GENETIC RELATIVES — THE CASE OF FISH CATHELICIDINS

Cathelicidins are one of the best recognized and studied families of host defense peptides. In a 1993 publication, Zanetti et al. (1993) identified a pro-sequence containing a cathelin-like domain as a common genetic feature of various neutrophil antimicrobials. Since then, many mammalian peptides in humans, mice, pigs, horses, rabbits, guinea pigs, cows, sheep, and goats containing the conserved cathelin domain at the N-terminal pre-pro region and assorted host defense peptides at the C-terminus have been identified. While the pre-pro region sequence is strongly conserved, the mature peptide region, released by proteolytic cleavage and removal of the pre-pro sequences, can vary enormously. For example, cathelicidins include the pig β-hairpin peptide protegrin, the related mouse and human α-helical peptides CRAMP and LL-37, the extended bovine peptide indolicidin and the bovine loop peptide bactenecin. We recently overviewed the broad range of activities and structures of these peptides (Powers *et al.*, 2003; Bowdish *et al.*, 2005).

At around the time that cathelicidins were identified, Shinnar *et al.* (1996) briefly reported on a 'new family of linear antimicrobial peptides from hagfish intestines'. The peptides were unusual, in the sense that they contained bromo-tryptophan and did not receive much attention for almost ten years, until a 2002 paper by Basanez who referred to them as 'cathelicidin antimicrobial peptides'. The assertion that the peptides were cathelicidins was not based on experimental data, but referred to unpublished data by Uzzel. Finally, in 2003, the basis of this statement was published (Uzzell *et al.*, 2003). In analyzing the clones from a hagfish intestine cDNA library, Uzzell was able to identify sequences that matched both the amino acid sequence of the previously known hagfish peptides and the sequence of the cathelin domain from mammalian peptides, especially bovine and goat. The consequence of discovering cathelicidins in the very primitive hagfish is that this gene family can now be truly stated to date back to the dawn of vertebrate evolution (Uzzell *et al.*, 2003).

This finding was further reinforced by the discovery of another fish cathelicidin *rt*CATH_1 in rainbow trout a year after the realisation that the hagfish peptides were cathelicidins (Chang *et al.*, 2005). Again, the pre-pro peptide combined the features of known cathelicidins and the novel *rt*CATH_1 active component. A follow-up study identified 3 more cathelicidins, one from rainbow trout and two from Atlantic salmon (Chang *et al.*, 2006). This further confirms that cathelicidins evolved early in vertebrate evolution and we can certainly expect to find more of these in fish.

As anticipated and already mentioned, fish cathelicidins exhibit identical genetic structure as mammalian cathelicidins with Exon 1 encoding a signal peptide, Exons 2 and 3 encoding the catheling domain, and Exon 4 being the variable region encoding diverse mature peptides. In addition, the fish cathelicidins described here were synthesized by their dicoverers and exhibited marked antimicrobial activity.

Mammalian cathelicidins are among the best-studied host defense peptides and exhibit a range of activities well beyond their antimicrobial properties. While these properties have not been studied in fish antimicrobial peptides, there are substantial implications to this finding and they will be discussed elsewhere in the chapter(Bowdish *et al.*, 2005).

STRUCTURAL RELATIVES — THE CASE OF HEPCIDINS

Unlike cathelicidins, which share their genetic structure between species but vary in the sequence of the mature peptide, hepcidins are related in their amino acid sequence (of the mature peptide) in addition to their genetic structure and the sequence of flanking regions.

Generally speaking, cysteine-containing peptides are well described. The best known among them are defensins. However, more recently, other groups of peptides, liver express antimicrobial peptides (LEAPs) and hepcidins, were isolated from mammalian species — the terms LEAPs and hepcidins have been used interchangeably by some authors.

Human hepcidin, an amphipathic 25 amino acid cationic peptide with 4 disulphide bridges, was first reported in 2001. While originally identified as an antimicrobial peptide, and with many of the features of such peptides, it seems to be more important as a master regulator of iron homeostasis in humans and other mammals (Ganz, 2003). The availability of expressed sequence tags libraries from fish immediately led to the discovery of related hepcidin sequences in various species of fish, for example bass (Shike *et al.*, 2002). Unlike cathelicidins, where the mammalian research has preceded the fish research, the story of fish hepcidins is unfolding parallel to the story of mammalian hepcidins.

However, one consequence of the fish hepcidin discovery arising from genetic information is that studies continue to concentrate on genetic organization and expression analysis of the genes, and have not ventured much into the realm of structure and function. This was true for zebrafish hepcidins (Shike *et al.*, 2004), rainbow trout hepcidins (Zhang *et al.*, 2004), red seabream hepcidins (Chen *et al.*, 2005), Japan sea bass hepcidins (Ren 2006), catfish hepcidins (Bao *et al.*, 2005), Japanese flounder hepcidin (Hirono *et al.*, 2005), as well as hepcidins from other fish species (Douglas *et al.*, 2002, 2003). In fact, the only fish hepcidin that has been described at the level of structure and function is bass hepcidin from the gills of hybrid striped bass (Lauth *et al.*, 2005).

The reason for this research focus is that genetic studies can be performed without the need for the actual peptide, which is difficult to manufacture and fold properly due to the need for the formation of adequate disulphide bonds. However, the genetic information is sufficient to obtain expression data through RT-PCR, real time PCR, or *in-situ* hybridization studies.

As for the previously mentioned bass hepcidins, their activities can be described as predictable (Lauth *et al.*, 2005). As expected, the report indicates that the bass and human hepcidins fold into almost-identical three-dimensional structures, share the disulfide-bonding pattern, and both contain a rare vicinal disulfide bond which is believed to be important for peptide function. The spectrum of antimicrobial activities exhibited by the bass hepcidin also corresponded to the spectrum previously described for human hepcidin (not the best antimicrobial activites *in vitro*). The authors argue that, given the structural and antimicrobial activity similarities to mammalian peptides, bass hepcidin should be expected to play a role in the hypoferremic response during inflammation, much like mammalian hepcidins.

This theory remains to be demonstrated, but it seems reasonable to anticipate that the activities identified for mammalian hepcidins will be similarly manifested for fish hepcidins. This expectation is not only based on the genetic and structural arguments made so far, but also on the fact that functional relationships have already been found between fish and non-fish peptides, as described below.

FUNCTIONS

Functionally, fish host defense peptides have not received nearly as much attention as their non-fish counterparts. The extent to which fish peptides have been studied is usually restricted to their *in vitro* bactericidal and/or bacteriostatic activity, which is comparable to non-fish peptides. These antimicrobial activities manifest themselves through either the direct attack of peptides on membranes and entry into bacteria and attack of cytoplasmic targets (Patrzykat *et al.*, 2002). Only recently have other aspects of fish host defense peptide activities been investigated.

Mammalian peptides have been reported to have both pro- and anti-inflammatory activities. For example, human peptide LL-37 induces so-called pro-inflammatory chemokines like IL-8 and MIP1α, while suppressing endotoxin induced TNF-α in monocytes (Mookherjee *et al.*, 2006). Conversely, human beta defensin-2 can activate pro-inflammatory mechanisms in dendritic cells through direct interaction on Toll-like receptor 4 (Biragyn *et al.*, 2002).

A recent manuscript (Chiou et al., 2005) described the pro-inflammatory effects of the fish peptide pleurocidin on trout macrophages, confirming that fish host defense peptides can modulate the immune system of their host. It was reported that the expression of two pro-inflammatory genes IL-1β and COX-2 was increased upon peptide treatment of the cultured cells. In addition, pleurocidin did not neutralize the pro-inflammatory effect of LPS on the same genes. These observations led the authors to propose that fish peptides should be evaluated for their potential to act as immune adjuvants in fish.

Another report (Chinchar et al., 2004) focused on the ability of piscine host defense peptides, piscidins, to protect ectothermic animals from viral infections. Comparable studies in mammals have been carried out over the past 20 years (Lehrer et al., 1985; Daher et al., 1986; Jenssen et al., 2006). Generally, these studies relied on the ability of peptides to directly inactivate viruses in vitro. In Daher's study, human neutrophil peptide-1 was shown to inactivate herpes simplex virus 1 and 2 but not cytomegalovirus. In the 1985 Lehrer study, the same pattern was observed; rabbit peptides MCP-1 and MCP-2 were shown to inactivate HSV-2, VSV and influenza A but not CMV. The antiviral properties of peptides gained prominence when it was shown that several human alpha-defensins possessed anti-HIV-1 properties (Zhang et al., 2002).

Piscidin-1N, -1H, -2 and -3 directly reduced the in vitro infectivity of channel catfish virus (Chinchar et al., 2004). However, in their discussion, the authors speculate that piscine host defense peptides may exert their antiviral activities in a two-punch model of blocking the virus infection by direct inactivation or, failing that, eliminating infection by modulating additional innate and adaptive responses. This course of thinking is indeed directly in line with some current research efforts in human health (Jenssen et al., 2006) and indicates how far the field of fish host defense peptides has come over the past 10 years.

While we would like to avoid the inevitably complicated discussion of the mode of action of host defense peptides—there are many venues where this is debated and the reader is directed to the numerous reviews and the Zlotkin chapter in this book to become familiar with the state of the art—a point should be made on the sophistication in the studies of the mode of action and structure of fish peptides. A recent study (Chekmenev et al., 2006) on the biological function of the previously mentioned piscidins employed ^{15}N solid state NMR, including 2D PISEMA (polarization inversion spin exchange at the magic angle) experiments to estimate peptide orientation in the membrane. In the conclusion, the authors themselves state that the research on piscidins was undertaken with the larger goal of providing 'insight about other species active at membranes including membrane proteins and fusion peptides'. This statement may well be indicative of the acceptance that fish peptides are gaining as model molecules.

COMMERCIAL POTENTIAL OF FISH PEPTIDES

When Magainin Pharmaceuticals Inc was developing magainin in 1988, the fact that it came from frogs was irrelevant. The properties of the peptide made it an attractive molecule for development and the company was formed and funded. Many more companies have been formed since then but today, almost 20 years later, none of the peptides that entered the commercial development path has obtained FDA approval (Hancock and Sahl, 2006). Names like Pexiganan, Iseganan and Neuprex are reminders of attempts made but as yet success remains limited. Indeed, many in the scientific and investment community have been asking for reasons for these failures. When easy-to-manufacture mammalian peptides were discovered, development efforts shifted away from non-mammalian peptides. There was a scientific rationale behind the shift—a belief that mammalian peptides may be better tolerated, and a non-scientific one. Due to the original failures, the investors were detecting higher level of risk and working on mammalian molecules gave them a greater level of comfort. However, as we approach the first potential success in this field (likely to be the Migenix peptide Omiganan that demonstrated statistically significant reduction in catheter colonisation and catheter associated tunnel infections in Phase IIIa clinical trials), it is worth reflecting on what are the issues that need to be considered to drive the field forward.

To date, the major limitations to development of peptides as commercial drugs have been the cost of goods, limited stability of peptides to proteolytic digestion and unknown toxicities. Each of these issues is indeed addressable (Hancock and Sahl, 2006). One major route forward is through the discovery of new and more effective peptides that can serve as building blocks for rational and/or semi-random design (Hilpert et al., 2005), and there is no doubt that marine species will assist in these efforts. The unique features of marine peptides, including unusual structures and post-translational modifications, may represent promising building blocks.

Today's investors benefit from 20 years of experience in financing peptide ventures. Whether the peptides come from mammals or fish, the prospects are evaluated based on the advantages and disadvantages of the particular molecule. If the peptide shows efficacy and low toxicity, it will then be subject to a standard list of enquiries. Is it cheap enough to make? What are the pharmacokinetic and pharmacodynamic properties of the molecule? Is it susceptible to proteolysis? Will it remain bioavailable at the site it needs to reach for efficacy? What exactly does it do and not do? If a fish peptide can be shown to address all of these queries, it will be as good a commercial opportunity as a non-fish peptide. But addressing these issues is not trivial, as many failed development efforts and even more failed financing efforts demonstrate. One exciting new avenue is created by the 'new' activities of these host defense peptides (Bowdish *et al.*, 2005) in that immunomodulation, which triggers host responses rather than killing directly, might be expected to require smaller amounts of peptides (and thus be cheaper).

While there is no doubt that host defense peptides are critical in protecting hosts from infections, a substantial amount of work is required to harness their power to provide effective therapeutants to the clinics. The sheer number and diversity of sequences of natural fish peptides that has become available after genetic discovery techniques were perfected is providing some of the answers. When we wrote our chapter on the potential of fish antimicrobial peptides 5 years ago (Patrzykat and Hancock, 2002), we predicted that the number of peptides from the sea would increase dramatically, as it indeed has. The next five years are likely to become a season for discovering patterns and large-scale screening for commercially useful members among the fish host defense peptides. Given the extent of current research on fish hepcidins, pleurocidins and piscidins, as well as our knowledge of the commercialisation efforts under way, there is a very good chance that a fish peptide will be in clinical trials within the next few years.

Another area in which we saw potential for fish peptides five years ago was for protecting fish from infections in aquaculture applications. Indeed, there has been a substantial effort devoted to developing peptides for fish disease applications and to developing transgenic animals expressing peptides for protection of disease. However, there have been no substantial commercial developments at the post-research stage. While the world aquaculture industry is growing, the high margin phase has given way to a high volume approach to the market. The inevitable consequence is that even in salmon farming, the producers are more focused on the bottom line than innovation. The ability of the aquaculture industry to absorb biotech innovations has, therefore, been somewhat impaired.

In addition, the controversy regarding the development of transgenic animals for food production has not yet been settled and many research organisations have opted against transferring related technologies. Finally, the cost and risk of developing a novel antimicrobial are so great that the rewards to the investors must also be proportionately high. The fish health market at this point probably does not offer as high a premium as the human health market and we are now only moderately optimistic that antimicrobial peptides will be developed specifically for use in controlling fish disease.

WHAT IS STILL LEFT IN THE OCEAN

Our discussion to this point has concentrated on describing the realm of fish peptides that have already been discovered. As indicated in the introduction to this chapter, we would like to offer the reader an opinion on the directions that fish peptide research might take over the next few years.

Tip of the Iceberg

When fish host defense peptides were first discovered, the rare discoveries of active compounds were exciting. Since then, the wealth of genetic information and technology is such that we can now say with reasonable confidence that there do not appear to be any cathelicidins (as an example) in some species, not just that we were unable to find them. As well-annotated genomes, or at least exhaustive EST libraries of more and more common species fill databases, the best chances of finding truly new compounds will present themselves as completely new organisms are discovered. And the oceans offer the greatest undiscovered variety of those in the sense that there are more than 70,000 fish species, of which only 1200 have been described; hence there is room to fill the void.

But even more importantly, when putative new peptides can be found by simply searching through databases, their mere identification will not constitute a great scientific leap. Understanding tissue-specific and disease stage-specific expression patterns, in vivo activities and entire organism impacts will be far more important and we believe that most studies identifying new peptides from fish in the future will require some or all of these to warrant scientific interest and publication.

This should contribute greatly to the body of knowledge about the patterns of occurrence, expression, and activity of host defense peptides. We expect to find further relationships between fish and non-fish peptides, much like the story of the cathelicidins where the porcine PR-39 and human LL-37 are now known to be related to hagfish and rainbow trout peptides.

Little is known about the innate, secondary and adaptive immune defenses in fish. The traditional knowledge of cell-mediated secondary defenses, antibodies, cytokines, chemokines and their receptors—which is available for mammalian species—is not available for fish as yet. Hence, the immunomodulatory effects of fish host defense peptides are almost certainly underappreciated. Based on genetics, we know that winter flounder possesses a large repertoire of pleurocidins. Our own studies have shown that only a portion of these is directly antimicrobial *in vitro*. Some of our further studies indicate a large array of other activities (based on the gene stimulation patterns shown in micro array studies with salmon head kidney cells). Till date, we have no appreciation of the role of most of the upregulated and downregulated genes or their contribution to flounder defenses. By combining the knowledge of peptides gained from fish and other organisms, we may gain valuable insight into the putative role of these genes. On the other side of the coin, many unanswered questions related to mammalian host defense peptides may find answers once the diversity of fish host defense peptides is adequately appreciated, discovered and studied.

Finally, better understanding of the role of host defense peptides in protecting from infections will then lead to better selections of compounds for commercial applications.

Acknowledgements

Hancock (REWH) is the recipient of a Canada Research Chair. His antimicrobial peptide research is supported by the Canadian Institutes of Health Research and the Advanced Foods and Materials Network while his immunomodulatory peptide research is supported by the Pathogenomics of Innate Immunity Program funded by Genome BC and Genome Prairie and by two grants from the Grand Challenges in Health Research program from the Foundation for the National Institutes of Health and the Gates Foundation.

References

Bao, B., E. Peatman, P. Li, C. He and Z. Liu. 2005. Catfish hepcidin gene is expressed in a wide range of tissues and exhibits tissue-specific upregulation after bacterial infection. *Developmental and Comparative Immunology* 29: 939-950.

Basanez, G., A.E. Shinnar and J. Zimmerberg. 2002. Interaction of hagfish cathelicidin antimicrobial peptides with model lipid membranes. *FEBS Letters* 532: 115-120.

Biragyn, A., P.A. Ruffini, C.A. Leifer, E. Klyushnenkova, A. Shakhov, O. Chertov, A.K. Shirakawa, J.M. Farber, D.M. Segal, J.J. Oppenheim and L.W. Kwak. 2002. Toll-like receptor 4-dependent activation of dendritic cells by beta-defensin 2. *Science* 298: 1025-1029.

Bowdish, D.M.E., D.J. Davidson and R.E.W. Hancock. 2005. A re-evaluation of the role of host defense peptides in mammalian immunity. *Current Protein Peptide Science* 6: 35-51.

Bowdish, D.M.E., D.J. Davidson, Y.E. Lau, K. Lee, M.G. Scott and R.E.W. Hancock. 2005. Impact of LL-37 on anti-infective immunity. *Journal of Leukocyte Biology* 77: 451-459.

Chang, C.I., O. Pleguezuelos, Y.A. Zhang, J. Zou and C.J. Secombes. 2005. Identification of a novel cathelicidin gene in the rainbow trout, *Oncorhynchus mykiss. Infection Immunology* 73: 5053-5064.

Chang, C.I., Y.A. Zhang, J. Zou, P. Nie and C.J. Secombes. 2006. Two cathelicidin genes are present in both rainbow trout (*Oncorhynchus mykiss*) and Atlantic salmon (*Salmo salar*). *Antimicrobial Agents Chemotherapy* 50: 185-195.

Chekmenev, E.Y., B.S. Vollmar, K.T. Forseth, M.N. Manion, S.M. Jones, T.J. Wagner, R.M. Endicott, B.P. Kyriss, L.M. Homem, M. Pate, J. He, J. Raines, P.L. Gor'kov, W.W. Brey, D.J. Mitchell, A.J. Auman, M.J. Ellard-Ivey, J. Blazyk and M. Cotton. 2006. Investigating molecular recognition and biological function at interfaces using piscidins, antimicrobial peptides from fish. *Biochimique et Biophysique Acta* 1758: 1359-1372.

Chen, S.L., M.Y. Xu, X.S. Ji, G.C. Yu and Y. Liu. 2005. Cloning, characterization, and expression analysis of hepcidin gene from red sea bream (*Chrysophrys major*). *Antimicrobial Agents Chemotherapy* 49: 1608-1612.

Chinchar, V.G., J. Wang, G. Murti, C. Carey and L. Rollins-Smith. 2001. Inactivation of frog virus 3 and channel catfish virus by esculentin-2P and ranatuerin-2P, two antimicrobial peptides isolated from frog skin. *Virology* 288: 351-357.

Chinchar, V.G., L. Bryan, U. Silphadaung, E. Noga, D. Wade and L. Rollins-Smith. 2004. Inactivation of viruses infecting ectothermic animals by amphibian and piscine antimicrobial peptides. *Virology* 323: 268-275.

Chiou, P.P., C.M. Lin, L. Perez and T.T. Chen. 2002. Effect of cecropin B and a synthetic analogue on propagation of fish viruses *in vitro. Marine Biotechnology* 4: 294-302.

Chiou, P., J. Khoo, N.C. Bols, S. Douglas and T.T. Chen. 2006. Effects of linear cationic alpha-helical antimicrobial peptides on immune-relevant genes in trout macrophages. *Developmental and Comparative Immunology* 30: 797-806.

Cole, A.M., P. Weis and G. Diamond. 1997. Isolation and characterization of pleurocidin, an antimicrobial peptide in the skin secretions of winter flounder. *Journal of Biological Chemistry* 272: 12008-12013.

Daher, K.A., M.E. Selsted and R.I. Lehrer. 1986. Direct inactivation of viruses by human granulocyte defensins. *Journal of Virology* 60: 1068-1074.

Douglas, S.E., J.W. Gallant, Z. Gong and C. Hew. 2001. Cloning and developmental expression of a family of pleurocidin-like antimicrobial peptides from winter flounder, *Pleuronectes americanus* (Walbaum). *Developmental and Comparative Immunology* 25: 137-147.

Douglas, S.E., A. Patrzykat, J. Pytyck and J.W. Gallant. 2003. Identification, structure and differential expression of novel pleurocidins clustered on the genome of the winter flounder, *Pseudopleuronectes americanus* (Walbaum). *European Journal of Biochemistry* 270: 3720-3730.

Douglas, S.E., J.W. Gallant, R.S. Liebscher, A. Dacanay and S.C. Tsoi. 2003. Identification and expression analysis of hepcidin-like antimicrobial peptides in bony fish. *Developmental and Comparative Immunology* 27: 589-601.

Ganz, T. 2003. Hepcidin, a key regulator of iron metabolism and mediator of anemia of inflammation. *Blood* 102: 783-788.

Hancock, R.E.W. and H.G. Sahl. 2006. Antimicrobial and host defense peptides as novel anti-infective therapeutic strategies. *Nature Biotechnology* 24: 1551-1557.

Hilpert, K., R. Volkmer-Engert, T. Walter and R.E.W. Hancock. 2005. High-throughput generation of small antibacterial peptides with improved activity. *Nature Biotechnology* 23: 1008-1012.

Hirono, I., J.Y. Hwang, Y. Ono, T. Kurobe, T. Ohira, R. Nozaki and T. Aoki. 2005. Two different types of hepcidins from the Japanese flounder *Paralichthys olivaceus*. FEBS *Journal* 272: 5257-5264.

Jenssen, H., P. Hamill and R.E.W. Hancock. 2006. Peptide antimicrobial agents. *Clinical Microbiological Reviews* 19: 491-511.

Jenssen, H., T.J. Gutteberg, O. Rekdal and T. Lejon. 2006. Prediction of activity, synthesis and biological testing of anti-HSV active peptides. *Chem Biol Drug Des.* 68: 58-66.

Jia, X., A. Patrzykat, R. Devlin, P.A. Ackerman, G.K. Iwama and R.E.W. Hancock. 2000. Antimicrobial peptides protect Coho salmon from *Vibrio anguillarum* infections. *Applied Environmental Microbiology* 66: 1928-1932.

Lauth, X., H. Shike, J.C. Burns, M.E. Westerman, V.E. Ostland, J.M. Carlberg, J.C. Van Olst, V. Nizet, S.W. Taylor, C. Shimizu and P. Bulet. 2002. Discovery and characterization of two isoforms of moronecidin, a novel antimicrobial peptide from hybrid striped bass. *Journal of Biological Chemistry* 277: 5030-5039.

Lauth, X., J.J. Babon, J.A. Stannard, S. Singh, V. Nizet, J.M. Carlberg, V.E. Ostland, M.W. Pennington, R.S. Norton and M.E. Westerman. 2005. Bass hepcidin synthesis, solution structure, antimicrobial activities and synergism, and in vivo hepatic response to bacterial infections. *Journal of Biological Chemistry* 280: 9272-9282.

Lehrer, R.I., K. Daher, T. Ganz and M.E. Selsted. 1985. Direct inactivation of viruses by MCP-1 and MCP-2, natural peptide antibiotics from rabbit leukocytes. *Journal of Virology* 54: 467-472.

Mookherjee, N., K.L. Brown, D.M.E. Bowdish, S. Doria, R. Falsafi, K. Hokamp, F.M. Roche, R. Mu, G.H. Doho, J. Pistolic, J. Powers, J. Bryan, F.S.L. Brinkman and R.E.W. Hancock. 2006. Modulation of the Toll-like receptor-mediated inflammatory response by the endogenous human host defense peptide LL-37. *Journal of Immunology* 176: 2455-2464.

Nizet, V., T. Ohtake, X. Lauth, J. Trowbridge, J. Rudisill, R.A. Dorschner, V. Pestonjamasp, J. Piraino, K. Huttner and R.L. Gallo. 2001. Innate antimicrobial peptide protects the skin from invasive bacterial infection. *Nature (London)* 414: 454-457.

Park, C.B., J.H. Lee, I.Y. Park, M.S. Kim and S.C. Kim. 1997. A novel antimicrobial peptide from the loach, *Misgurnus anguillicaudatus*. *FEBS Letters* 411: 173-178.

Patrzykat, A. and R.E.W. Hancock. 2002. Antimicrobial peptides for fish disease control. In: *Recent Advances in Marine Biotechnology* (Seafood safety and human health), M. Fingerman and R. Nagabhushananam (eds.). Oxford and IBH Publishing Co., New Delhi, Vol. 7, pp. 141-155.

Patrzykat, A. and S.E. Douglas. 2003. Gone gene fishing: How to catch novel marine antimicrobials. *Trends in Biotechnology* 21: 362-369.

Patrzykat, A. and S.E. Douglas. 2005. Antimicrobial peptides: cooperative approaches to protection. *Protein Peptide Letters* 12: 19-25.

Patrzykat, A., J.W. Gallant, J.K. Seo, J. Pytyck and S.E. Douglas. 2003. Novel antimicrobial peptides derived from flatfish genes. *Antimicrobial Agents Chemotherapy* 47: 2464-2470.

Patrzykat, A., C.L. Friedrich, L. Zhang, V. Mendoza and R.E.W. Hancock. 2002. Sublethal concentrations of pleurocidin-derived antimicrobial peptides inhibit macromolecular synthesis in *Escherichia coli*. *Antimicrobial Agents Chemotherapy* 46: 605-614.

Powers, J-P.S. and R.E.W. Hancock. 2003. The relationship between peptide structure and antibacterial activity. *Peptides* 24: 1681-1691.

Ren, H.L., K.J. Wang, H.L. Zhou and M. Yang. 2006. Cloning and organisation analysis of a hepcidin-like gene and cDNA from Japan sea bass, *Lateolabrax japonicus*. *Fish and Shellfish Immunology* 21: 221-227.

Salzman, N.H., D. Ghosh, K.M. Huttner, Y. Paterson and C.L. Bevins. 2003. Protection against enteric salmonellosis in transgenic mice expressing a human intestinal defensin. *Nature (London)* 422: 522-526.

Shai, Y., J. Fox, C. Caratsch, Y.L. Shih, C. Edwards and P. Lazarovici. 1988. Sequencing and synthesis of pardaxin, a polypeptide from the Red Sea Moses sole with ionophore activity. *FEBS Letters* 242: 161-166.

Shike, H., C. Shimizu, X. Lauth and J.C. Burns. 2004. Organization and expression analysis of the zebrafish hepcidin gene, an antimicrobial peptide gene conserved among vertebrates. *Developmental and Comparative Immunology* 28: 747-754.

Shike, H., X. Lauth, M.E. Westerman, V.E. Ostland, J.M. Carlberg, J.C. Van Olst, C. Shimizu, P. Bulet and J.C. Burns. 2002. Bass hepcidin is a novel antimicrobial peptide induced by bacterial challenge. *European Journal of Biochemistry* 269: 2232-2237.

Shinnar, A.E., T. Uzzell, M.N. Rao, E. Spooner, W.S. Lane and M.A. Zasloff. 1996. Peptides: Chemistry and Biology. Proceedings of the 14th American Peptide Symposium, P. Kauyama and R. Hodges (eds.), pp. 189-191. Mayflower Scientific, Leiden.

Sogin, M.L., H.G. Morrison, J.A. Huber, D.M. Welch, S.M. Huse, P.R. Neal, J.M. Arrieta and G.J. Herndl. 2006. Microbial diversity in the deep sea and the underexplored 'rare biosphere'. *Proceedings of the National Academy of Sciences of the United States of America* 103: 12115-12120.

Uzzell, T., E.D. Stolzenberg, A.E. Shinnar and M. Zasloff. 2003. Hagfish intestinal antimicrobial peptides are ancient cathelicidins. *Peptides* 24: 1655-1667.

Welling, M.M., P.S. Hiemstra, M.T. van den Barselaar, A. Paulusma-Annema, P.H. Nibbering, E.K. Pauwels and W. Calame. 1998. Antibacterial activity of human neutrophil defensins in experimental infections in mice is accompanied by increased leukocyte accumulation. *Journal of Clinical Investigations* 102: 1583-1590.

Zanetti, M., G. Del Sal, P. Storici, C. Schneider and D. Romeo. 1993. The cDNA of the neutrophil antibiotic Bac5 predicts a pro-sequence homologous to a cysteine proteinase inhibitor that is common to other neutrophil antibiotics. *Journal of Biological Chemistry* 268: 522-526.

Zhang, L., W. Yu, T. He, J. Yu, R.E. Caffrey, E.A. Dalmasso, S. Fu, T. Pham, J. Mei, J.J. Ho, W. Zhang, P. Lopez and D.D. Ho. 2002. Contribution of human alpha-defensin 1, 2, and 3 to the anti-HIV-1 activity of CD8 antiviral factor. *Science* 298: 995-1000.

Zhang, Y.A., J. Zou, C.I. Chang and C.J. Secombes. 2004. Discovery and characterization of two types of liver-expressed antimicrobial peptide 2 (LEAP-2) genes in rainbow trout. *Veterinary Immunology and Immunopathology* 101: 259-269.

Viral Immune Defences in Fish

A. Estepa[1,*], C. Tafalla[2] and J.M. Coll[3]

FISH VIRUSES AND FISH VIRAL DISEASES

As effective biosecurity measures to maintain the health status of fish stocks have increased in the past decades and bacterial diseases have been partially managed, viral diseases have emerged as serious problems to the fish aquaculture industry. Several major viral diseases such as Infectious Pancreatic Necrosis (IPN), Infectious Haematopoietic Necrosis (IHN), Viral Haemorrhagic Septicaemia (VHS), Spring Viraemia of Carp (SVC), Infectious Salmon Anaemia (ISA), Channel Catfish virus Disease (CCVD), etc., are a cause of severe losses in worldwide fish farming. Moreover, the fact that most of the viral fish diseases are notifiable to the OIE (Office International des Epizooties) indicates the importance of fish viruses worldwide.

There are no specific agents for the treatment of any of these viral diseases. Consequently, the use of preventive measures, including

Authors' addresses: [1]UMH, IBMC, Miguel Hernandez University, 03202, Elche, Spain.
[2]INIA, CISA Valdeolmos-28130 Madrid, Spain.
[3]INIA, Dept Biotecnología-Crt. Coruña Km 7-28040 Madrid, Spain.
*Corresponding author: E-mail: aestepa@umh.es

vaccination, seems as the most adequate method to control these viral agents in aquacultured fish (Biering *et al.*, 2005; Sommerset *et al.*, 2005). Despite extensive research over the years, the development of cheap and effective vaccines for the prevention of diseases caused by viruses in fish has proven to be a difficult task. As a result, only a few commercial vaccines are available for use against viral infections. For instance, in China and Japan, an inactivated vaccine and a recombinant protein vaccine against spring viraemia of the carp virus (SVCV), another fish rhabdovirus and red seabream iridovirus have been licensed, respectively (Sommerset *et al.*, 2005). In both Chile and Norway, different variants of a polyvalent vaccine against IPNV consisting in the recombinant VP2 viral protein are commercialized by different companies. One live attenuated VHSV vaccine is licensed in Germany, although its effectiveness is really limited. Most recently, the first DNA vaccine for fish has been approved for use in Canada against IHNV.

In this context, the knowledge of immune system function becomes essential for viral disease prevention strategies such as the development of vaccines and selection for increased disease resistance.

VIRAL HAEMORRHAGIC SEPTICAEMIA VIRUS (VHSV) AND OTHER FISH RHABDOVIRUSES

Rhabdoviruses constitute one of the largest groups of viruses isolated from teleost fish and are responsible for great losses in aquaculture production since they not only affect fish at the early stages of development, as most other viruses that infect fish, but can also produce a high percentage of mortality in adult fish of high economic value. Among rhabdovirus, VHSV, the causative agent of viral haemorrhagic septicaemia (VHS) disease, is one of the most important viral diseases of salmonid fish in aquaculture (Olesen, 1998; Skall *et al.*, 2005). In addition, VHSV has been also isolated from an increasing number of free-living marine fish species. So far, it has been isolated from at least 48 fish species from the Northern hemisphere, including North America, Asia and Europe, and fifteen different species including herring, sprat, cod, Norway pout and flatfish from northern European waters (Skall *et al.*, 2005).

VHSV is a negative-stranded RNA virus, and as all rhabdovirus, is an enveloped virus presenting a bullet-shaped morphology (Fig. 3.1A). Its genome consists of a negative single-stranded RNA molecule of ~11.1 kilobases (Kb) (GenBanK Accession number Y18263) (Schutze *et al.*,

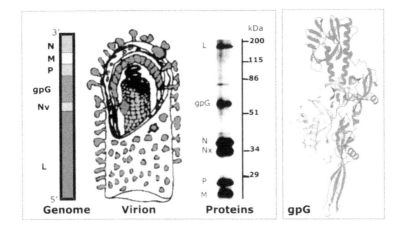

Fig. 3.1 Schemes of the genome and particles of VHSV and proteins from purified VHSV virions (A) and the gpG tridimensional structure (gpG) (B). A. Location of the N (nucleoprotein), M (membrane protein), P (phosphorilated protein), gpG (glycoprotein G), Nv (non-viral protein) and L (polymerase) genes in the negative RNA genome and particle of VHSV. Proteins present in purified VHSV virions separated by SDS-PAGE and visualized by Western blotting with an anti-VHSV polyclonal Ab. Numbers indicate molecular weights in kDa. B. Tridimensional structure of the gpG at physiological pH (yellow) and at the fusion pH 6 (green) modelled after the gpG from VSV (Roche *et al.*, 2006, 2007).

1999). VHSV, together with other piscine rhabdovirus such as IHNV, hirame rhabdovirus (HIRRV) (Essbauer and Ahne, 2001) and snakehead rhabdovirus, have been placed into the newly recognized *Novirhabdovirus* genus. This is due to the presence of an additional small gene encoding a non-structural Nv protein in their genome, a gene not present in other rhabdovirus (Schutze *et al.*, 1996; Walker and Tordo, 2000) (Fig. 3.1A). Thus, the VHSV genome codes for 6 different proteins, 5 of which are structural proteins (L, N, P, M and G proteins) and one is a non-structural protein (Nv protein). Inside the virus particle (Fig. 3.1A), the RNA genome is tightly packed by the nucleocapsid protein N. The viral RNA-dependent RNA polymerase, L, composed of L, is associated with the nucleocapsid core and P proteins, to form the replication complex. The viral envelope is a lipid bilayer derived from the host cell containing approximately 400 trimeric transmembrane spikes consisting of the single viral glycoprotein G (gpG). Actually, the sequence variation of the gpG from 74 isolates of VHSV have been published (Einer-Jensen *et al.*, 2004), their phylogenetics estimated (Thiery *et al.*, 2002) and the main sequence-fusion relationships studied (Estepa and Coll, 1996; Nunez *et al.*, 1998; Estepa *et al.*, 2001; Rocha *et al.*, 2004). The matrix protein, M, is localized inside the viral envelope between the membrane and the nucleocapsid. The non-structural protein Nv is only present in piscine novirhabdovirus (Basurco and Benmansour, 1995; Schutze *et al.*, 1996, 1999; Essbauer and Ahne, 2001), whose gene is localized between the G and L genes (3′N-P-M-G-NV-L 5′). It seems to play a role in virus pathogenicity in trout not yet fully characterized (Thoulouze *et al.*, 2004). After binding to its cellular receptor/s, VHSV enters into the host cells by endocytosis. The virions are then transported along endocytic pathway towards lysosomes. Beyond early endosomes but prior to lysosomes, the acid pH triggers the fusion of the viral envelope with endosomal membranes, releasing the nucleocapsid into the cytosol, where transcription and replication of the viral genome occurs. Finally, new virus particles are assembled and released in a process known as budding. Both the virus cell attachment and fusion are mediated by VHSV-gpG. Moreover, VHSV-gpG protein is the only viral protein able to induce neutralizing antibodies (Ab) in fish (Boudinot *et al.*, 1998; Lorenzen, 1998).

Most studies concerning VHSV have been performed in rainbow trout, since it is highly susceptible to the virus and also an important species in levels of production worldwide. From these rainbow trout studies, as well as from other performed in other susceptible species, it is well known that the transmission of this virus takes place horizontally throughout the water at temperatures lower than 15°C. Although previous immunohistochemistry studies had proposed the gut as an entry tissue (Helmick *et al.*, 1995), a recombinant IHNV expressing a luciferase gene obtained though reverse genetics recently pointed to the fin bases, where the virus persisted for at least 3 weeks, as the main site of entry of rhabdoviruses into the trout body (Harmache *et al.*, 2006).

Once inside the host, rhabdoviruses spread throughout the body and produce a systemic infection affecting mainly lymphoid organs such as head kidney and spleen, and most organs and tissues in later stages as shown by many different *in vivo* and *in vitro* studies (Estepa and Coll, 1991; Estepa *et al.*, 1993, 1994; Dorson, 1994; DeKinkelin and LeBerre, 1977; Tafalla *et al.*, 1998; Tafalla and Novoa, 2001). It has been demonstrated that VHSV replicates in head kidney and blood leukocytes, the monocytes/macrophages being one of susceptible populations. However, differences in the percentage of cells that support viral replication have been observed between different studies (Estepa *et al.*, 1992; Tafalla *et al.*, 1998). The dissemination within the host is thought to be via blood circulation, although may be not exclusively. Replication in endothelial cells results in characteristic petechial haemorrhages in muscle and internal organs, while external signs of disease include lethargy, darkened body, exophthalmia, pale gills and external haemorrhages as well in the skin and fins. The affected fish are slow and lethargic. In chronic stages dark discolouration and abnormal swimming behaviour may be observed (Wolf, 1988). Mortality depends on the age of the fish but it may be up to 100% in fry, although often less in older fish, typically from 30% to 70% (Skall *et al.*, 2005).

Despite many previous efforts (DeKinkelin *et al.*, 1995; Leong *et al.*, 1995) only DNA vaccination seems to be effective against VHSV. The DNA vaccine against VHSV, as well as the DNA vaccines against other rhabdovirus, is based on the plasmid in which the rhabdovirus-gpG gene is inserted under the control of cytomegalovirus (CMV) promoter. None of the other rhabdoviral genes has proven to be useful for the induction of immunity when delivered as DNA vaccines (Corbeil *et al.*, 2000; Lorenzen, 2005). These plasmids, when injected intramuscularly, induce a long-term specific immunity which is preceded by a strong early non-specific protective response (Lorenzen, 2000). Although not available as yet for VHSV, an IHNV DNA vaccine was approved in 2005 by Vical-Aqua Health Ltd. of Canada, a company related to Novartis (APEX-IHN). In Europe, until other problems such as security due to the viral promoter are not as well resolved, its commercialization seems difficult.

FISH DEFENCES AGAINST VIRAL INFECTIONS: THE TROUT/VHSV MODEL

It is a well-known fact that aquatic environments contain very high concentrations of pathogenic organisms and, therefore, fish live in intimate contact with high concentrations of bacteria, viruses and parasites. Regarding virus, it was recently estimated that about 10^{10} virus particles/l exist in aquatic habitats (Wilhelm, 1999; Tort, 2004). Taking in account that a picornavirus, for example, is capable of producing 10^4 progeny per cell, after only 3-4 replication cycles, the yield of virus will be high enough to infect all of the cells in an animal (Leong, 1997). However, under normal conditions, fish maintain a healthy state by defending themselves against virus or any other potential invaders through a complex network of defence mechanisms. Since fish represent the earliest class of vertebrates in which both innate and acquired immune mechanisms are present, this defence network includes structural barriers, antiviral cytokines and serum factors as well as the hallmark components of the adaptive immune response (T- and B-cell receptors and the major histocompatibility complex (MHC) molecules) (Du Pasquier et al., 1998; Plouffe et al., 2005).

Although many aspects of the fish immune response against virus infections have been widely analyzed, some of the basic parameters determining the balance between virus and fish immunity are yet to be understood completely. In particular, although the role of antigen, the bases of protective immune response and the nature of immunological memory have been studied during the past few years, these issues still remain controversial. These topics are worthy of further efforts because they impinge directly upon improved concepts for vaccines and adoptive immunotherapy.

NON-SPECIFIC DEFENCE MECHANISMS

Importantly, the non-specific or innate immune system of fish rather than the adaptive system appears to play a more central role in the response to infections (Tort, 2004). The reason is basically the intrinsic inefficiency of the acquired immune response of fish due to its evolutionary status and poikilothermic nature (Magnadottir, 2006), which results in a limited Ab repertoire, affinity maturation and memory and a slow lymphocyte proliferation. The acquired immune response of fish is, therefore, sluggish (up to 12 weeks), as compared to the antigen-independent (Tort, 2004), instant and relatively temperature-independent innate immune response (Du Pasquier, 1982; Ellis, 2001; Magnadottir, 2006). Concerning viral infections, it is particularly evident from studies on VHSV in rainbow trout that innate or non-specific defences may play a significant role in resistance to viral diseases (Ellis, 2001).

The innate defences against viruses in fish comprise a wide repertoire of biological actions in which both cellular and humoral components like the activity of macrophages and cytotoxic cells, complement, interferon (IFN) and antimicrobial peptides are implicated. As a whole, these actions are seemingly driven by germline-encoded pattern-recognition receptors capable of recognizing virus-associated molecular patterns, such as nucleic acid or viral surface glycoproteins. Since the molecular bases involved in the recognition of virus-associated molecular patterns by fish receptors have been poorly studied, no references to this process that initiates the fish antiviral responses will be made.

THE INTERFERON SYSTEM

The first and foremost line of defence against viruses is the type I Interferon system that encompasses in mammals at least 8 subclasses, including the classical IFN-α/βs. Type I IFNs are pH-resistant cytokines, which are produced by almost all cell types in response to a viral infection. The importance of type I IFNs in innate antiviral responses is underlined by the fact that mice lacking IFN-α/β receptor show a marked increase in susceptibility to a wide variety of different viruses (Muller *et al.*, 1994). Typically, IFN-α/β genes are induced rapidly during virus infection. Then, the IFN secreted by virus-infected cells triggers the up-regulation of several hundred genes, the so-called interferon-stimulated genes (ISG), some of which encode products directly responsible for inhibiting viral replication. Some of the ISG-encoding proteins with antiviral activity are the dsRNA-dependent protein kinase R (PKR), 2′,5′-oligoadenylate synthetase (OAS) and the GTP-ase Mx genes (Stark *et al.*, 1998; Samuel, 2001). Only the gene for OAS, which is a very important IFN-induced protein in higher vertebrates, is yet to be cloned in fish (Robertsen, 2006).

The fact that VHSV induces an IFN response was already indirectly demonstrated during the 1970s *in vivo* (DeKinkelin and Dorson, 1973) as well as *in vitro* in established cell lines (de Sena and Rio, 1975) and leukocytes using both live and inactivated virus (Rogel-Gaillard *et al.*, 1993; Congleton and Sun, 1996; O'Farrell *et al.*, 2002; Thorgaard *et al.*, 2002). The IFN found in the trout serum had 26 kDa (Dorson *et al.*, 1975) and its production was shown to occur very rapidly after VHSV infection (within 2 days in rainbow trout injected with VHSV (Dorson, 1994)) and in very young fish (rainbow trout fry of less than 0.2 g, 600 degree days; (Boudinot *et al.*, 1998). Thus, IFN-mediated antiviral defence mechanisms are able to respond during the early stages of a viral infection and this information has led many authors to believe that IFN responses provide some degree of protection until the specific immune defences are able to respond. Moreover, IFN synthesis increased *in vivo* with water temperature (11-15°C) and VHSV virulence (Dorson and DeKinkelin, 1974). It was proportionally synthesized with respect to viraemia to reach its maximal titres 4-5 days after infection (DeKinkelin *et al.*, 1977) and, as expected, protected not only against VHSV but also against heterologous viruses such as IHNV and IPNV (DeKinkelin and Dorson, 1973; Tengelsen *et al.*, 1991; Rogel-Gaillard *et al.*, 1993; Snegaroff, 1993; Congleton and Sun, 1996).

In the past, type I IFN-like sequences have began to become available in several fish species including zebrafish *Danio rerio* (Altmann *et al.*, 2003), catfish *Ictalurus punctatus* (Long *et al.*, 2004), Atlantic salmon *Salmo salar* (Robertsen, 2003), puffer fish *Fugu rubripes* (Zou *et al.*, 2005) and, recently, in rainbow trout, where three different genes have been identified (Zou *et al.*, 2007). All three rainbow trout genes are up-regulated *in vivo* in response to VHSV, but antiviral activity against this virus has only been demonstrated for two of them. The IFN designated as rtIFN3, which seems to act through another receptor, does not seem to have antiviral activity against VHSV (Zou *et al.*, 2007).

Among the antiviral protein expressed from the ISG, Mx proteins have been the most thoroughly studied in different fish species. Mx proteins comprise a family of large GTPases with homology to the dyamin superfamily (Haller *et al.*, 1998, 2007; Haller and Kochs, 2002). Mx genes have been cloned in many different species, including rainbow trout, where three different isoforms (Mx1, Mx2 and Mx3) have been characterized through PCR amplification, cloning and sequencing the mRNA, induced that was in the rainbow trout cell line RTG-2 upon IHNV infection (Trobridge and Leong, 1995; Trobridge *et al.*, 1997). Mx1 and Mx3 are very similar but Mx2 has differential cysteins and a nuclear localization signal (Leong, 1998), whereas the other two are localized in the cytoplasm. Antiviral activity has been demonstrated, but only for some teleost Mx proteins. Thus, inhibition of VHSV and HIRRV infectivity was clearly demonstrated in Japanese flounder cells stably expressing flounder Mx protein (Caipang *et al.*, 2003, 2005) but not antiviral activity was observed for any of the three trout Mx proteins against VHSV in salmon cell transfected with plasmids coding for the different Mx isoforms (Trobridge *et al.*, 1997). However, as the antiviral activity of Mx proteins against some virus appears to be cell type-specific (Haller *et al.*, 2007) and the activity of trout Mx proteins was evaluated in salmon cells, more experiments must be performed before it can be stated that rainbow trout Mx proteins do not possess an antiviral capacity. In any case, it has been demonstrated that the expression of the different trout Mx genes is induced in response to VHSV and Poly I:C both *in vivo* and *in vitro* in different cell types (Tafalla *et al.*, 2007), where it was observed that the differential expression of the different isoforms is more linked to the cell type than to the type of stimulus that triggered the expression. Moreover, Mx genes are known to be up-regulated in response to DNA vaccination against VHSV and are even thought to play a role in the early protection conferred by the vaccine since some correlation between Mx3 gene expression and protection has been observed in DNA vaccinated rainbow trout (McLauchlan *et al.*, 2003).

The presence of PKR in rainbow trout is suggested by the demonstration of increased eIF2 phosphorylation in rainbow trout cells after treatment with polyribocytidylic acid (Poly I:C) or infection with IPNV (Garner *et al.*, 2003). Moreover, the use of a specific inhibitor of PKR, 2-aminopurine (2-AP) in fish cells and its consequent effects also demonstrate the presence of PKR. It has been shown that 2-AP down-regulates the expression of Mx genes (DeWitte-Orr *et al.*, 2007; Tafalla *et al.*, 2008). In the rainbow trout monocyte cell line RTS11, in which VHSV is unable to complete its replication cycle, it has been demonstrated that the antiviral effects against the virus induced by Poly I:C such as Mx expression are mediated by PKR (Tafalla *et al.*, 2008), since the expression of N mRNA was significantly inhibited in cells pre-treated with Poly I:C and when cells were pre-incubated with Poly I:C in the presence of 2-AP, the levels of N mRNA were restored. This demonstrated that Poly I:C can limit viral transcription through an antiviral mechanism dependent of PKR.

INNATE CELL-MEDIATED CYTOTOXICITY (CMC)

In mammals, innate cell-mediated cytotoxicity (CMC) reactions that lead to the lysis of virus infected cells in the early stages of infection are well characterized and differentiated from adaptive CMC performed by T lymphocytes and they are known to be mainly executed by natural killer (NK) cells. Both non-specific cytotoxic cells (NCCs) and NK-like cells as functional equivalent to mammalian NK cells have been identified in several fish species (Evans *et al.*, 1984, 1987, 1990; Fischer *et al.*, 2006; Utke *et al.*, 2008), but there are only a few functional studies determining the mechanism of action against virus-infected target cells by NK-like effector cells (Yoshinaga K, 1994; Hogan *et al.*, 1996), in part due to the lack of tools for studying these processes in detail. In fact, neither Abs for the clear discrimination of NK cells according to the mammalian CD nomenclature nor genes homologous to NK cell receptors in higher vertebrates, such as Fcg receptor(R)III (CD16), CD56, CD158 (KIR) or CD161c (NK1.1), have been reported for fish (Fischer *et al.*, 2006; Utke *et al.*, 2007).

In response to VHSV infection, both innate and adaptive cell-mediated immune response represented by NK-like cells and T cytotoxic cells have been recently described (Utke et al., 2007). In this work and regarding innate CMC, peripheral blood leucocytes (PBL) isolated from VHSV-infected rainbow trout killed xenogeneic MHC class I-mismatched VHSV-infected cells (carp EPC cells infected with VHSV). When compared to PBL from uninfected control fish, PBL from the infected fish showed a higher transcriptional level of the natural killer cell enhancement factor (NKEF)-like gene (Zhang et al., 2001) as measured by real-time RT-PCR. To date, NKEF, which is involved in NK-cell regulation in mammals, is the only marker that can be used to obtain information on NK cell activation in rainbow trout. Unexpectedly, the NK-like cell-mediated cytotoxicity observed against VHSV-infected EPC cells was found later during infection than CTL-like responses against VHSV-infected MHC class I-matched target cells, something in contradiction with the generally accepted rule that innate immune mechanisms represent the first line of defence after viral infections. Therefore, more studies on CMC in rainbow trout are needed to clarify this point.

MACROPHAGE-MEDIATED RESPONSES

Immune functions carried out by macrophages are thought to be of particular importance in the resistance to viral infections. Macrophages can limit viral dissemination within the host by two different mechanisms that have been named as either extrinsic or intrinsic activity (Stohlman et al., 1982). Extrinsic antiviral activity is the ability of macrophages to inhibit viral replication in another susceptible cell line. This can be performed through the action of many different factors, such as IFN production, or the liberation of reactive oxygen and nitrogen metabolites (Croen, 1993; Tafalla et al., 1999). On the other hand, intrinsic antiviral activity is defined as the permissiveness or non-permissiveness of macrophages themselves to support viral replication. These two mechanisms, generally independent (Stohlman et al., 1982), are of great importance in determining the outcome of a viral infection.

Previous reports demonstrate that rainbow trout (*Oncorhynchus mykiss*) primary cultures of macrophages are susceptible to VHSV (Estepa et al., 1992; Tafalla et al., 1998). However, many differences were found between these two studies. Certain studies (Estepa et al., 1992; Tafalla *et al.*, 1998) infected unfractioned cell kidney cells in fibrin clots and found total cell lysis after one week of infection, whereas in Tafalla *et al.* (1998), the total blood leucocytes or head kidney macrophages did not show an apparent cytophatic effect and, although the viral titre increased with time in cultures indicating susceptibility, the percentage of macrophages that were, in fact, supporting the infection was very low as determined by immunoflourescence. These different results could be due to many factors, but in the case of immune cells, probably one of the most important factors in determining the outcome of a viral infection is the differentiation/ activation state (McCullough *et al.*, 1999). Recent experiments performed in our group demonstrate that the established monocyte-like cell line from rainbow trout RTS11 is not susceptible to VHSV replication, since although there is a transcription of viral genes, the translation of viral proteins is interrupted (Tafalla *et al.*, 2008).

Different studies concerning the effect of VHSV on macrophage functions have also been performed in turbot (*Scophthalmus maximus*), another susceptible species. In this species, it was demonstrated that VHSV does not have a significant effect on the respiratory burst capacity of macrophages *in vitro* (Tafalla *et al.*, 1998), although *in vivo* the viral infection produced a significant up-regulation of this function through the activity of an immune factor liberated to serum by the infection (Tafalla and Novoa, 2001). This factor was postulated to be IFN-γ. As well, different studies have been performed concerning the role of nitric oxide (NO) production in the defence against VHSV in this same specie (Tafalla *et al.*, 2001). It was demonstrated that NO is capable of decreasing the infectivity of VHSV (Tafalla *et al.*, 1999). Therefore, it seems that as observed for many mammalian viruses, NO production also plays an important role in antiviral defence in fish. In rainbow trout, the inducible NO synthase (iNOS) has been cloned and sequenced (Wang *et al.*, 2001), and as expected from the results obtained in turbot, VHSV *in vivo* infection up-regulated its levels of expression (Tafalla *et al.*, 2005). This iNOS up-regulation was observed in the spleen, head kidney and liver of rainbow trout intraperitoneal injected with VHSV mostly at day 7 post infection.

INDUCTION OF OTHER CYTOKINES

In addition to IFN, upon the encounter with a virus, all immune cells within the host secrete a great number of cytokines, which are responsible for the triggering of the non-specific immune response and also act as mediators of the specific defence mechanisms. However, very little is known about the specific role of these molecules in fish antiviral defences and fish virus resistance so far.

Genes encoding pro-inflammatory cytokines such as interleukin 1 β (IL-1 β), tumour necrosis factor α (TNF-α), transforming growth factor β (TGF-β) and IL-8 are known to be up-regulated in response to VHSV at early times post-infection. IL-1 β was mainly induced on haematopoietic organs such as the spleen and head kidney (Tafalla et al., 2005), as occurred in response to IHNV (Purcell et al., 2004). Although its role in antiviral defence in unknown, it has been demonstrated that the in vivo administration of IL-1 β-derived peptides confers resistance to VHSV in rainbow trout 2 days post-administration (Peddie et al., 2003), thus indicating a role in defence more decisive for the outcome of the infection that just triggering the immune response. Two different TNF-α molecules have been identified in rainbow trout (Zou et al., 2003), although differences in regulation and functionality have not been thoroughly studied.

Transcription of IL-8, a cytokine belongs to the CXC family of chemokines (Laing et al., 2002) that can be catalogued within the pro-inflammatory cytokines as well as within chemokines (cytokines with chemotactic activity) was also induced in response to VHSV (Tafalla et al., 2005). As previously demonstrated for IHNV (Purcell et al., 2004), Tafalla et al. (2005) show that in the spleen, IL-8 expression was strongly induced in response to the virus at days 1 and 2 post-infection. In the head kidney, although the results were not significant, an increased transcription in response to VHSV was also observed in most individuals 1 day post-infection. Therefore, since a strong IL-8 expression was induced in lymphoid organs in response to the virus in vivo, it may be possible that in vivo IL-8 expression is not only induced directly by the virus but through other factors or cytokines produced by cell types other than macrophages. This is confirmed by the fact that VHSV in vitro does not significantly stimulate head kidney macrophages for IL-8 production (Tafalla et al., 2005). In mammals, IL-8 is known to be induced by pro-inflammatory cytokines such as IL-1, IL-6 or TNF-α (Grignani and Maiolo, 2000). In trout head kidney leukocytes, it has been demonstrated that IL-8 expression can be induced by a combination of LPS and TNF-α (Sangrador-Vegas et al., 2002). When subtractive suppressive hybridization was performed with VHSV-infected rainbow trout leukocytes, an homologue to a human CXC chemokine and to other chemo-attractant molecules were obtained (O'Farrell *et al.*, 2002). All these results give weight to the hypothesis that chemokines play an essential role in viral defence, as can be concluded from the fact that many viruses have created different strategies to inactivate chemokines in the host (Liston and McColl, 2003).

Although mainly inhibitory, it is known that TGF-β—at early stages of infection—can facilitate CD8+ CT responses such as differentiation (Suda and Zlotnik, 1992) and IL-2 secretion (Swain *et al.*, 1991). Although it is unknown whether all these functions are true for fish TGF-β, it was demonstrated that bovine TGF-β inhibited the respiratory burst of rainbow trout macrophages (Jang *et al.*, 1995). Therefore, it may be possible that the induction of TGF-β that takes place in response to VHSV immediately after infection, mostly in the spleen, allows the virus to enter into macrophages, as it is known that VHSV replicates in rainbow trout macrophages (Estepa *et al.*, 1992). We still do not know whether this TGF-α induction in response to VHSV at the early stages of the infection is beneficial or detrimental for the host.

IDENTIFICATION OF NEW EARLY GENES INDUCED BY VHSV BY USING MICROARRAYS

Although some genes directly induced by VHSV infection, for example, vig-1 and vig-2 (Boudinot *et al.*, 1999, 2001b), have been identified using mRNA differential display methodology, with the use of microarrays, a new technology is available to study new genes involved in the innate response to rhabdoviruses. To design which trout sequences could be included in the microarrays, orthologous genes related to rhabdoviral resistance or immune-related genes mapped in other species might be first selected and compared with trout EST sequences. This compared analysis will help identify the existence of other possible candidate sequences not yet identified or mapped in the trout genome, but present in the genome maps of other fish or mammalian species. The cDNA obtained from tissue mRNA and/or oligo probes defined from EST trout sequences could be also used randomly to design microarrays. The experiments should include parallel non-infected healthy tissues versus infected tissues under different conditions. From the comparison between the transcriptomes obtained from healthy and infected samples, the genes included in the microarrays would be classified in up-, down- and non-regulated by rhabdoviral infections. Actually, more than 300,000 ESTs from over 175 salmonid cDNA libraries derived from a wide variety of tissues and different developmental stages (von Schalburg et al., 2005), are deposited in Genbank.

To our knowledge, there are only a few examples on the use of cDNA microarrays to study immunity to VHSV. A cDNA microarray was performed in Japanese flounder (Paralychthys olivaceous) following injection of a VHSV DNA vaccine based on gpG (Byon et al., 2005, 2006; Yasuike et al., 2007). The cDNA chip used in this study contained a total of 779 clones consisting in 228 immune-related genes and 551 unknown genes. The gene expression profiles were compared between gpG and empty vector injected groups. The greatest number of genes (16.6%) with changed expression levels were observed at 3 days after injection. Of those, 91.4% were up-regulated (31% known and 60.4% unknown). Up-regulated genes include genes related to the non-specific immune response such as Kupffer cell receptor, TNF-α, MIP1-α, IL1 receptor, coagulation factor XIII, CC chemokine receptor, Mx, etc., transcription factors such as IF-induced protein, TAP2 protein, caspase-10d, etc., and even a few genes related to the late specific antibody response such as the CD20 receptor and B cell adhesion molecule. A number of unknown genes were also up-regulated. One such gene mRNA increased a maximum of 56-fold 3 days after infection. A promising area of new research, therefore, is to characterize those highly up-regulated unknown new genes. These first studies demonstrate that the microarray technology has opened a new way to analyze the expression profiles induced by rhabdovirus infections and/or immunizations and to discover new immune-related genes that will help us to gain further insights into the molecular mechanisms of immunity to rhabdoviruses.

SPECIFIC DEFENCE MECHANISMS

As a group, fish are in the baseline of vertebrate 'radiation' (Schluter, 1999) and their specific immune system anticipates the most sophisticated mammalian repertoire of specific immune responses (Tort, 2004). Fish above the level of the agnatha display typical vertebrate adaptive immune responses characterized by the presence of immunoglobulins (Ig), T-cell receptors, major histocompatibility (MHC) complexes and recombination activator genes (RAG 1 and RAG 2) (Watts, 2001). The genes for T-cell receptor α and β polypeptides have been cloned and sequenced several fish species such as rainbow trout (Partula et al., 1994, 1995, 1996) and channel catfish (Wilson, 1998). RAG 1 and 2 from rainbow trout (Hansen and Kaattari, 1995, 1996) and zebra fish (*Danio rerio*) (Greenhalgh, 1995; Willett, 1997 #4941; Willett *et al.*, 1997) have been cloned and sequenced and MHC I and II genes have been extensively characterised (Flajnik, 1999). In humans, MHC I and II genes are linked, but in teleosts they are found on separate chromosomes (Flajnik, 1999). Nevertheless, it is clear that fish adaptive immune responses are less developed than those of higher vertebrates. For example, in comparison with mammals, the piscine specific humoral responses are generally considered to be less efficient due to limited Ig isotype diversity and a poor anamnestic response (Pilstrom, 1996). Regarding fish Ig isotype diversity, although only two classes of Ig had been described in teleosts; IgM and IgD (Pilstrom, 1996; Wilson *et al.*, 1997; Hordvik, 2002), other isotype, IgT, have been recently discovered in trout (Boshra *et al.*, 2005).

Although our knowledge is still limited to have a full understanding of the reasons for those differences, one major dissimilarity between higher vertebrates and fish is that piscine immune response is severely affected by environmental temperature (Bly and Clem, 1991, 1992; Bly, 1997). The specific immune response is particularly affected since, at non-permissive temperatures (low ambient/water temperature), the T-dependent specific immune response is compromised mostly due to the non-adaptive lipid composition of T-cell membranes (Bly, 1994). In contrast, memory T-cells and macrophages are less affected.

On the other hand, farmed fish may be more affected by temperature fluctuations than wild fish because, due to confinement, they are unable to thermoregulate by moving away from the adverse temperatures (Watts, 2001).

THE ROLE OF VIRUS-SPECIFIC ANTIBODIES IN PROTECTION

The existence of a specific humoral immune response, which by definition requires B, T-helper and antigen-presenting cell collaboration, has been demonstrated in all teleost species so far studied (Kaattari, 1992; Watts, 2001) because the presence of specific Abs against viruses, bacteria, helminths and protozoa are presented in serum as well as in mucus, bile and eggs (Lobb and Clem, 1981; Romboult, 1993; Yousif *et al.*, 1993). However, the role of specific Abs in protection against infectious agents is not always evident in fish.

The fish Ab response to virus has been characterized in detail for VHSV and IHNV rhabdovirus (Lorenzen *et al.*, 1999, 1999b). Against VHSV and IHNV, the protective role of the virus-specific Abs seems unquestionable since the transference of sera or purified Abs from rainbow trout surviving infection with IHNV or VHSV or sera from vaccinated fish to naïve fish protects them against an infection with virulent-virus (*in vivo* passive immunization assays) (Amend and Smith, 1974; DeKinkelin *et al.*, 1977; Bernard *et al.*, 1983, 1985; Olesen and Vestergard-Jorgensen, 1986; Lorenzen *et al.*, 1999b). Although the precise mechanism/s involved in the protection by passive immunization are still not well known, *in vivo* protection correlated with the presence of *in vitro* neutralizing activity in those sera (Bernard *et al.*, 1983, 1985; LaPatra, 1993). On the contrary, there is more ambiguity about a role of the virus-specific Abs in ongoing infection. Trout anti-VHSV Abs peak 6–10 weeks after VHSV natural infection (Olesen, 1986; Olesen *et al.*, 1991) or 8-20 weeks after natural IHNV infection (Hattenberg-Baudouy *et al.*, 1989) at optimum temperature. In both cases, Abs peaked after mortality had ceazed (maximal mortalities occur ~ 1 week after natural infection). Therefore, Abs produced as response to viral infection appear too late to play any role in protection of non-immunised fish against acute disease. One possibility is that these Ab might have a protective effect during the later stages of a disease outbreak and may allow survivors to eliminate or suppress residual virus (Lorenzen, 1999).

The protective effect of MAbs to the different rhabdoviral proteins was also tested in passive immunization experiments. No protection was observed in fingerling trout injected with MAbs to the N, M1 and M2 proteins and protection was observed in two (neutralizing and non-neutralizing) out of three gpG-specific MAbs (Lorenzen *et al.*, 1990). Therefore, Abs induced by the gpG can be protective but not always and *in vitro* non-neutralizing Abs can also be protective *in vivo*. More recent passive immunizations using sera from trout immunised with plasmids encoding gpG from VHSV or IHNV, have confirmed their *in vivo* protection (Boudinot *et al.*, 1998).

As variability exists between rhabdovirus isolates, an important aspect concerns whether protection occurs across variability. No differences were detected in cross-neutralization assays of sera from trout hyper-immunized by injection with five IHNV electropherotypes (Basurco *et al.*, 1993). Similarly, trout antiserum produced against one isolate of IHNV was capable of *in vitro* neutralizing isolates from 3-10 different antigenic groups and protected against all variants *in vivo*. Furthermore, sera from trout resistant to infection with a VHSV serotype were capable of conferring resistance to other serotypes (DeKinkelin and Bearzotti, 1981; Basurco and Coll, 1992). Immunization with the gpG gene of an isolate of VHSV protected against challenge with two serologically different VHSV isolates (LaPatra, 1993; LaPatra *et al.*, 1994a, b; Lorenzen *et al.*, 1999a). All these cross-protection studies suggest that in the case of trout rhabdoviruses, a single vaccine might be efficacious against most of the antigenic variants.

In the case of rhabdovirus infections, the temperature is a critical factor in the development of virus-specific Ab in fish. At temperatures lower than 15°C, the optimal rhabdoviral *in vitro* replication temperature, outbreaks of VHSV cause massive mortalities but there is also the development of neutralizing Abs (Lorenzen, 1999). However, at higher temperatures, lower mortality and absence of virus-specific Ab are observed, at least after infection with no highly-virulent VHSV (Lorenzen *et al.*, 1999b) Probably, other defence mechanisms non-related to the specific immune response are implicated in virus clearance at higher temperatures since long lasting immunity is not established in these circumstances. To date, the factors that determine the outcome of a primary infection in non-immunized fish and their inter-relationships have not been determined (Lorenzen, 1999). In this context, to determine the specific early immune response-related genes directly implicated in the outcome of an infection would constitute an interesting task of research.

NEUTRALIZING VIRUS-SPECIFIC ANTIBODIES

Initial attempts to demonstrate the development of a neutralizing antibody response in trout surviving IHNV or VHSV infections had limited success (Jørgensen, 1971; de Kinkelin, 1977; Lorenzen *et al.*, 1999b) because, as later demonstrated, only when including complement in the *in vitro* assays, trout serum neutralizing activity could be detected (Dorson and Torchy, 1979; Olesen and Vestergard-Jorgensen, 1986). The inhibitory effect of EDTA/EGTA on the complement activity indicated that a similar process to complement activation in mammals (classical pathway) is involved in this complement-dependent neutralization of VHSV and IHNV, but attempts to demonstrate involvement of C3 have not been successful and the full mechanism of the role of complement is still unknown (Lorenzen *et al.*, 1999b; Ellis, 2001). The complement-dependent neutralization mechanism may be related to the enveloped nature of the rhabdovirus since neutralization by trout serum to non-enveloped viruses could be detected without the use of complement. Future studies will have to address whether viral neutralisation requires the action of the lytic pathway (i.e., assembly of the membrane attack complex), or whether C3/C4 fixation on the surface of the virus is sufficient for its neutralization (Boshra *et al.*, 2006). In addition, since teleosts contain multiple C3 isoforms, it will be of interest to determine which particular isoforms are involved in the neutralization of viruses (Boshra *et al.*, 2006).

Involvement of trout Ig in the neutralization of VHSV and IHNV has been well documented. Thus, the macroglobulin fraction of trout serum was used for neutralization tests (Bernard *et al.*, 1985) and rabbit anti-trout Ig (Olesen and Vestergard-Jorgensen, 1986; Hattenberg-Baudouy *et al.*, 1989) or MAbs anti-trout Ig (Lorenzen, 1998) inhibited neutralization.

Neutralization titers varied among individuals from the same population around 100 (dilution of the serum that causes 50% reduction in the number of rhabdoviral plaques obtained *in vitro*). Trout with titers ~100 could not be re-infected experimentally with VHSV (DeKinkelin *et al.*, 1995). Occasionally, titers of about 1000 can be found among natural survivors to the disease and those titers can even be increased 10-fold by further immunization by repeated injections. The detectable levels of *in vitro* neutralising Abs after infection lasts during 4-6 months (Olesen and Vestergard-Jorgensen, 1986; Noonan *et al.*, 1995) and there was no increment of their titer after re-infection (Cossarini-Dunier, 1985; Olesen *et al.*, 1991; Traxler *et al.*, 1999). The low sensitivity of the neutralizing assays continues to be a limiting factor to accurately estimate the neutralizing activity of those sera.

The trout Abs, which neutralize VSHV or IHNV, only recognize their gpG (Engelking and Leong, 1989; Lorenzen *et al.*, 1990, 1993b; Olesen *et al.*, 1991; Xu *et al.*, 1991; Bearzotti *et al.*, 1995; Huang *et al.*, 1996). It was also demonstrated by injection of the gpG gene that the gpG alone is able to induce a neutralizing Ab response in trout (Boudinot *et al.*, 1998; Lorenzen, 1998).

One month after injection of the gpG gene, the protection was specifically restricted to the homologous rhabdovirus (Kanellos *et al.*, 1999; Traxler *et al.*, 1999). The switching time from non-specific to specific immune responses depends on the size of the trout, the dosage of

VHSV/vaccine and the temperature (McLauchlan *et al.*, 2003). The neutralizing Abs induced by injection of the gpG gene of VHSV were detected during 6 months but protection lasted longer than 9 months (McLauchlan *et al.*, 2003), suggesting that either: (i) the *in vitro* technique did not detect all neutralizing Abs; (ii) non-neutralizing Abs mediated protection *in vivo* as it happened with some MAbs (Lorenzen *et al.*, 1990); or (iii) there were cellular mechanisms (i.e., cytotoxic) involved in protection.

Affinity maturation due to somatic hypermutation of the V genes (genes coding the variable part of each Ig chains) is a well-known mammal mechanism to enhance their Ab response. Following immunization of trout with the T-cell dependent antigen FITC-KLH, the Ab response after ~ 30 days shifted to a 2-3-fold higher affinity at ~ 90 days (Cain *et al.*, 2002), a much lower increase than those seen in mammals. In the case of natural infections of VHSV, it is well known that: (i) *In vitro* neutralizing Abs can only be detected in 54% of the survivor trout (Olesen *et al.*, 1991); (ii) The majority of the survivor trout endure a second VHSV infection (Basurco and Coll, 1992), thus allowing for the genetic selection of trout strains with a > 90% survival to the VHSV infection (Dorson *et al.*, 1995); (iii) After the second infection, there is no detectable increase in the levels of the neutralizing Abs (Olesen *et al.*, 1991); (iv) The injection of trout with recombinant gpG proteins produced in *E. coli* (Lorenzen *et al.*, 1993a; Estepa *et al.*, 1994), yeast (Estepa *et al.*, 1994) and/or baculovirus (Koener and Leong, 1990) have not obtained good protection levels despite good correlations between *in vitro* neutralization titers and *in vivo* protection obtained by injection with recombinant gpG fragments of IHNV (Xu *et al.*, 1991) or VSHV (Lorenzen *et al.*, 1993a; Estepa *et al.*, 1994); and (v) Protection > 95% have been obtained by intramuscular injection of the gpG gene of rhabdoviruses with low levels of neutralizing Abs (Anderson *et al.*, 1996a, b; Lorenzen, 2000, 2008; Fernandez-Alonso *et al.*, 2001; LaPatra *et al.*, 2001; Lorenzen *et al.*, 2001; McLauchlan *et al.*, 2003; Lorenzen and LaPatra, 2005).

VHSV-gpG REGIONS IMPLICATED IN THE INDUCTION OF NEUTRALIZING VIRUS-SPECIFIC ANTIBODIES

As above indicated, VHSV-gpG protein is the only viral protein able to induce a neutralizing Ab response in trout (Boudinot *et al.*,

1998; Lorenzen, 1998) and, accordingly, neutralizing Ab only recognize VHSV-gpG (Engelking and Leong, 1989; Lorenzen *et al.*, 1990, 1993a; Olesen *et al.*, 1991; Xu *et al.*, 1991; Bearzotti *et al.*, 1995; Huang *et al.*, 1996). Specifically, these neutralizing Abs seem to recognize discontinuous conformational epitopes rather than lineal continuous epitopes on VHSV gpG (Lorenzen *et al.*, 1990, 1993a). Attempts to map the main Ab epitopes (B-cell epitopes) on the VHSV gpG by the use of overlapping 15-mer synthetic peptides showed that of 3 neutralizing MAbs none could be mapped and highly neutralizing trout sera only significantly recognized a few peptides (Fig. 3.2) (Fernandez-Alonso *et al.*, 1998b). Conformational (discontinuous) B-cell epitopes of IHNV and VHSV may, thus, be more immunogenic than linear (continuous) epitopes in trout or, alternatively, the antigenicity of B-cell epitopes might be more easily lost in immunoblotting assays.

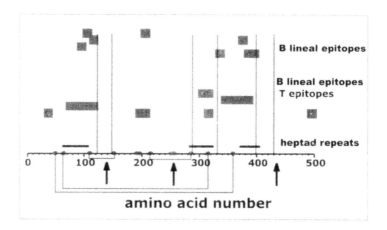

Fig. 3.2   Relative positions of B- and T-cell epitopes and the VHSV gpG structural features. Red solid circles, Cysteines connected by horizontal lines meaning its pairing by disulphide bridges (Einer-Jensen *et al.*, 1998). Vertical arrows, positions 140 and 433 where the simultaneous epitopes of neutralizing MAb C10 have been mapped (Bearzotti *et al.*, 1995) and 253 where the neutralizing MAb 3F1A12 was mapped. Vertical lines, locations of the α-helixes in the corresponding VSV tridimensional structure of the protein G of VSV at low pH (Roche *et al.*, 2006). Blue horizontal thin lines, non-cannonical hydrophobic amino acids forming 4-5 contigouos heptad repeats abcdefg (Coll, 1995b). Green horizontal black wide lines, B-cell lineal epitopes mapped by pepscan ELISA in 6 trout sera showing main recognized peptides 99-113 (6 trout), 199-213 (6 trout) and other peptides (1-4 trout) (Fernandez-Alonso *et al.*, 1998b) and T-cell epitopes mapped by pepscan lymphoproliferation in 12 VHSV-surviving trout showing main recognized peptides 299-323 (7 trout), 339-393 (4 trout) and other peptides (1-3 trout) (Lorenzo *et al.*, 1995d).

Most of the epitopes recognized in gpG by neutralizing MAbs are conformation-dependent and some discontinuous, although few neutralizing MAbs are yet available to reach definitive conclusions (Olesen *et al.*, 1993; Huang *et al.*, 1994; Bearzotti *et al.*, 1995). Thus, there are only 2 well-characterized VHSV conformational neutralizing MAbs: C10 (Bearzotti *et al.*, 1994, 1995) and 3F1A12 (Lorenzen, pers. comm.). An attempt to obtain more neutralising MAbs to VHSV among those MAbs selected by FACS screening (to maintain gpG conformation during screening), failed to obtain any neutralising MAbs among 25 reacting with gpG by FACS and ELISA (unpubl.). The difficulties to obtain neutralising MAbs could be due to an easy loss of conformation of the native gpG with the elevated temperature of the mice (37°C) (Lorenzen *et al.*, 1990; Coll, 1995a).

By sequencing MAb resistant mutants, the neutralizing MAb C10 was mapped simultaneously to positions 140 y 433 (Bearzotti *et al.*, 1995) and the 3F1A12 to position 253 (Lorenzen, pers. comm.) (Fig. 3.2). On the other hand, neutralization escape mutants selected by the use of IHNV-neutralizing MAbs were fully neutralized by sera from trout immunized with the wild-type IHNV strain. Additionally, attempts to isolate mutants escaping the neutralising activity of immune trout sera led to mutants mapping at sites distant from those identified by the MAbs (Roberti *et al.*, 1991; Winton *et al.*, 1998).

The neutralization *in vitro* assays are not only time consuming, labor intensive, low sensitivity and require sterile conditions but are also restricted to the detection of neutralizing Abs which are gpG conformation-dependent Abs (Lorenzen *et al.*, 1993a, 1999b; Fernandez-Alonso, 1999). On the other hand, non-neutralizing Abs (those directed towards lineal epitopes) can also induce *in vivo* protection and persist longer than neutralizing Abs (Lorenzen *et al.*, 1990). Furthermore, *in vitro* VHSV neutralizing Abs do not always correlate with its protection properties *in vivo* (LeFrancois, 1984; Lorenzen *et al.*, 1990; Lorenzen, 1999) and, as shown very recently, their detection by *in vitro* neutralization is highly dependent on the VHSV isolate used (Fregeneda-Grandes and Olesen, 2007). Therefore, assays to detect trout Abs directed towards non-neutralizing lineal epitopes on the gpG of VHSV (such as ELISA) could be a complement to the neutralizing Ab assays.

By using MAbs with specificity to trout IgM (DeLuca *et al.*, 1983; Sanchez and Dominguez, 1991; Warr, 1996), the response of serum Abs to

VHSV could be estimated not only by *in vitro* neutralizing Abs assays (Jorgensen *et al.*, 1991; Sanz and Coll, 1992a; Lorenzo *et al.*, 1996) but also by ELISA using captured VHSV (Olesen *et al.*, 1991), yeast recombinant gpG (Sanz and Coll, 1992b) or recombinant gpG fragments obtained in *E. coli* (Rocha *et al.*, 2002) as solid-phases. The solid-phase VHSV ELISA, although able to detect trout Abs to both conformational and lineal epitopes, suffers from high backgrounds and false positives (Olesen *et al.*, 1991). To detect anti-VHSV Abs in trout sera, it would be most convenient to have an ELISA method with higher sensibility and lower backgrounds.

Previous attempts to detect anti-VHSV Abs by ELISA using purified VHSV as solid-phase also had high backgrounds and involved the preparation of large amounts of purified VHSV. To increase the number of gpG epitopes per well, recombinant fragments of the gpG were used as solid-phase. Linearized recombinant G4 (aa 9-443) produced in yeast after destroying the inter molecular disulphide bonds of inclusion bodies (Estepa *et al.*, 1994, 1996) and frg11 (aa 56-110), a shorter fragment recognized by 40% of the trout anti-VHSV Abs on the pepscan peptides of the gpG (Fig. 3.2) (Fernandez-Alonso *et al.*, 1998b; Rocha *et al.*, 2002) were used to develop a higher sensitivity ELISA. The ELISA made with frg11 is presently being improved to detect lineal anti-VHSV Abs with higher sensibility and lower backgrounds (data not published).

The most recent elucidation of the gpG structure at physiological (Roche *et al.*, 2007) and low pH (Roche *et al.*, 2006) (Fig. 3.1B), suggested the reason of why the frg11 appeared to be so immunologically relevant to the trout immunological system: the frg11 goes though the exterior of the molecule from its top to bottom ~ 100 angstrom (by using the accepted homology alignment between the corresponding sequences of VSV and VHSV). Furthermore, recombinant frg11 was recognised by VHSV immunised trout serum in Western blots at low pH (data not published).

An explanation for the existence of VHSV resistant trout without *in vitro* detectable neutralizing Abs (Jorgensen *et al.*, 1991) could be the presence of lineal yet *in vivo* protective Abs. The existence of anti-VHSV Abs with *in vivo* protective activity despite the absence of *in vitro* neutralizing activity was demonstrated long ago (Lorenzen *et al.*, 1990). Because the anti-frg11 Abs seem to be abundant in trout serum, those could be a good candidates to further develop ELISA diagnostic methods

(Rocha *et al.*, 2002) with enough sensitivity to detect survivor trout carriers of VHSV.

Furthermore, the present ELISA methods to detect trout anti-VHSV Abs, however, rely only on the trout anti-IgM MAbs developed to date (DeLuca *et al.*, 1983; Sanchez *et al.*, 1989, 1993; Sanchez and Dominguez, 1991). However, although predominantly tetrameric, trout IgM exhibits further structural heterogeneities due to both different disulfide polymerization and/or halfmeric (1H+1L chains) subunits present in mucus (Bromage *et al.*, 2006). This redox diversity is not related to isotypic differences since single C genes (genes coding the constant part of each of the Ig chains) can generate all redox diversity (Ledford *et al.*, 1993) and it can be observed among all Ab clonotypes (Kaattari, 1998), in contrast to mammalian structural diversity (isotype). At least two more transcriptionally active Ig genes have been detected in other fish (Bromage *et al.*, 2006) and other isotype IgT have been recently discovered in trout (Boshra *et al.*, 2005). There are no definitive studies on the possible significance of those other isotypes and/or redox molecular species of trout IgM on the anti-rhabdoviral Ab response. To make these studies possible, however, more specific MAbs will be required.

SPECIFIC CELL-MEDIATED CYTOTOXICITY (CMC)

In mammals, specific cell-mediated cytotoxicity responses are executed by CD8+ cytotoxic T lymphocytes (CTLs) and it has been demonstrated that CTL responses provide a major defence mechanism for elimination of virus-infected cells (Zinkernagel and Doherty, 1979; Oldstone, 1987; Somamoto *et al.*, 2002). Furthermore, CTL activity has been able to confer complete protection in some cases, even in the absence of an antibody response (Lukacher, 1984; Bevan, 1989; Somamoto *et al.*, 2002).

The results of some studies strongly suggest that CTL are present in fish (Manning and London, 1996; Fischer *et al.*, 1998; Hasegawa *et al.*, 1998; Stuge, 2000; Nakanishi *et al.*, 2002) but the lack of specific surface markers of the CD nomenclature has not allowed the appropriate characterization of those cells in fish so far. Therefore, fish CTL and fish CTL responses are being mostly characterized at genetic level because homologous sequences of mammalian immunologically relevant molecules such as, MHC class I, TCR and CD8, have become available during the past few years. However, for most of these genes with homologous sequences to mammalian genes the function of the corresponding proteins has yet to be shown (Fischer et al., 2006). Other genes, as the genes involved in peptide loading of MHC class I molecules, β2-microglobuline (β 2m) (Shum et al., 1996; Rodrigues et al., 1998), low molecular mass protein and transporter associated with antigen processing, have also been detected in rainbow trout (Fischer et al., 2006). Moreover, as shown by a monoclonal Ab (mAb) directed against the recombinant Onmy-UBA*501 protein, rainbow trout classical MHC class I molecules are expressed in similar cell types as mammalian classical MHC class I molecules (Dijkstra et al., 2003).

Specific CMC against virus-infected autologous cells has been reported in catfish (Hogan et al., 1996) and in ginbuna crucian carp (Somamoto, 2000). However, the role of specific CMC in the antiviral defence against VHSV as well as against other fish virus are not well documented so far because the absence of appropriate MHC class I compatible effector/target cell systems for the establishment of specific CMC assays in susceptible fish (Utke et al., 2007, 2008). Since it has been recently discovered that the MHC class I sequence Onmy-UBA*501 (GenBank accession number AF287488) is shared by the clonal rainbow trout strain C25 and the rainbow trout gonad cell line RTG-2 (Dijkstra et al., 2003) an MHC class I restricted cytotoxicity assay using the combination of these clonal fish and VHSV-infected RTG-2 cells has been able established (Utke et al., 2007, 2008). By using this system, Utke et al. (2007, 2008) have demonstrated that PBL isolated from low dose viral haemorrhagic septicaemia virus (VHSV)-infected rainbow trout killed MHC class I-matched VHSV-infected cells and that those PBL showed a higher transcriptional level of the CD8α gene which is a typical marker for mammalian cytotoxic T cells. In addition, those studies also shown that VHSV-gpG protein was a more potent trigger of cytotoxic cells than the VHSV-N protein since leucocytes from fish DNA immunized against the N protein of VHSV kill only MHC class I compatible infected cells, while DNA immunization against the VHSV G-protein yielded leucocytes killing both, MHC class I compatible and incompatible virally infected cells (Utke et al., 2008). As recognized by the authors of these works, the relative importance and potential interdependence of humoral and cellular mechanisms for protection of rainbow trout against VHSV now remains to be determined.

INDUCTION OF T CELLULAR MEMORY RESPONSES

In addition to protection provided by virus-specific Abs, long-term protection to VHSV infection might also be mediated by T cellular memory. Until recently, most of the studies on this topic were restricted to *in vitro* estimation of leucocyte proliferative responses (lymphoproliferation assays) performed by adding polyclonal mitogens (Chilmonczyk, 1978a; Estepa and Coll, 1992a, b), whole inactivated rhabdovirus (Chilmonczyk, 1978b), isolated rhabdoviral proteins, recombinant rhabdoviral protein fragments (Estepa *et al.*, 1994) or pepscan peptides derived from the gpG of VHSV and covering their whole amino acid sequence (Lorenzo *et al.*, 1995a, c). Stimulation of *in vitro* lymphoproliferative responses (Nakanishi *et al.*, 1999) resulted only when the leucocytes were obtained from trout surviving VHSV infection but not when obtained from healthy trout (Estepa *et al.*, 1994).

On mammal/virus models leucocyte proliferative responses occur after presentation of a limited number of short viral protein peptides in the membrane of the host infected cells in the MHC context, a mechanism reinforced in anamnestic responses. Leucocytes from most of the survivor trout of VHSV infections were capable of *in vitro* proliferation when cultured in the presence of short synthetic peptides designed from the gpG or N cDNA derived protein sequences of VHSV (Fernandez-Alonso *et al.*, 1995a, 1998a). The recognition of each of the 15-mer peptides of the gpG varied largely within individuals from the outbred trout population. Thus, T cell epitopes mapped by pepscan lymphoproliferation in 12 trout showed peptides 299-323 (7 trout), 339-393 (4 trout) and other peptides (1-3 trout) to stimulate proliferation (Lorenzo *et al.*, 1995b) (Fig. 3.2). In contrast, no significant proliferative responses were obtained for the above-mentioned peptides when leucocytes were obtained from either non-infected or genetically VHSV-resistant trout.

Head kidney cultures obtained from trout resistant to VHSV infections could be maintained during more than a year, retaining the capacity of gpG antigen-dependent lymphoproliferation when incubated with autologous adherent cells (mostly macrophages) treated with gpG. The proliferating long-term haematopoietic cell lines have the morphology of lymphocytes, cell surface β-TcR staining, and expression of α and β-chain TcR mRNA sequences and secreted non-specific immunostimulating molecules (Estepa *et al.*, 1996, 1999; Estepa and Coll, 1997). Because the *in vitro* cell immunological memory to VHSV exposure lasted during more than a year (Estepa *et al.*, 1994; Lorenzo *et al.*, 1995b) in contrast with the 4-6 months of the neutralizing Abs, lymphoproliferation could be perhaps used for diagnostic purposes.

Identification and separation of T cell subsets is critical for the continuation of the study of lymphoproliferative responses (cytotoxic or helper) to rhabdoviral antigens. Although initial attempts begun in catfish (Clem *et al.*, 1996) and trout TCR genes were identified by PCR (Partula *et al.*, 1994, 1995, 1996), production of MAbs to T cell markers have met with difficulties (Nakanishi *et al.*, 1999). For instance, after developing MAbs to trout head kidney melanomacrophages, all the obtained MAbs reacted not only with monocyte/macrophages but also with lymphocytes and granulocytes. Similarly, immunohistochemistry of gut, gill, liver, spleen, head kidney, and endothelial tissues showed similar patterns of staining with the different MAbs.

On the other hand, other assays are beginning to be used to study cellular memory to rhabdoviral antigens. Upregulation of MHC class II expression (another sign of T-cell activation) was observed in trout immunized with the gpG gene (Boudinot *et al.*, 1998) and the gpG was shown to be the target of most of the public anti-VHSV T cell response, suggesting that T helper cells probably contribute to the Ab response (Boudinot *et al.*, 2004). VHSV infection induced modifications of the TCRβ repertoire from polyclonal to oligoclonal as studied by espectratyping (methodology that delivers a global view of the TCRβ repertoires by showing the size distribution of part of the V region of the TcR) (Bernard *et al.*, 2006a). Specific VßJß rearrangements were amplified among spleen T cells in response to injection with the gpG gene from VHSV (Boudinot *et al.*, 2001a). Sequencing of cloned VßJß PCR products corresponding to spectratypes with reduced number of peaks (oligoclonal) identified recurrent sequences corresponding to the expanded clones. Interestingly, the sequence SSGDSYSE (amino acids in single letter code), was the most expanded in the spleen public T cell response to the VHSV gpG (Boudinot *et al.*, 2004). It was also amplified in gut intraepithelial lymphocytes (IELs) from VHSV infected trout (Bernard *et al.*, 2006b).

Despite all the studies mentioned above, the role of trout cellular immune memory in protection against rhabdovirus infections remains to be fully characterized until lymphocyte subpopulation MAb markers can be developed.

MUCOSAL IMMUNITY

The gut, final gastrointestinal tract, gills and skin represent the major interfaces between the trout and their water environment. Because of the permanent exposure to antigens, the lymphocytes existing in these body surfaces should be implicated in some kind of early immune response.

The gut-associated lymphoid tissue (GALT) of teleosts contains only intra-epithelial lymphocytes (IELs) scattered throughout the mucosa but no specialized structures similar to mammalian Peyer's patches. IELs in between gut epithelial cells have been observed in trout, carp (*Cyprinus carpio*) (Rombout *et al.*, 1993) and sea bass (*Dicentrarchus labrax*).

IELs prepared from the gut of sea bass expressed TCRβ transcripts (Scapigliati *et al.*, 2000) and a ~ 90% of leukocytes isolated from the carp intestine were Ig-negative lymphoid cells (Rombout *et al.*, 1997). More recently, the rearing of germfree zebrafish revealed an evolutionarily conserved gut innate response to bacteria (Rawls *et al.*, 2004). IELs purified from trout gut epithelium constituted an homogeneous population of small round cells with typical T lymphocyte morphology, no IgM transcripts and no IgM + cells, as estimated by flow cytometry (Bernard *et al.*, 2006a). In contrast, trout IELs expressed mRNA coding for the homologs of T cell markers CD8, CD4, CD28, CD3ε, TCRξ, TCRγ, and TCRβ, as did trout thymocytes and spleen leukocytes (Bernard *et al.*, 2006b). All these genes displayed high similarity with their respective mammalian counterparts and most likely are true orthologs. Taken together, these observations suggested that trout IELs were mostly T cells as in mammals.

Bell-shaped CDR3 TCRβ spectratypes (polyclonal) (Boudinot *et al.*, 2001a) with 6-10 peaks (amino acids) for all VβJβ combinations were observed in IELs from either young or adult trout, indicating a polyclonal TCRβ repertoire as in pronephros and spleen and contrary to mammals. IELs and spleen T cells could not be distinguished by either morphological or phenotypic characteristics but their TCRβ repertoire changed from polyclonal to oligoclonal in VHSV-infected trout (Bernard *et al.*, 2006a).

Further studies will be necessary to fully elucidate the origin and functions of trout IELs, which would provide interesting clues about the evolution of mucosal immunity and may improve the efficiency of possible oral vaccines to rhabdovirus.

Mucosal immunity in the final gastrointestinal tract and the skin epithelial cells and mucus was studied during IHNV infection after injection and waterborne routes (Cain et al., 1996). A moderate infiltration of lymphocytes was observed in the skin but specific Abs could not be detected in mucus by ELISA and were detected by neutralization only 1 day after infection and with very low titers. In contrast, serum neutralizing Abs appeared in survivor trout 21-28 days after infection. This study confirmed the early presence of rhabdovirus in the skin mucus (Helmick et al., 1995; Harmache et al., 2006) but with none or little associated pathology, suggesting that innate mechanisms of rhabdoviral resistance may be important as a first line of defence in skin/mucus.

However because neutralizing Ab titers were low and the ELISA used only detected one isotype of trout IgM (DeLuca et al., 1983), it is possible that some Ab response might have been not detected in the studies mentioned above due to low sensibility. Thus, in mammals, pentameric IgM is found in blood, dimeric IgA in the secretions and mucosa, monomeric IgE in the epidermis, and monomeric IgG in plasma and lymph. In trout, while the structure of the majority of induced anti-TNP Abs in serum, mucus, egg and ovarian fluid were tetrameric, the degree of polymerisation varied within individuals and halfmeric molecules consisting in 1H+1L chain appeared in mucus (Bromage et al., 2006). Furthermore, purified serum and mucus Ig from non-immunised trout showed different protein banding patterns by SDS–PAGE under reducing conditions, suggesting that mucosal Ab responses in trout may consist of heterogeneous forms of Ig differing from serum IgM (Cain, 2000). On the other hand, serum IgM was rapidly degraded when added to gut mucus in salmon (Hatten et al., 2001) and no estimations have been yet made on these activities under rhabdoviral infections.

The gills are an important site of inducible isoform of nitric oxide synthase (iNOS) when trout were injected with Renibacterium (Hong et al., 2003), suggesting that the gills might be also important not only as a point of entry of pathogens but also as a tissue capable of mounting an immune response. To our knowledge, there are no studies on the possible involvement of gill immunity on rhabdoviral infections.

There are histological and biochemical differences between the skin and mucus of trout and different salmonid species with different susceptibilities to the same pathogens, which suggest their importance to disease resistance. Susceptibility to rhabdovirus might depend on some of those intestinal/skin/mucus innate parameters yet to be studied.

GENETICS OF RESISTANCE TO RHABDOVIRAL INFECTION

Susceptibility to rhabdoviral infections may depend not only on different non-specific or specific mechanisms but also on other individual epigenetic characteristics, as suggested by the wide variation on mortality kinetics observed among individual trout belonging to genetically homogeneous clones (Ristow *et al.*, 2000; Quillet, 2007 #4165). However, there are strong evidences that genetics traits are involved in resistance of rainbow trout to rhabdoviral infections (Yamamoto, 1991; Dorson *et al.*, 1995).

Recently, the resistance to VHSV bath infection of nine rainbow trout homozygous clones produced from a genetically diverse population by using gynogenesis-based strategies (doubled haploid populations) was analysed (Quillet *et al.*, 2007). The results of this experiment showed a large variability in susceptibility to VHSV among to different clones since some clones were > 95% resistant to VHSV, while others were 0% resistant to waterborne infection, the natural route of diseases (Quillet *et al.*, 2007). The variability of resistance among homozygous clones was consistent with previous selection breeding procedures to improve to > 90% their resistance to rhabdovirus (Yamamoto, 1991; Dorson *et al.*, 1995; Slierendrecht *et al.*, 2001).

Susceptibility to IHNV was also variable among those homozygous clones, confirming previous studies with IHNV (Yamamoto, 1991; Trobridge *et al.*, 2000) and correlated with the susceptibility to VHSV, suggesting the existence of common non-specific mechanisms of resistance. Accordingly, an absence of correlation between rhabdovirus resistance and MHC haplotype was demonstrated (Slierendrecht *et al.*, 2001).

Regarding the non-specific mechanism underlying the resistance to VHSV and IHNV waterborne, a barrier mechanism of resistance is proposed since VHSV was seldom detected in resistant clones after a waterborne-challenge (Quillet *et al.*, 2007). The existence of such a 'barrier' mechanism in trout was also supported by the lack of rhabdovirus growth on fin tissue obtained from resistant families or heterozygous clones from the same strain (Dorson and Torchy, 1993; Quillet *et al.*, 2001). A previous work with IHN-resistant trout hybrids showed that resistance to waterborne-challenge correlated best with the lack of entry of rhabdovirus into the trout body while resistance to injection-challenge correlated best with production of neutralising Abs (LaPatra *et al.*, 1996).

Up to now, trout with increased resistance to rhabdoviruses have been produced by selective breeding and homozygous trout clones with opposite susceptibility to rhabdoviruses have been produced in one single generation. The trout clones obtained with extreme phenotypes (full resistance versus full susceptibility), can be used now for further genetic (search for QTL and candidate genes) and physiological (gene expression profiling by microarrays) studies to identify novel antiviral pathways and genetic (innate) factors involved in resistance to rhabdoviruses.

CONCLUSIONS

Most of the innate immune genes up- or downregulated upon rhabdoviral infections remain to be characterised in detail. Furthermore, each of these genes have isoforms and present individual sequence variability which are only beginning to be studied. In some of these aspects, trout populations with increased resistance (after selective breeding) or clones with >95% resistance or susceptibility to rhabdoviruses are now available to search for QTL, candidate new genes and for gene expression profiling by microarray analysis. These studies will contribute to identify new antiviral innate genes involved in resistance to rhabdoviruses.

Neutralization by trout Abs of VHSV and IHNV *in vitro* is dependant on complement; however, although this was discovered long ago, their mechanism of neutralization remains to be characterized in detail. The analysis of the specificity of anti-VHSV trout Abs has been complicated by a difficulties in their binding to rhabdoviral proteins by immunoblotting, while other assays, have demonstrated that trout can produce specific and functional Abs. Fractionation of trout sera with different levels of neutralizing Abs by affinity columns made by solid-phase gpG recombinant fragments could be a novel way to further characterize the Ab response between lineal and conformation-dependent Abs.

New anti-trout Ig MAbs will also be required to detect other isotypes and/or redox molecular species of trout IgM, so that their possible significance during the anti-viral Ab response could be studied, especially in mucosas. Similarly, the role of trout cellular immune memory in protection against rhabdovirus infections would remain to be fully characterised until lymphocyte subpopulation-specific MAb markers could be developed.

A challenging task for future research is also the identification of the parameters that determine the outcome of an infection with virulent rhabdovirus in naïve trout at low temperatures, i.e., whether the trout die or survive and become immune. Most probably some of the responses are to be derived from the deeper study of mucosal immunity. In addition, the identification of the receptors on the surface of susceptible cells will be of interest.

A better understanding of the determinants of trout immunity to rhabdoviruses could be one of the first steps towards the effective prevention of their infections. Till date, salmonid rhabdoviruses have been important research objectives due to their negative economic impact on aquacultured species. In the future, they might produce new tools for the basic study of the fish immune system (Lorenzen *et al.*, 2002).

In this context, the study of the fish immune systems and closer look to the relationship between pathogens and their hosts will be of benefit to the design of more potent vaccines in fish and anti-viral therapeutic agents, and to the identification of new targets for preventive actions in different cultured aquatic species.

Acknowledgements

This work was supported by the Spanish MEC projects AGL2004-07404-CO2/ACU, AGL2008-03519-C04 and Consolider ingenio 2010, CSD2007-02.

References

Altmann, S.M., M.T. Mellon, D.L. Distel and C.H. Kim. 2003. Molecular and functional analysis of an interferon gene from the zebrafish, *Danio rerio*. *Journal of Virology* 77: 1992-2002.

Amend, D.F. and L. Smith. 1974. Pathophysiology of infectious hematopoietic necrosis virus disease in rainbow trout (*Salmo gairdneri*): early changes in blood and aspects of the immune response after injection of IHN virus. *Journal of the Fisheries Research Board Canada* 31: 1371-1378.

Anderson, E.D., D.V. Mourich, S.C. Fahrenkrug, S. LaPatra, J. Shepherd and J.A. Leong. 1996a. Genetic immunization of rainbow trout (*Oncorhynchus mykiss*) against infectious hematopoietic necrosis virus. *Molecular Marine Biology and Biotechnology* 5: 114-122.

Anderson, E.D., D.V. Mourich and J.C. Leong. 1996b. Gene expression in rainbow trout (*Oncorhynchus mykiss*) following intramuscular injection of DNA. *Molecular Marine Biology and Biotechnology* 5: 105-113.

Basurco, B. and J.M. Coll. 1992. *In vitro* studies and *in vivo* immunisation with the first viral haemorrhagic septicaemia viruses isolated in Spain compared to international reference serotypes. *Research Veterinary Science* 53: 93-97.

Basurco, B. and A. Benmansour. 1995. Distant strains of the fish rhabdovirus VHSV mantain a sixth functional cistron which codes for a nonstructural protein of unknown function. *Virology* 212: 741-745.

Basurco, B., S. Yun and R.P. Hedrick. 1993. Comparison of selected strains of infectious hematopoietic necrosis virus (IHNV) using neutralizing trout antisera. *Diseases of Aquatic Organisms* 15: 229-233.

Bearzotti, M., A.F. Monnier, P. Vende, J. Grosclaude, P. DeKinkelin and A. Benmansour. 1994. Molecular aspect of antigenicity and pathogenicity of viral hemorrhagic septicemia virus (VHSV), a fish rhabdovirus. *IX International Conference on Negative Strand Viruses Estoril Portugal* Oct. 2-7, 242.

Bearzotti, M., A.F. Monnier, P. Vende, J. Grosclaude, P. DeKinkelin and A. Benmansour. 1995. The glycoprotein of viral hemorrhagic septicemia virus (VHSV): antigenicity and role in virulence. *Veterinary Research* 26: 413-422.

Bernard, D., B. Riteau, J.D. Hansen, R.B. Phillips, F. Michel, P. Boudinot and A. Benmansour. 2006a. Costimulatory Receptors in a Teleost Fish: Typical CD28, Elusive CTLA4. *Journal of Immunology* 176: 4191-4200.

Bernard, D., A. Six, L. Rigottier-Gois, S. Messiaen, S. Chilmonczyk, E. Quillet, P. Boudinot and A. Benmansour. 2006b. Phenotypic and functional similarity of gut intraepithelial and systemic T cells in a teleost fish. *Journal of Immunology* 176: 3942-3949.

Bernard, J., M.B. LeBerre and P. DeKinkelin. 1983. Viral haemorrhagic septicemia of rainbow trout: relation between the G polypeptide and antibody production of fish after infection with the F25 attenuated variant. *Infection and Immunity* 39: 7-14.

Bernard, J., M.B. Le Berre and P. De Kinkelin. 1985. Viral haemorrhagic septicaemia in rainbow trout: attempt to relate interferon production, antibody synthesis and structure of the virus with the mechanism of virulence. *Annals of Institute Pasteur/ Virology* E136, 13-26.

Bevan, M.J. 1989. Stimulating killer cells. *Nature* (London) 342: 478-479.

Biering, E., S. Villoing, I. Sommerset and K.E. Christie. 2005. Update on viral vaccines for fish. *Developmental Biology* (Basel) 121: 97-113.

Bly, J.E. and L.W. Clem. 1991. Temperature-mediated processes in teleost immunity: *In vitro* immunosuppression induced by *in vivo* low temperature in channel catfish. *Veterinary Immunology and Immunopathology* 28: 365-377.

Bly, J.E. and L.W. Clem. 1992. Temperature and teleost immune functions. *Fish and Shellfish Immunology* 2: 159-171.

Bly, J.E. and L.W. Clem. 1994. Temperature adaptation of lymphocyte function in fish. In: *Temperature Adaptation of Biological Membranes*, A.R. Cossins (ed.). Portland Press, London, pp. 169-184.

Bly, J.E., S.M.A. Quiniou and L.W. Clem. 1997. Environmental effects on fish immune mechanisms. In: *Fish Vaccinology*, R. Gudding, A. Lillehaug, P.J. Midtlyng and P.J. Brown (eds.). S. Karger Basel, pp. 33-43.

Boshra, H., J. Li and J.J.O. Sunyer. 2006. Recent advances on the complement system of teleost fish. *Fish and Shellfish Immunology* 20: 239-262.

Boshra, H., T. Wang, L. Hove-Madsen, J. Hansen, J. Li, A. Matlapudi, C.J. Secombes, L. Tort and J.O. Sunyer. 2005. Characterization of a C3a receptor in rainbow trout and Xenopus: The first identification of C3a receptors in nonmammalian species. *Journal of Immunology* 175: 2427-2437.

Boudinot, P., S. Boubekeur and A. Benmansour. 2001a. Rhabdovirus Infection Induces Public and Private T Cell Responses in Teleost Fish. *Journal of Immunology* 167: 6202-6209.

Boudinot, P., S. Salhi, M. Blanco and A. Benmansour. 2001b. Viral haemorrhagic septicaemia virus induces vig-2, a new interferon-responsive gene in rainbow trout. *Fish and Shellfish Immunology* 11: 383-397.

Boudinot, P., M. Blanco, P. de Kinkelin and A. Benmansour. 1998. Combined DNA immunization with the glycoprotein gene of viral hemorrhagic septicemia virus and infectious hematopoietic necrosis virus induces double-specific protective immunity and nonspecific response in rainbow trout. *Virology* 249: 297-306.

Boudinot, P., P. Massin, M. Blanco, S. Riffault and A. Benmansour. 1999. vig-1, a new fish gene induced by the rhabdovirus glycoprotein, has a virus-induced homologue in humans and shares conserved motifs with the MoaA family. *Journal of Virology* 73: 1846-1852.

Boudinot, P., D. Bernard, S. Boubekeur, M.I. Thoulouze, M. Bremont and A. Benmansour. 2004. The glycoprotein of a fish rhabdovirus profiles the virus-specific T-cell repertoire in rainbow trout. *Journal of General Virology* 85: 3099-3108.

Bromage, E.S., J. Ye and S.L. Kaattari. 2006. Antibody structural variation in rainbow trout fluids. *Comparative Biochemistry and Physiology* B143: 61-69.

Byon, J.Y., T. Ohira, I. Hirono and T. Aoki. 2005. Use of a cDNA microarray to study immunity against viral hemorrhagic septicemia (VHS) in Japanese flounder (*Paralichthys olivaceus*) following DNA vaccination. *Fish and Shellfish Immunology* 18: 135-147.

Byon, J.Y., T. Ohira, I. Hirono and T. Aoki. 2006. Comparative immune responses in Japanese flounder, *Paralichthys olivaceus* after vaccination with viral hemorrhagic septicemia virus (VHSV) recombinant glycoprotein and DNA vaccine using a microarray analysis. *Vaccine* 24: 921-930.

Cain, K., D.R. Jones and R.L. Raison. 2000. Characterisation of mucosal and systemic immune responses in rainbow trout (*Oncorhynchus mykiss*) using surface plasmon resonance. *Fish and Shellfish Immunology* 10: 651-656.

Cain, K.D., D.R. Jones and R.L. Raison. 2002. Antibody-antigen kinetics following immunization of rainbow trout (*Oncorhynchus mykiss*) with a T-cell dependent antigen. *Developmental and Comparative Immunology* 26: 181-190.

Cain, K.D., S.E. LaPatra, T.J. Baldwin, B. Shewmaker, J. Jones and S. Ristow. 1996. Characterization of mucosal immunity in rainbow trout *Oncorhynchus mykiss* challenged with infectious hematopoietic necrosis virus: Identification of antiviral activity. *Diseases of Aquatic Organisms* 27: 161-172.

Caipang, C.M.A., I. Hirono and T. Aoki. 2003. In vitro inhibition of fish rhabdoviruses by Japanese flounder, *Paralichthys olivaceus* Mx. *Virology* 317: 373-382.

Caipang, C.M.A., I. Hirono and T. Aoki. 2005. Induction of antiviral state in fish cells by Japanese flounder, *Paralichthys olivaceus*, interferon regulatory factor-1. *Fish and Shellfish Immunology* 19: 79-91.

Clem, L.W., J.E. Bly, M. Wilson, V.G. Chinchar, T. Stuge, K. Barker, C. Luft, M. Rycyzyn, R.J. Hogan, T. vanLopik and N.W. Miller. 1996. Fish immunology: The utility of immortalized lymphoid cells — A mini review. *Veterinary Immunology and Immunopathology* 54: 137-144.

Coll, J.M. 1995a. The glycoprotein G of rhabdoviruses. *Archives of Virology* 140: 827-851.

Coll, J.M. 1995b. Heptad-repeat sequences in the glycoprotein of rhabdoviruses. *Virus Genes* 10: 107-114.

Congleton, J. and B. Sun. 1996. Interferon-like activity produced by anterior kidney leucocytes of rainbow trout stimulated in vitro by infectious hematopoietic necrosis virus or poly I:C. *Diseases of Aquatic Organisms* 25: 185-195.

Corbeil, S., G. Kurath and S.E. LaPatra. 2000. Fish DNA vaccine against infectious hematopoietic necrosis virus: Efficacy of various routes of immunisation. *Fish and Shellfish Immunology* 10: 711-723.

Cossarini-Dunier, M. 1985. Effect of different adjuvants on the humoral immune response of rainbow trout. *Developmental and Comparative Immunology* 9: 141-146.

Croen, K.D. 1993. Evidence for antiviral effect of nitric oxide. Inhibition of herpes simplex virus type 1 replication. *Journal of Clinical Investigation* 91: 2446-2452.

Chilmonczyk, S. 1978a. In vitro stimulation by mitogens of peripheral blood lymphocytes from rainbow trout (*Salmo gairdneri*). *Annals of Immunology* 129: 3-12.

Chilmonczyk, S. 1978b. Stimulation specifique des lymphocytes de truites arc-en-ciel (*Salmo gairdneri*) resistantes a la septicemie hemorragique virale. *Comptes-Rendus de l' Académie des Sciences*, Paris 287: 387-389.

DeKinkelin, P. and M. Dorson. 1973. Interferon production in Rainbow trout (*Salmo gairdneri*) experimentally infected with Egtved virus. *Journal of General Virology* 19: 125-127.

DeKinkelin, P. and M. LeBerre. 1977. Isolament d'un rhabdovirus pathogéne de la truite fario (*Salmo trutta*, L. 1766). *Comptes-Rendus de l' Académie des Sciences*, Paris 284: 101-104.

DeKinkelin, P. and M. Bearzotti. 1981. Immunization of rainbow trout against viral haemorrhagic septicaemia (VHS) with a thermoresistant variant of the virus. *Developments Biological Standards* 49: 431-439.

De Kinkelin, P., A.M. Baudouy and M. Le Berre. 1977. Immunologie. Réaction de la truite fario (*Salmo trutta*, L. 1766) et arc-en-ciel (*Salmo gairdneri* Richardson,1836) à l'infection par un nouveau rhabdovirus. *Comptes-Rendus de l' Académie des Sciences*, Paris 284: 401-404.

DeKinkelin, P., J.P. Geraed, M. Dorson and M. Le Berre. 1977. Viral Haemorrhagic Septicaemia: Demonstration of a protective immune response following natural infection. *Fish Health News* 6: 43-45.

DeKinkelin, P., M. Bearzotti, J. Castric, P. Nougayrede, F. Lecocq-Xhonneux and M. Thiry. 1995. Eighteen years of vaccination against viral haemorrhagic septicaemia in France. *Veterinary Research* 26: 379-387.

DeLuca, D., M. Wilson and G.W. Warr. 1983. Lymphocyte heterogeneity in the trout, *Salmo gairdneri*, defined with monoclonal antibodies to IgM. *European Journal of Immunology* 13: 546-551.

de Sena, J. and G.J. Rio. 1975. Partial purification and characterization of RTG-2 fish cell interferon. *Infection and Immunity* 11: 815-822.

DeWitte-Orr, S.J., J.A. Leong and N.C. Bols. 2007. Induction of antiviral genes, Mx and vig-1, by dsRNA and Chum salmon reovirus in rainbow trout monocyte/macrophage and fibroblast cell lines. *Fish and Shellfish Immunology* 23: 670-682.

Dijkstra, J.M., B. Kollner, K. Aoyagi, Y. Sawamoto, A. Kuroda, M. Ototake, T. Nakanishi and U. Fischer. 2003. The rainbow trout classical MHC class I molecule Onmy-UBA*501 is expressed in similar cell types as mammalian classical MHC class I molecules. *Fish and Shellfish Immunology* 14: 1-23.

Dorson, M. and P. DeKinkelin. 1974. Mortalitet production d'interferon circulan chez la truite arc-en-ciel apres infection experimentale avec le virus d'Egtved:Influence de la temperature. *Annals Recherches Veterinary* 5: 365-372.

Dorson, M. and C. Torchy. 1979. Complement dependent neutralization of Egtved virus by trout antibodies. *Journal of Fish Diseases* 2: 345-347.

Dorson, M. and C. Torchy. 1993. Viral haemorrhagic septicaemia virus replication in external tissue excised from rainbow trout, *Onchorynchus mykiss* (Walbaum), and hybrids of different susceptibilities. *Journal of Fish Diseases* 16: 403-408.

Dorson, M., A. Barde and P. De Kinkelin. 1975. Egtved virus induced Rainbow trout serum interferon: some physicochemical properties. *Annals Microbiologie* (Institute Pasteur) 126: 485-489.

Dorson, M., C. Torchy and P. De Kinkelin. 1994. Viral haemorrhagic septicaemia virus multiplication and interferon production in rainbow trout and in rainbow trout × brook trout hybrids. *Fish and Shellfish Immunology* 4.

Dorson, M., E. Quillet, M.G. Hollebecq, C. Torhy and B. Chevassus. 1995. Selection of rainbow trout resistant to viral haemorrhagic septicaemia virus and transmission of resistance by gynogenesis. *Veterinary Research* 26: 361-368.

Du Pasquier, L. 1982. Antibody diversity in lower vertebrates—Why is it so restricted? *Nature (London)* 296: 311-313.

Du Pasquier, L., M. Wilson, A.S. Greenberg and M.F. Flajnik. 1998. Somatic mutation in ectothermic vertebrates: musings on selection and origins. *Current Topics in Microbiology and Immunology* 229: 199-216.

Einer-Jensen, K., T.N. Krogh, P. Roepstorff and N. Lorenzen. 1998. Characterization of Intramolecular Disulfide Bonds and Secondary Modifications of the Glycoprotein from Viral Hemorrhagic Septicemia Virus, a Fish Rhabdovirus. *Journal of Virology* 72: 10189-10196.

Einer-Jensen, K., P. Ahrens, R. Forsberg and N. Lorenzen. 2004. Evolution of the fish rhabdovirus viral haemorrhagic septicaemia virus. *Journal of General Virology* 85: 1167-1179.

Ellis, A.E. 2001. Innate host defense mechanisms of fish against viruses and bacteria. *Developmental and Comparative Immunology* 25: 827-839.

Engelking, H. and J.C. Leong. 1989. The glycoprotein of Infectious Hematopoietic Necrosis Virus eluciting antibody and protective responses. *Virus Research* 13: 213-230.

Essbauer, S. and W. Ahne. 2001. Viruses of lower vertebrates. *Journal of Veterinary Medicine* B 48: 403-475.

Estepa, A. and J.M. Coll. 1991. Infection of mitogen stimulated colonies from trout kidney cell cultures with salmonid viruses. *Journal of Fish Diseases* 14: 555-562.

Estepa, A. and J.M. Coll. 1992a. In vitro immunostimulants for optimal responses of kidney cells from healthy trout and from trout surviving viral haemorrhagic septicaemia virus disease. *Fish and Shellfish Immunology* 2: 53-68.

Estepa, A. and J.M. Coll. 1992b. Mitogen-induced proliferation of trout kidney leucocytes by one-step culture in fibrin clots. *Veterinary Immunology and Immunopathology* 32: 165-177.

Estepa, A. and J.M. Coll. 1996. Pepscan mapping and fusion-related properties of the major phosphatidylserine-binding domain of the glycoprotein of viral hemorrhagic septicemia virus, a salmonid rhabdovirus. *Virology* 216: 60-70.

Estepa, A. and J.M. Coll. 1997. An *in vitro* method to obtain T-lymphocyte like cells from the trout. *Journal of Immunological Methods* 202: 77-83.

Estepa, A., D. Frias and J.M. Coll. 1992. Susceptibility of trout kidney macrophages to viral haemorrhagic septicaemia virus. *Viral Immunology* 5: 283-292.

Estepa, A., D. Frías and J.M. Coll. 1993. *In vitro* susceptibility of rainbow trout fin cell lines to viral haemorrhagic septicaemia virus. *Diseases of Aquatic Organisms* 15: 35-39.

Estepa, A., M. Thiry and J.M. Coll. 1994. Recombinant protein fragments from haemorrhagic septicaemia rhabdovirus stimulate trout leucocyte anamnestic in vitro responses. *Journal of General Virology* 75: 1329-1338.

Estepa, A., F. Alvarez, A. Villena and J.M. Coll. 1996. Morphology of antigen-dependent haematopoietic cells from trout surviving rhabdoviral infections. *Bulletin of European Association of Fish Pathologists* 16: 203-207.

Estepa, A., F. Alvarez, A. Ezquerra and J.M. Coll. 1999. Viral-antigen dependence and T-cell receptor expression in leucocytes from rhabdovirus immunized trout. An *in vitro* model to study fish anti-viral responses. *Veterinary Inmunology and Inmunopathology* 68: 73-89.

Estepa, A., A. Rocha, L. Pérez, J.A. Encinar, E. Nuñez, A. Fernandez, J.M. Gonzalez Ros, F. Gavilanes and J.M. Coll. 2001. A protein fragment from the salmonid VHS rhabdovirus induces cell-to-cell fusion and membrane phosphatidylserine translocation at low pH. *Journal of Biological Chemistry* 276: 46268-46275.

Evans, D.L., E.E. Smith and F.C. Brown. 1987. Nonspecific cytotoxic cells in fish (*Ictalurus punctatus*) VI. Flow cytometric analysis. *Developmental and Comparative Immunology* 11: 95-104.

Evans, D.L., R.L. Carlson, S.S. Graves and K.T. Hogan. 1984. Nonspecific cytotoxic cells in fish (*Ictalurus punctatus*) VI. Target cell binding and recycling capacity. *Developmental and Comparative Immunology* 8: 823-829.

Evans, D.L., D.T. Harris, D.L. Staton and L.J. Friedman. 1990. Pathways of signal transduction in teleost nonspecific cytotoxic cells. *Developmental and Comparative Immunology* 14: 295-304.

Fernandez-Alonso, M., A. Rocha and J.M. Coll. 2001. DNA vaccination by immersion and ultrasound to trout viral haemorrhagic septicaemia virus. *Vaccine* 19: 3067-3075.

Fernandez-Alonso, M., F. Alvarez, A. Estepa, R. Blasco and J.M. Coll. 1999. A model to study fish DNA immersion vaccination by using the green fluorescent protein. *Journal of Fish Diseaeses* 22: 237-241.

Fernandez-Alonso, M., G. Lorenzo, L. Perez, R. Bullido, A. Estepa, N. Lorenzen and J.M. Coll. 1998a. Mapping of linear antibody epitopes of the glycoprotein of VHSV, a salmonid rhabdovirus. *Diseases of Aquatic Organisms* 34: 167-176.

Fernandez-Alonso, M., G. Lorenzo, L. Perez, R. Bullido, A. Estepa, N. Lorenzen and J.M. Coll. 1998b. Mapping of the lineal antibody epitopes of the glycoprotein of VHSV, a salmonid rhabdovirus. *Diseases of Aquatic Organisms* 34: 167-176.

Fischer, U., M. Ototake and T. Nakanishi. 1998. In vitro cell-mediated cytotoxicity against allogeneic erythrocytes in ginbuna crucian carp and goldfish using a non-radioactive assay. *Developmental and Comparative Immunology* 22: 195-206.

Fischer, U., K. Utke, T. Somamoto, B. Kollner, M. Ototake and T. Nakanishi. 2006. Cytotoxic activities of fish leucocytes. *Fish and Shellfish Immunology* 20: 209-226.

Flajnik, M., Y. Ohta, C. Namikama-Yamada and M. Nonaka. 1999. Insight into the primordial MHC from studies in ectothermic vertebrates. *Immunology Reviews* 167: 59-67.

Fregeneda-Grandes, J.M. and N.J. Olesen. 2007. Detection of rainbow trout antibodies against viral haemorrhagic septicaemia virus (VHSV) by neutralisation test is highly dependent on the virus isolate used. *Diseases of Aquatic Organisms* 74: 151-158.

Garner, J.N., B. Joshi and R. Jagus. 2003. Characterization of rainbow trout and zebrafish eukaryotic initiation factor 2[alpha] and its response to endoplasmic reticulum stress and IPNV infection. *Developmental and Comparative Immunology* 27: 217-231.

Greenhalgh, P.A.S., L. 1995. Recombination activator gene 1 (Rag 1) in zebrafish and shark. *Immunogenetics* 41: 54-55.

Grignani, G. and A. Maiolo. 2000. Cytokines and hemostasis. *Haematologica* 85: 967-972.

Haller, O. and G. Kochs. 2002. Interferon-Induced Mx Proteins: Dynamin-Like GTPases with Antiviral Activity. *Traffic* 3: 710-717.

Haller, O., M. Frese and G. Kochs. 1998. Mx proteins: Mediators of innate resistance to RNA viruses. *Research in Veterinary Science* 17: 220-230.

Haller, O., P. Staeheli and G. Kochs. 2007. Interferon-induced Mx proteins in antiviral host defense. *Biochimie* 89: 812-818.

Hansen, J.D. and S.L. Kaattari. 1995. The recombination activation gene 1 (RAG1) of rainbow trout (*Oncorhynchus mykiss*): Cloning, expression, and phylogenetic analysis. *Immunogenetics* 42: 188-195.

Hansen, J.D. and S.L. Kaattari. 1996. The recombination activating gene 2 (RAG2) of the rainbow trout *Oncorhynchus mykiss*. *Immunogenetics* 44: 203-211.

Harmache, A., M. LeBerre, S. Droineau, M. Giovannini and M. Bremont. 2006. Bioluminescence imaging of live infected salmonids reveals that the fin bases are the major portal of entry for Novirhabdovirus. *Journal of Virology* 80: 3655-3659.

Hasegawa, S., C. Nakayasu, T. Yoshitomi, T. Nakanishi and N. Okamoto. 1998. Specific cell-mediated cytotoxicity against an allogeneic target cell line in isogeneic ginbuna crucian carp. *Fish and Shellfish Immunology* 8: 303-313.

Hatten, F., A. Fredriksen, I. Hordvik and C. Endresen. 2001. Presence of IgM in cutaneous mucus, but not in gut mucus of Atlantic salmon, *Salmo salar*. Serum IgM is rapidly degraded when added to gut mucus. *Fish and Shellfish Immunology* 11: 257-268.

Hattenberg-Baudouy, A.M., M. Danton, G. Merle, C. Torchy and P. De Kinkelin. 1989. Serological evidence of infectious hematopoietic necrosis in rainbow trout from a French outbreak of disease. *Journal of Aquatic Animal Health* 1: 126-134.

Helmick, C.M., J.F. Bailey, S. LaPatra and S. Ristow. 1995. The esophagus/cardiac stomach region: site of attachment and internalization of infectious hematopoietic necrosis virus in challenged juvenile rainbow trout *Oncorynchus mykiss* and coho salmon *O. kisutch*. *Diseases of Aquatic Organisms* 23: 189-199.

Hogan, R.J., T.B. Stuge, W. Clem, N.W. Miller and V.G. Chinchar. 1996. Anti-viral cytotoxic cells in the channel catfish (*Ictalurus punctatus*). *Developmental and Comparative Immunology* 20: 115-127.

Hong, S., S. Peddie, A.J. Campos-Perez, J. Zou and C.J. Secombes. 2003. The effect of intraperitoneally administered recombinant IL-1beta on immune parameters and resistance to *Aeromonas salmonicida* in the rainbow trout (*Oncorhynchus mykiss*). *Developmental and Comparative Immunology* 27: 801-812.

Hordvik, I. 2002. Identification of a novel immunoglobulin [delta] transcript and comparative analysis of the genes encoding IgD in Atlantic salmon and Atlantic halibut. *Molecular Immunology* 39: 85-91.

Huang, C., M.S. Chien, M. Landolt and J. Winton. 1994. Characterization of the infectious hematopoietic necrosis virus glycoprotein using neutralizing monoclonal antibodies. *Diseases of Aquatic Organisms* 18: 29-35.

Huang, C., M.S. Chien, M. Landolt, W. Batts and J. Winton. 1996. Mapping the neutralizing epitopes on the glycoprotein of infectious haematopoietic necrosis virus, a fish rhabdovirus. *Journal of General Virology* 77: 3033-3040.

Jang, S.I., L.J. Hardie and C.J. Secombes. 1995. Elevation of rainbow trout *Oncorhynchus mykiss* macrophage respiratory burst activity with macrophage-derived supernatants. *Journal of Leukocyte Biology* 57: 943-947.

Jørgensen, P.E.V. 1971. Egtved virus: demonstration of neutralizing antibodies in serum from artificially infected rainbow trout. *Journal of the Fisheries Research Board of Canada* 28: 875-877.

Jorgensen, P.E.V., N.J. Olesen, N. Lorenzen, J.R. Winton and S. Ristow. 1991. Infectious haematopoietic necrosis (IHN) and viral haemorragic septicaemia (VHS): detection of trout antibodies to the causative viruses by means of plaque neutralization, immunofluorescence, and enzyme-linked immunosorbent assay. *Journal of Aquatic Animal Health* 3: 100-108.

Kaattari, S. 1992. Fish B lymphocytes: defining their from and function. *Annual Review Fish Diseases* 2: 161-180.

Kaattari, S., D.A. Evans and J.V. Klemer. 1998. Varied redox forms of teleost IgM: an alternative to isotypic diversity? *Immunological Reviews* 166: 133-142.

Kanellos, T., I.D. Sylvester, C.R. Howard and P.H. Russell. 1999. DNA is as effective as protein at inducing antibody in fish. *Vaccine* 17: 965-972.

Koener, J.F. and J.C. Leong. 1990. Expression of the glycoprotein gene from a fish rhabdovirus by using baculovirus vectors. *Journal of Virology* 64: 428-430.

Laing, K.J., J.J. Zou, T. Wang, N. Bols, I. Hirono, T. Aoki and C.J. Secombes. 2002. Identification and analysis of an interleukin 8-like molecule in rainbow trout *Oncorhynchus mykiss*. *Developmental and Comparative Immunology* 26: 433-444.

LaPatra, S.E., K.A. Lauda and G.R. Jones. 1994a. Antigenic variants of infectious hematopoietic necrosis virus and implications for vaccine development. *Diseases of Aquatic Organisms* 20: 119-126.

LaPatra, S.E., K.A. Lauda, G.R. Jones, S.C. Walker and W.D. Shewmaker. 1994b. Development of passive immunotherapy for control of infectious hematopoietic necrosis. *Diseases of Aquatic Organisms* 20: 1-6.

LaPatra, S.E., T. Turner, K.A. Lauda, S.C. Walker and G.R. Jones. 1993. Characterization of the humoral response of rainbow trout to infectious hematopoietic necrosis virus. *Journal of Aquatic Animal Health* 5: 165-171.

LaPatra, S.E., K.A. Lauda, R.G. Jones, W.D. Shewmaker, J.M. Grff and D. Routledge. 1996. Susceptibility and humoral response of brown trout × lake trout hybrids to infectious hematopoietic necrosis virus: a model for examining disease resistance mechanisms. *Aquaculture* 146: 179-188.

LaPatra, S.E., S. Corbeil, G.R. Jones, W.D. Shewmaker, N. Lorenzen, E.D. Anderson and G. Kurath. 2001. Protection of rainbow trout against infectious hematopoietic necrosis virus four days after specific or semi-specific DNA vaccination. *Vaccine* 19: 4011-4019.

Ledford, B.E., B.G. Magor, D.L. Middleton, R.L. Miller, M.R. Wilson, N.W. Miller, L.W. Clem and G.W. Warr. 1993. Expression of a mouse-channel catfish chimeric IgM molecule in a mouse myeloma cell. *Molecular Immunology* 30: 1405-1417.

LeFrancois, L. 1984. Protection against lethal viral infection by neutralizing and nonneutralizing monoclonal antibodies: aistinct mechanisms action in vivo. *Journal of Virology* 51: 208-214.

Leong, J.A.C.B., L., P.C. Chiou, B. Drolet, M. Johnson, C. Kim, D. Mourich, K. Suzuki and G. Trobridge. 1997. Immune response to viral diseases in fish. *Developmental and Comparative Immunology* 21: 223-223.

Leong, J.C., L. Bootland, E. Anderson, P.W. Chiou, B. Drolet, C. Kim, H. Lorz, D. Mourich, P. Ormonde, L. Perez and G. Trobridge. 1995. Viral vaccines for aquaculture. *Journal of Marine Biotechnology* 3: 16-23.

Leong J.C., T.G.D., C.H. Kim, M. Johnson and B. Simon. 1998. Interferon-inducible Mx proteins in fish. *Immunological Reviews* 166: 349-363.

Liston, A. and S. McColl. 2003. Subversion of the chemokine world by microbial pathogens. *Bioessays* 25: 478-488.

Lobb, C.J. and L.W. Clem. 1981. Phylogeny of immunoglobulin structure and function X. Humoral immunoglobulins of the sheepseed (*Archosargus probatocephalus*). *Developmental and Comparative Immunology* 5: 271-282.

Long, S., M. Wilson, E. Bengten, L. Bryan, L.W. Clem, N.W. Miller and V.G. Chinchar. 2004. Identification of a cDNA encoding channel catfish interferon. *Developmental and Comparative Immunology* 28: 97-111.

Lorenzen, N. and S.E. LaPatra. 1999. Immunity to rhabdoviruses in rainbow trout: the antibody response. *Fish and Shellfish Immunology* 9.

Lorenzen, N. and S.E. LaPatra. 2005. DNA vaccines for aquacultured fish. *Research in Veterinary Science* 24: 201-213.

Lorenzen, N., N.J. Olesen and P.E. Jorgensen. 1990. Neutralization of Egtved virus pathogenicity to cell cultures and fish by monoclonal antibodies to the viral G protein. *Journal of General Virology* 71: 561-567.

Lorenzen, N., E. Lorenzen and K. Einer-Jensen. 2001. Immunity to viral haemorrhagic septicaemia (VHS) following DNA vaccination of rainbow trout at an early life-stage. *Fish and Shellfish Immunology* 11: 585-591.

Lorenzen, N., E. Lorenzen, K. Einer-Jensen and S.E. LaPatra. 2002. DNA vaccines as a tool for analysing the protective immune response against rhabdoviruses in rainbow trout. *Fish and Shellfish Immunology* 12: 439-453.

Lorenzen, N., E. Lorenzen, K. Einer-Jensen, J. Heppell and H.L. Davis. 1999a. Genetic vaccination of rainbow trout against viral haemorrhagic septicaemia virus: small amounts of plasmid DNA protect against a heterologous serotype. *Virus Research* 63: 19-25.

Lorenzen, N., N.J. Olesen and C. Koch. 1999b. Immunity to VHS virus in rainbow trout. *Aquaculture* 172: 41-61.

Lorenzen, E., K. Einer-Jensen, T. Martinussen, S.E. LaPatra and N. Lorenzen. 2000. DNA vaccination of rainbow trout against viral hemorrhagic septicemia virus: a dose-response and time-course study. *Journal of Aquatic Animal Health* 12: 167-180.

Lorenzen, N., N.J. Olesen, P.E. Jorgensen, M. Etzerodt, T.L. Holtet and H.C. Thogersen. 1993a. Molecular cloning and expression in *Escherichia coli* of the glycoprotein gene of VHS virus, and immunization of rainbow trout with the recombinant protein. *Journal of General Virology* 74: 623-630.

Lorenzen, N., N.J. Olesen, P.E. Vestergaard-Jorgensen, M. Etzerodt, T.L. Holtet and M.C. Thorgersen. 1993b. Molecular cloning and expression in *Escherichia coli* of the glycoprotein gene of VHS virus and immunization of rainbow trout with the recombinant protein. *Journal of General Virology* 74: 623-630.

Lorenzen, N., E. Lorenzen, K. Einer-Jensen, J. Heppell, T. Wu and H. Davis. 1998. Protective immunity to VHS in rainbow trout (*Oncorhynchus mykiss*, Walbaum) following DNA vaccination. *Fish and Shellfish Immunology* 8: 261-270.

Lorenzo, G., A. Estepa and J.M. Coll. 1996. Fast neutralization/immunoperoxidase assay for viral haemorrhagic septicaemia with anti-nucleoprotein monoclonal antibody. *Journal of Virological Methods* 58: 1-6.

Lorenzo, G., A. Estepa, S. Chilmonczyk and J.M. Coll. 1995a. Mapping of the G and N regions of viral haemorrhagic septicaemia virus (VHSV) inducing lymphoproliferation by pepscan. *Veterinary Research* 26: 521-525.

Lorenzo, G.A., A. Estepa, S. Chilmonczyk and J.M. Coll. 1995b. Different peptides from hemorrhagic septicemia rhabdoviral proteins stimulate leucocyte proliferation with individual fish variation. *Virology* 212: 348-355.

Lorenzo, G.A., A. Estepa, S. Chilmonczyk and J.M. Coll. 1995c. Mapping of the G-Regions and N-Regions of Viral Hemorrhagic Septicemia Virus (VHSV) Inducing Lymphoproliferation by Pepscan. *Veterinary Research* 26: 521-525.

Lorenzo, G.A., A. Estepa, S. Chilmonczyk and J.M. Coll. 1995d. Mapping of the G and N regions of viral haemorrhagic septicaemia virus (VHSV) inducing lymphoproliferation by pepscan. *Veterinary Research* 26: 521-525.

Lukacher, A.E., V.L. Braciale and T.J. Braciale. 1984. In vivo effector function of influenza virus-specific cytotoxic T lymphocyte clones is highly specific. *Journal of Experimental Medicine* 160: 814-826.

Magnadottir, B. 2006. Innate immunity of fish (overview). *Fish and Shellfish Immunology* 20: 137-151.

Manning, M.J. and T. Nakanishi. 1996. The specific immune system: Cellular defenses. In: *Fish Physiology. The Fish Immune System*, G. Iwama and T. Nakanishi (eds.). Academic Press, London, Vol. 15, pp. 159-205.

McCullough, K.C., S. Basta, S. Knotig, H. Gerber, R. Schaffner, Y.B. Kim, A. Saalmuller and A. Summerfield. 1999. Intermediate stages in monocyte-macrophage differentiation modulate phenotype and susceptibility to virus infection. *Immunology* 98: 203-212.

McLauchlan, P.E., B. Collet, E. Ingerslev, C.J. Secombes, N. Lorenzen and A.E. Ellis. 2003. DNA vaccination against viral haemorrhagic septicaemia (VHS) in rainbow trout: Size, dose, route of injection and duration of protection-early protection correlates with Mx expression. *Fish and Shellfish Immunology* 15: 39-50.

Muller, U., U. Steinhoff, L.F. Reis, S. Hemmi, J. Pavlovic, R.M. Zinkernagel and M. Aguet. 1994. Functional role of type I and type II interferons in antiviral defense. *Science* 264: 1918-1921.

Nakanishi, T., K. Aoyagi, C. Xia, J.M. Dijkstra and M. Ototake. 1999. Specific cell-mediated immunity in fish. *Veterinary Immunology and Immunopathology* 72: 101-109.

Nakanishi, T., U. Fischer, J.M. Dijkstra, S. Hasegawa, T. Somamoto, N. Okamoto and M. Ototake. 2002. Cytotoxic T cell function in fish. *Developmental and Comparative Immunology* 26: 131-139.

Noonan, B., P.J. Enzmann and T.J. Trust. 1995. Recombinant infectious necrosis virus and viral hemorrhagic septicemia virus glycoprotein epitopes expressed in *Aeromonas salmonicida* induce protective immunity in rainbow trout (*Oncorhynchus mykiss*). *Applied Environmental Microbiology* 61: 3586-3591.

Nunez, E., A.M. Fernandez, A. Estepa, J.M. Gonzalez-Ros, F. Gavilanes and J.M. Coll. 1998. Phospholipid interactions of a peptide from the fusion-related domain of the glycoprotein of VHSV, a fish rhabdovirus. *Virology* 243: 322-330.

O'Farrell, C., N. Vaghefi, M. Cantonnet, B. Buteau, P. Boudinot and A. Benmansour. 2002. Survey of transcript expression in rainbow trout leukocytes reveals a major contribution of interferon-responsive genes in the early response to a rhabdovirus infection. *Journal of Virology* 76: 8040-8049.

Oldstone, M.B. 1987. Immunotherapy for virus infection. *Current Top Microbiology and Immunology* 134: 211-229.

Olesen, N.J. 1986. Quantification of serum immunoglobulin in rainbow trout *Salmo gairdneri* under various environmental conditions. *Diseases of Aquatic Organisms* 1: 183-186.

Olesen, N. 1998. Sanitation of viral haemorrhagic septicaemia (VHS). *Journal of Applied Ichthyology* 14: 173-177.

Olesen, N.J. and P.E. Vestergard-Jorgensen. 1986. Detection of neutalizing antibody to Egtved virus in Rainbow trout (*Salmo gairdneri*) by plaque neutralization test with complement addition. *Journal of Applied Ichthyology* 2: 33-41.

Olesen, N.J., N. Lorenzen and P.E. Vestergaard-Jorgensen. 1991. Detection of rainbow trout antibody to Egtved virus by enzyme-linked immunosorbent assay (ELISA), immunofluorescence (IF), and plaque neutralization tests (50% PNT). *Diseases of Aquatic Organisms* 10: 31-38.

Olesen, N.J., N. Lorenzen and P.E.V. Jorgensen. 1993. Serological differences among isolates of viral haemorrhagic septicemia virus detected by neutralizing monoclonal and polyclonal antibodies. *Diseases of Aquatic Organisms* 16: 163-170.

Partula, S., J.S. Fellah, A. DeGuerra and J. Charlemagne. 1994. Identification of cDNA clones encoding the T-cell receptor beta chain in the rainbow trout (*Oncorhynchus mykiss*). *Concepts Rendues Academy Sciences Paris* 317: 765-770.

Partula, S., A. DeGuerra, J.S. Fellah and J. Charlemagna. 1995. Structure and diversity of the T cell antigen receptor beta-chain in a teleost fish. *Journal of Immunology* 155: 699-706.

Partula, S., A. DeGuerra, J.S. Fellah and J. Charlemagne. 1996. Structure and diversity of the TCR alpha-chain in a teleost fish. *Journal of Immunology* 157: 207-212.

Peddie, S., P.E. McLauchlan, A.E. Ellis and C.J. Secombes. 2003. Effect of intraperitoneally administered IL-1beta-derived peptides on resistance to viral haemorrhagic septicaemia in rainbow trout *Oncorhynchus mykiss*. *Diseases of Aquatic Organisms* 56: 195-200.

Pilstrom, L. and E. Bengten. 1996. Immunoglobulin in fish — Genes, expression and structure. *Fish and Shellfish Immunology* 4: 243-262.

Plouffe, D.A., P.C. Hanington, J.G. Walsh, E.C. Wilson and M. Belosevic. 2005. Comparison of select innate immune mechanisms of fish and mammals. *Xenotransplantation* 12: 266-277.

Purcell, M.K., G. Kurath, K.A. Garver, R.P. Herwig and J.R. Winton. 2004. Quantitative expression profiling of immune response genes in rainbow trout following infectious haematopoietic necrosis virus (IHNV) infection or DNA vaccination. *Fish and Shellfish Immunology* 17: 447-462.

Quillet, E., M. Dorson, G. Aubard and C. Torhy. 2001. In vitro viral haemorrhagic septicaemia virus replication in excised fins of rainbow trout: correlation with resistance to waterborne challenge and genetic variation. *Diseases of Aquatic Organisms* 45: 171-182.

Quillet, E., M. Dorson, S. Le Guillou, A. Benmansour and P. Boudinot. 2007. Wide range of susceptibility to rhabdoviruses in homozygous clones of rainbow trout. *Fish and Shellfish Immunology* 22: 510-519.

Rawls, J.F., B.S. Samuel and J.I. Gordon. 2004. Gnotobiotic zebrafish reveal evolutionarily conserved responses to the gut microbiota. *Proceedings of the National Academy of Sciences of the United States of America* 101: 4596-4601.

Ristow, S.S., S.E. LaPatra, R. Dixon, C.R. Pedrow, W.D. Shewmaker, J.W. Park and G.H. Thorgaard. 2000. Responses of cloned rainbow trout *Oncorhynchus mykiss* to an attenuated strain of infectious hematopoietic necrosis virus. *Diseases of Aquatic Organisms* 42: 163-172.

Roberti, K.A., J.R. Winton and J.S. Rohovec. 1991. Variants of infectious hematopoietic necrosis virus selected with glycoprotein-specific monoclonal antibodies. *Proceedings Second International Symposium on Viruses of Lower Vertebrates*, Oregon State University of Oregon, USA, July 29-31, pp. 33-42.

Robertsen, B. 2006. The interferon system of teleost fish. *Fish and Shellfish Immunology* 20: 172-191.

Robertsen, B., V. Bergan, Røkenes Torunn, Larsen Rannveig and A. Albuquerque. 2003. Atlantic Salmon Interferon Genes: Cloning, Sequence Analysis, Expression, and Biological Activity. *Journal of Interferon and Cytokine Research* 23: 601-612.

Rocha, A., S. Ruiz, C. Tafalla and J.M. Coll. 2004. Conformation and fusion defective mutants in the hypothetical phospholipid-binding and fusion peptides of the protein G of viral haemorrhagic septicemia salmonid rhabdovirus. *Journal of Virology* 78: 9115-9122.

Rocha, A., M. Fernandez-Alonso, V. Mas, L. Perez, A. Estepa and J.M. Coll. 2002. Antibody response to a fragment of the protein G of VHS rhabdovirus in immunised trout. *Veterinary Immunology and Immunopathology* 86: 89-99.

Roche, S., S. Bressanelli, F.A. Rey and Y. Gaudin. 2006. Crystal structure of the low-pH form of the vesicular stomatitis virus glycoprotein G. *Science* 313: 187-191.

Roche, S., F.A. Rey, Y. Gaudin and S. Bressanelli. 2007. Structure of the prefusion form of the vesicular stomatitis virus glycoprotein G. *Science* 315: 843-848.

Rodrigues, P. N., B. Dixon, J. Roelofs, J.H. Rombout, E. Egberts, B. Pohajdak and R.J. Stet. 1998. Expression and temperature-dependent regulation of the beta2-microglobulin (Cyca-B2m) gene in a cold-blooded vertebrate, the common carp (*Cyprinus carpio* L.). *Developmental Immunology* 5: 263-275.

Rogel-Gaillard, C., S. Chilmonczyk and P. DeKinkelin. 1993. *In vitro* induction of interferon-like activity from rainbow trout leucocytes stimulated by Egtved virus. *Fish and Shellfish Immunology* 3: 382-394.

Rombout, J.H.W.M., N. Taverne, M. van de Kamp and A.J. Taverne-Thiele. 1993. Differences in mucus and serum immunoglobulin of carp (*Cyprinus carpio* L.). *Developmental and Comparative Immunology* 17: 309-317.

Rombout, J.H.W.M., J.W. VandeWal, N. Companjen, N. Taverne and J.J. Taverne-Thiele. 1997. Characterization of a T cell lineage marker in carp (*Cyprinus carpio* L.). *Developmental and Comparative Immunology* 21: 35-46.

Samuel, C.E. 2001. Antiviral Actions of Interferons. *Clinical Microbiology Rev* 14: 778-809.

Sanchez, C. and J. Dominguez. 1991. Trout Immunoglobulin populations differing in light chains revealed by monoclonal antibodies. *Molecular Immunology* 28: 1271-1277.

Sanchez, C., J. Dominguez and J.M. Coll. 1989. Immunoglobulin heterogeneity in the rainbow trout, *Salmo gairdneri* Richardson. *Journal of Fish Diseases* 12: 459-465.

Sanchez, C., M. Babin, J. Tomillo, F.M. Obeira and J. Dominguez. 1993. Quantification of low levels of rainbow trout immunoglobulin by enzyme immunoassay using two monoclonal antibodies. *Veterinary Immunology and Immunopathology* 36: 64-74.

Sangrador-Vegas, A., J.B. Lennington and T.J. Smith. 2002. Molecular cloning of an IL-8-like CXC chemokine and tissue factor in rainbow trout (*Oncorhynchus mykiss*) by use of suppression subtractive hybridization. *Cytokine* 17: 66-70.

Sanz, F. and J.M. Coll. 1992a. Detection of viral haemorrhagic septicemia virus by direct immunoperoxidase with selected anti-nucleoprotein monoclonal antibody. *Bulletin of European Association of Fish Pathologists* 12: 116-119.

Sanz, F.A. and J.M. Coll. 1992b. Detection of hemorrhagic virus of samonid fishes by use of an enzyme-linked immunosorbent assay containing high sodium chloride concentration and two concompetitive monoclonal antibodies against early viral nucleoproteins. *American Journal of Veterinary Research* 53: 897-903.

Scapigliati, G., N. Romano, L. Abelli, S. Meloni, A.G. Ficca, F. Buonocore, S. Bird and C.J. Secombes. 2000. Immunopurification of T-cells from sea bass *Dicentrarchus labrax* (L.). *Fish and Shellfish Immunology* 10: 329-341.

Schluter, S.F., R.M. Bernstein and J.J. Marchalonis. 1999. "Big Bang" emergence of the combinatorial immune system. *Developmental and Comparative Immunology* 23: 107-111.

Schutze, H., E. Mundt and T.C. Mettenleiter. 1999. Complete genomic sequence of viral hemorrhagic septicemia virus, a fish rhabdovirus. *Virus Genes* 19: 59-65.

Schutze, H., P.J. Enzmann, E. Mundt and T.C. Mettenleiter. 1996. Identification of the non-virion (NV) protein of fish rhabdoviruses viral haemorrhagic septicaemia virus and infectious haematopoietic necrosis virus. *Journal of General Virology* 77: 1259-1263.

Shum, B.P., K. Azumi, S. Zhang, S.R. Kehrer, R.L. Raison, H.W. Detrich and P. Parham. 1996. Unexpected beta2-microglobulin sequence diversity in individual rainbow trout. *Proceedings of the National Academy of Sciences of the United States of America* 93: 2779-2784.

Skall, H.F., N.J. Olesen and S. Mellergaard. 2005. Viral haemorrhagic septicaemia virus in marine fish and its implications for fish farming—A review. *Journal of Fish Diseases* 28: 509-529.

Slierendrecht, W.J., N.J. Olesen, H.R. Juul-Madsen, N. Lorenzen, M. Henryon, P. Berg, J. Sondergaard and C. Koch. 2001. Rainbow trout offspring with different resistance to viral haemorrhagic septicaemia. *Fish and Shellfish Immunology* 11: 155-167.

Snegaroff, J. 1993. Induction of interferon synthesis in rainbow trout leucocytes by various homeotherm viruses. *Fish and Shellfish Immunology* 3: 191-198.

Somamoto, T., T. Nakanishi and N. Okamoto. 2000. Specific cell-mediated cytotoxicity against a virus-infected syngeneic cell line in ginbuna crucian carp. *Developmental and Comparative Immunology* 24: 633-640.

Somamoto, T., T. Nakanishi and N. Okamoto. 2002. Role of Specific Cell-Mediated Cytotoxicity in Protecting Fish from Viral Infections. *Virology* 297: 120-127.

Sommerset, I., B. Krossoy, E. Biering and P. Frost. 2005. Vaccines for fish in aquaculture. *Expert Reviews in Vaccines* 4: 89-101.

Stark, G.R., I.M. Kerr, B.R.G. Williams, R.H. Silverman and R.D. Schreiber. 1998. How cells respond to interferons? *Annual Review of Biochemistry* 67: 227-264.

Stohlman, S.A., J.G. Woodward and J.A. Frelinger. 1982. Macrophage antiviral activity: Extrinsic versus intrinsic activity. *Infection and Immunity* 36: 672-677.

Stuge, T. B., M.R. Wilson, H. Zhou, K.S. Barker, E. Bengten, G. Chinchar, N.W. Miller and L.W. Clem. 2000. Development and analysis of various clonal alloantigen-dependent cytotoxic cell lines from channel catfish. *Journal of Immunology* 164: 2971-2977.

Suda, T. and A. Zlotnik. 1992. In vitro induction of CD8 expression on thymic pre-T cells. I. Transforming growth factor-beta and tumor necrosis factor-alpha induce CD8 expression on CD8- thymic subsets including the CD25+CD3-CD4-CD8- pre-T cell subset. *Journal of Immunology* 148: 1737-1745.

Swain, S.L., G. Huston, S. Tonkonogy and A. Weinberg. 1991. Transforming growth factor-beta and IL-4 cause helper T cell precursors to develop into distinct effector helper cells that differ in lymphokine secretion pattern and cell surface phenotype. *Journal of Immunology* 147: 2991-3000.

Tafalla, C. and B. Novoa. 2001. Respiratory burst of turbot (*Scophthalmus maximus*) macrophages in response to experimental infection with viral haemorrhagic septicaemia virus (VHSV). *Fish and Shellfish Immunology* 11: 727-734.

Tafalla, C., A. Figueras and B. Novoa. 1998. In vitro interaction of viral haemorrhagic septicaemia virus and leukocytes from trout (*Oncorhynchus mykiss*) and turbot (*Scophthalmus maximus*). *Veterinary Immunology and Immunopathology* 62: 359-366.

Tafalla, C., A. Figueras and B. Novoa. 1999. Role of nitric oxide on the replication of viral haemorrhagic septicaemia virus (VHSV), a fish rhabdovirus. *Veterinary Immunology and Immunopathology* 72: 249-256.

Tafalla, C., A. Figueras and B. Novoa. 2001. Viral hemorrhagic septicemia virus alters turbot *Scophthalmus maximus* macrophage nitric oxide production. *Diseases of Aquatic Organisms* 47: 101-107.

Tafalla, C., J. Coll and C.J. Secombes. 2005. Expression of genes related to the early immune response in rainbow trout (*Oncorhynchus mykiss*) after viral haemorrhagic septicaemia virus (VHSV) infection. *Developmental and Comparative Immunology* 29: 615-626.

Tafalla, C., V. Chico, L. Perez, J.M. Coll and A. Estepa. 2007. In vitro and in vivo differential expression of rainbow trout (*Oncorhynchus mykiss*) Mx isoforms in response to viral haemorrhagic septicaemia virus (VHSV) G gene, poly I:C and VHSV. *Fish and Shellfish Immunology* 23: 210-221.

Tafalla, C., E. Sanchez, N. Lorenzen, S.J. Dewitte-Orr and N.C. Bols. 2008. Effects of viral hemorrhagic septicemia virus (VHSV) on the rainbow trout (*Oncorhynchus mykiss*) monocyte cell line RTS-11. *Molecular Immunology* 45: 1439-1448.

Tengelsen, L.A., G.D. Trobidge and J.C. Leong. 1991. Characterization of an inducible B-interferon-like antiviral activity in salmonids. *Proceedings of the Second International Symposium of Viruses of Lower Vertebrates* - Corvallis, pp. 219-226.

Thiery, R., C. de Boisseson, J. Jeffroy, J. Castric, P. de Kinkelin and A. Benmansour. 2002. Phylogenetic analysis of viral haemorrhagic septicaemia virus (VHSV) isolates from France (1971-1999). *Diseases of Aquatic Organisms* 52: 29-37.

Thorgaard, G.H., G.S. Bailey, D. Williams, D.R. Buhler, S.L. Kaattari, S.S. Ristow, J.D. Hansen, J.R. Winton, J.L. Bartholomew and J.J. Nagler. 2002. Status and opportunities for genomics research with rainbow trout. *Comparative Biochemistry and Physiology* B133: 609-646.

Thoulouze, M.I., E. Bouguyon, C. Carpentier and M. Bremont. 2004. Essential role of the NV protein of Novirhabdovirus for pathogenicity in rainbow trout. *Journal of Virology* 78: 4098-4107.

Tort, L., J.C. Balasch and S. MacKenzie. 2004. Fish health challenge after stress. Indicators of immunocompetence. *Contributions to Science* 2: 443-454.

Traxler, G.S., E. Anderson, S.E. LaPatra, J. Richard, B. Shewmaker and G. Kurath. 1999. Naked DNA vaccination of Atlantic salmon (*Salmo salar*) against IHNV. *Diseases of Aquatic Organisms* 38: 183-190.

Trobridge, G.D. and J.C. Leong. 1995. Characterization of a rainbow trout Mx Gene. *Journal of Interferon and Cytokine Research* 15: 691-702.

Trobridge, G.D., P.P. Chiou and J.C. Leong. 1997. Cloning of the rainbow trout (*Oncorhynchus mykiss*) Mx2 and Mx3 cDNAs and characterization of trout Mx protein expression in salmon cells. *Journal of Virology* 71: 5304-5311.

Trobridge, G.D., S.E. LaPatra, C.H. Kim and J.C. Leong. 2000. Mx mRNA expression and RFLP analysis of rainbow trout *Oncorhynchus mykiss* genetic crosses selected for susceptibility or resistance to IHNV. *Diseases of Aquatic Organisms* 40: 1-7.

Utke, K., S. Bergmann, N. Lorenzen, B. Kollner, M. Ototake and U. Fischer. 2007. Cell-mediated cytotoxicity in rainbow trout, *Oncorhynchus mykiss*, infected with viral haemorrhagic septicaemia virus. *Fish and Shellfish Immunology* 22: 182-196.

Utke, K., H. Kock, H. Schuetze, S.M. Bergmann, N. Lorenzen, K. Einer-Jensen, B. Kollner, R.A. Dalmo, T. Vesely, M. Ototake and U. Fischer. 2008. Cell-mediated immune responses in rainbow trout after DNA immunization against the viral hemorrhagic septicemia virus. *Developmental and Comparative Immunology* 32: 239-252.

von Schalburg, K., M. Rise, G. Cooper, G. Brown, A.R. Gibbs, C. Nelson, W. Davidson and B. Koop. 2005. Fish and chips: Various methodologies demonstrate utility of a 16,006-gene salmonid microarray. *BMC Genomics* 6: 126.

Walker, P.J., A. Benmansour, R. Dietzgen, R.X. Fang, A.O. Jackson, G. Kurath, J.C. Leong, S. Nadin-Davies, R.B. Tesh and N. Tordo. 2000. Family Rhabdoviridae. In: *Virus Taxonomy Classification and Nomenclature of Viruses*, M.H.V. Van Regenmortel, C.M. Fauquet, D.H.L. Bishop, E.B. Carstens, M.K. Estes, S.M. Lemon, J. Maniloff, M.A. Mayo, D.J. McGeoch, C.R. Pringle and R.B. Wickner (eds.). Academic Press, San Diego, pp. 563-583.

Wang, T., M. Ward, P. Grabowski and C.J. Secombes. 2001. Molecular cloning, gene organization and expression of rainbow trout (*Oncorhynchus mykiss*) inducible nitric oxide synthase (iNOS) gene. *Biochemical Journal* 358: 747-755.

Warr, G. 1996. Adaptative immunity in fish cells. *International Symposium on Fish Vaccinology Oslo*, June, 5-7.

Watts, M., B.L. Munday and C.M. Burke. 2001. Immune tesponses of teleost fish. *Australian Veterinary Journal* 79: 570-574.

Wilhelm, S.W. and C.A. Suttle. 1999. Viruses and nutrient cycles in the sea. *Bioscience* 49: 781-788.

Wilson, M., Z.H., E. Bengten, *et al.* 1998. T-cell receptors in channel catfish: structure and expression of TCR alpha and beta genes. *Molecular Immunology* 35: 545-557.

Wilson, M., E. Bengten, N.W. Miller, L.W. Clem, L. Du Pasquier and G.W. Warr. 1997. A novel chimeric Ig heavy chain from a teleost fish shares similarities to IgD. *Proceedings of the National Academy of Sciences of the United States of America* 94: 4593-4597.

Willett, C.E., A.G. Zapata, N. Hopkins and L.A. Steiner. 1997. Expression of zebrafish rag genes during early development identifies the thymus. *Developmental Biology* 182: 331-341.

Winton, J.R., C.K. Arakawa, C.N. Lannan and J.L. Fryer. 1998. Neutralizing monoclonal antibodies recognize antigenic variants among isolates of infectious hematopoietic necrosis. *Diseases of Aquatic Organisms* 4: 199-204.

Wolf, K. 1988. *Fish Virus and Fish Viral Diseases.* Cornell University Press, Ithaca, pp. 476.

Xu, L., D.V. Mourich, H.M. Engelking, S. Ristow, J. Arnzen and J.C. Leong. 1991. Epitope mapping and characterization of the infectious hematopoietic necrosis virus glycoprotein, using fusion proteins synthesized in *Escherichia coli. Journal of Virology* 65: 1611-1615.

Yamamoto, T., I. Sanyo, M. Kohara and H. Thara. 1991. Estimation of the heritability for resistance to infectious hematopoietic necrosis in rainbow trout. *Bulletin of the Japanese Society of Scientific Fisheries* 57: 1519-1522.

Yasuike, M., H. Kondo, I. Hirono and T. Aoki. 2007. Difference in Japanese flounder, *Paralichthys olivaceus* gene expression profile following hirame rhabdovirus (HIRRV) G and N protein DNA vaccination. *Fish and Shellfish Immunology* 23: 531-541.

Yoshinaga, K.O.N., O. Kurata and Y. Ikeda. 1994. Individual variation of natural killer activity of rainbow trout leukocytes against IPN virus infected and uninfected RTG-2 cells. *Fish Pathology* 29: 1-4.

Yousif, A., L.J. Albright and T.P.T. Evelyn. 1993. Immunological evidence for the presence of an IgM-like immunoglobulin in the eggs of coho salmon *Oncorhynchus kisutch. Diseases of Aquatic Organisms* 23: 109-114.

Zhang, H., J.P. Evenhuis, G.H. Thorgaard and S.S. Ristow. 2001. Cloning, characterization and genomic structure of the natural killer cell enhancement factor (NKEF)-like gene from homozygous clones of rainbow trout (*Oncorhynchus mykiss*). *Developmental and Comparative Immunology* 25: 25-35.

Zinkernagel, R.M. and P.C. Doherty. 1979. MHC-restricted cytotoxic T cells: studies on the biological role of polymorphic major transplantation antigens determining T-cell restriction-specificity, function, and responsiveness. *Advances in Immunology* 27: 51-177.

Zou, J., S. Peddie, G. Scapigliati, Y. Zhang, N.C. Bols, A.E. Ellis and C.J. Secombes. 2003. Functional characterisation of the recombinant tumor necrosis factors in rainbow trout, *Oncorhynchus mykiss. Developmental and Comparative Immunology* 27: 813-822.

Zou, J., C. Tafalla, J. Truckle and C.J. Secombes. 2007. Identification of a second group of type I IFNs in fish sheds light on IFN evolution in vertebrates. *Journal of Immunology* 179: 3859-3871.

Zou, J., A. Carrington, B. Collet, J.M. Dijkstra, Y. Yoshiura, N. Bols and C.J. Secombes. 2005. Identification and bioactivities of IFN-gamma in rainbow trout *Oncorhynchus mykiss*: The first Th1-type cytokine characterized functionally in fish. *Journal of Immunology* 175: 2484-2494.

Vaccination Strategies to Prevent Streptococcal Infections in Cultured Fish

Jesús L. Romalde[1,*], Beatriz Magariños[1],
Carmen Ravelo[2] and Alicia E. Toranzo[1]

INTRODUCTION

Nowadays, infections caused by Gram-positive cocci have to be considered as re-emerging fish diseases. Although streptococcosis outbreaks have been occurring for four decades in Japanese farms culturing rainbow trout (*Oncorhynchus mykiss*) and yellowtail (*Seriola quinqueradiata*) (Ringø and Gatesoupe, 1998), this disease has been described in other cultured fish species throughout the world, such as hybrid tilapia (*Oreochromis aureus*× *O. niloticus*) and striped bass (*Morone saxatilis*) in North America, or rainbow trout in South Africa and Australia (Bragg and Broere, 1986; Carson *et al.*, 1993; Ghittino, 1999).

Authors' addresses: [1] Departamento de Microbiología y Parasitología, C1BUS-Facultad de Biología. Universidad de Santiago de Compostela, 15782, Santiago de Compostela, Spain.
[2] Laboratorio de Ictiopatología, Estación de Investigaciones Hidrobiológicas de Guayana, Fundación La Salle de C.N. 8051, Ciudad Guayana, Venezuela.
Corresponding author: E-mail: jesus.romalde@usc.es

In recent years, new genera and species of Gram-positive cocci, including streptococci, lactococci and vagococci, have been isolated from diseased fish in Europe and the Mediterranean basin (Eldar *et al.*, 1996), and later on in other parts of the world (Bromage *et al.*, 1999; Shoemaker *et al.*, 2001; Agnew and Barnes, 2007). All these agents produced similar clinical signs in their hosts and, therefore, streptococcosis or 'pop-eye' disease of fish can be considered a complex of similar diseases caused by different taxa of Gram-positive cocci. These infections now constitute the most important diseases affecting farmed finfish, namely rainbow trout, yellowtail and tilapia, with estimated global economic losses of more than US$150 million. In addition, streptococcal episodes have been also detected in wild fish (Baya *et al.*, 1990; Colorni *et al.*, 2002, 2003) throughout the world.

Several attempts have been made to develop appropriate vaccination programmes for fish streptococcosis (Iida *et al.*, 1981; Sakai *et al.*, 1987; Carson and Munday, 1990; Ghittino *et al.*, 1995a, b; Toranzo *et al.*, 1995b; Akhlaghi *et al.*, 1996; Romalde *et al.*, 1996, 1999b; Bercovier *et al.*, 1997; Eldar *et al.*, 1997b). Considerable variability in the protection achieved has been observed, depending on the fish and bacterial species, the formulation of the vaccine, the route of administration, the age of the fish, as well as the use of immunostimulants. However, due to the failure of chemotherapy in most streptococcal outbreaks, vaccination remains the only possible approach to control this disease.

HOST RANGE AND GEOGRAPHIC DISTRIBUTION

The first description of streptococcal infection causing fish mortalities is in 1956 (Hoshina *et al.*, 1958), which affected populations of farmed rainbow trout in Japan with high mortality levels (0.3% per day). Since then, the disease has increased its host range as well as its geographical distribution. In fact, severe mortalities of rainbow trout were also described in the USA, Australia, South Africa, Israel, and several European countries including Spain, France and Italy. The disease may have occurred sporadically in Great Britain and Norway (Austin and Austin, 2007).

Outbreaks in yellowtail (Kusuda *et al.*, 1976; Kitao *et al.*, 1979), Coho salmon (*Oncorhynchus kisutch*) (Atsuta *et al.*, 1990), jacopever (*Sebastes schelegeli*) (Sakai *et al.*, 1986), Japanese eel (*Anguilla japonica*) (Kusuda *et al.*, 1978), ayu (*Plecoglossus altivelis*), tilapia (*Oreochromis* spp.) (Kitao *et al.*, 1981), and Japanese flounder (*Paralichthys olivaceus*) (Nakatsugawa,

1983) have been reported in Japan. In the USA, there are evidence of the disease in a variety of fish species including rainbow trout, sea trout (*Cynoscion regalis*), silver trout (*Cynoscion nothus*), Atlantic croaker (*Micropogon undulatus*), blue fish (*Pomatomus saltatrix*), golden shiner (*Notemigonous chrysoleuca*), menhaden (*Brevoortia patronius*), striped bass and striped mullet (*Mugil cephalus*), among others (Robinson and Meyer, 1966; Plumb *et al.*, 1974; Baya *et al.*, 1990; Ringø and Gatesoupe, 1998; Austin and Austin, 2007). In the Mediterranean area, apart from being widely recognized in rainbow trout, streptococcosis has been described in turbot (*Scophthalmus maximus*) in Spain, in sturgeon (*Acipenser naccarii*) in Italy; and in tilapia, striped mullet, striped bass, seabass (*Dicentrarchus labrax*), and gilthead seabream (*Sparus aurata*) in Israel (Toranzo *et al.*, 1994; Salati *et al.*, 1996; Ghittino, 1999; Romalde and Toranzo, 1999).

CLINICAL SIGNS OF DISEASE

The typical gross pathology observed in fish streptococcosis, apart from elevated rates of mortality (up to 50%), include external signs such as anorexia, loss of orientation, lethargy, reduced appetite and erratic swimming. Uni- or bilateral exophthalmia (Figs. 4.1, 4.2) is frequent with intra-ocular haemorrhage and clouding of the eye. In many cases abdominal distension, darkening of the skin and haemorrhage around the opercula and anus are also observed (Kusuda *et al.*, 1991; Eldar *et al.*, 1994; Nieto *et al.*, 1995; Stoffregen *et al.*, 1996; Michel *et al.*, 1997; Eldar and Ghittino, 1999). Internally, the principal organs affected are the spleen, liver and brain and, to a lesser extent, the kidney, gut and heart (Austin and Austin, 2007). The spleen may be enlarged and necrotized and the liver is generally pale with areas of focal necrosis. The intestine usually contains fluid and focal areas of haemorrhage. The abdominal cavity may contain varying amounts of exudate, which may be purulent or contain blood. Acute meningitis is often observed, consisting of a yellowish exudate covering the brain surface and often containing numerous bacterial cells (Kitao, 1993; Múzquiz *et al.*, 1999; Romalde and Toranzo, 1999; Austin and Austin, 2007).

While the general clinical picture is relatively consistent, some variation in the clinical signs of fish streptococcosis have been described depending on which fish species are affected, the stage of infection, and the aetiological agent (Michel *et al.*, 1997; Eldar and Ghittino, 1999). Eldar and Ghittino (1999) pointed out that in rainbow trout, *L. garvieae*

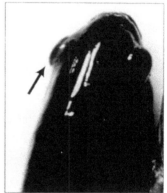

Fig. 4.1 Gross external symptoms of streptococcal infections: pronounced bilateral exophthalmia with haemorrhages in the periocular area (arrows) in rainbow trout suffering infection with *L. garvieae*.

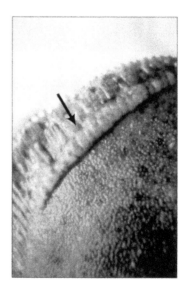

Fig. 4.2 Gross external symptoms of streptococcal infections: exophthalmia and accumulation of purulent material in the base of fins (arrows) in turbot affected by *S. parauberis*.

produces a hyper acute systemic disease, whereas *S. iniae* causes more specific lesions as part of a slower illness course. The main difference between both pathological processes, that could be indicative of the aetiological agent, is termed 'oculo-splanchnic dissociation' and consists in a severe serositis, sometimes extended to the myocardium, restricted to *L. garvieae* infected fish.

The acute course of disease associated with *L. garvieae* and *S. iniae* are markedly different from the chronic condition caused by *V. salmoninarum* where hyperaemia, tegumentary lesions and a proliferative response in the cardiovascular system are commonly observed (Michel *et al.*, 1997; Eldar and Ghittino, 1999). Other clinical signs associated with *V. salmoninarum* include impaired swimming, unilateral exophthalmia, haemorrhages in the eyes and on gills, and enlargement of the liver and spleen (Michel *et al.*, 1997).

IDENTIFICATION OF FISH PATHOGENIC STREPTOCOCCI

The presence of typical clinical symptoms and the demonstration of Gram-positive cocci from the internal organs, such as kidney, brain, etc., constitute a presumptive diagnosis. Gram-positive cocci can be isolated on standard general-purpose media but growth is enhanced by the addition of blood to a final concentration of 5% (v/v) (Frerichs, 1993). Several useful media in the recovery of the pathogen from diseased fish tissues include Nutrient agar (Kusuda *et al.*, 1991), Todd-Hewitt broth (Kitao, 1993), Brain Heart Infusion (BHI) agar (Eldar *et al.*, 1994), 5% [v/v] defibrinated sheep blood agar (Domenech *et al.*, 1996), Trypticase Soy agar (Michel *et al.*, 1997), or Columbia agar (Austin and Austin, 2007). Media can be supplemented with 1% (w/v) sodium chloride (Austin and Robertson, 1993). A selective procedure for *Streptococcus* spp. was described by Bragg *et al.* (1989), consisting of an enrichment step in nutrient broth supplemented with naladixic acid (100 µg/ml), oxolinic acid (160 µg/ml) or sodium azide (200 µg/ml) followed by plating the enriched samples onto tetrazolium agar. More recently, Nguyen and Kanai (1999) developed two selective media for the identification of *Streptococcus iniae* from Japanese flounder (*Paralichthys olivaceus*). Both media were based on BHI agar supplemented with horse blood, which served as a source of micronutrients and facilitated the identification of haemolytic isolates. The first media contained thallium acetate and oxolinic acid (TAOA) while the other colistin sulphate and oxolinic acid (CSOA). These two media may also be of some use in the differentiating *S. iniae* isolates from other fish species.

Small, translucent colonies, 1-2 mm in diameter develop after incubation at 15-37°C for 48-72 h, but may require up to 7 days to develop fully. In Gram-stained preparations, cells appear as cocci or ovoid forms in pairs or chains. Presumptive identification is made on the basis of a few characters, including cellular morphology, Voges-Proskauer reaction, type of haemolysis, growth on bile (40%)-aesculin agar, growth at 10 and 45°C at pH 9.6, production of H_2S, hydrolysis of sodium hippurate, starch hydrolysis, and presence of specific enzymes like arginine dihydrolase and pyrrolidonilarylamidase, among others. However, it should be emphasized that identification of species is difficult, as the isolates are identified only at genus level in numerous occasions (Elliot and Facklam, 1996; Ravelo et al., 2001).

After preliminary biochemical identification, the use of serological analysis can be useful to determine the streptococcal species involved in a particular outbreak. Serological confirmation may be performed by a variety of methods such as slide agglutination (Kitao, 1982) or fluorescent antibody staining (Kawahara and Kusuda, 1987). An indirect fluorescent antibody procedure has also been used to identify Streptococcus sp. from pure cultures and smears from experimentally and natural diseased salmonid fish (Bragg, 1988). More recently, Japanese authors have developed a rapid flow cytometry based method that proved to be useful to detect the pathogen in mixed cultures (Endo et al., 1998). While some isolates have been identified as Lancefield serogroup B or D, the majority of the fish pathogenic strains have proved to be not typable by the conventional Lancefield grouping system (Frerichs, 1993). For this reason and also because a number of different bacterial species are implicated with 'streptococcosis' in fish, it appears that the Lancefield scheme is of limited usefulness in the identification of the aetiological agents of fish streptococcosis.

The application of molecular techniques—developed in the last 15 years—to the diagnosis of fish streptococcosis has been a great help to clarify the aetiology of the disease as well as to correctly identified the causal organisms of the outbreaks (Romalde and Toranzo, 2002). There are a number of reports describing specific PCR protocols for S. iniae or L. garvieae (Goh et al., 1996, 1997, 1998, 2000; Berridge et al., 1998; Zlotking et al., 1998a, b; Aoki et al., 2000), which have facilitated the detection of these pathogens from fish tissues or the environment. On the other hand, sequencing of the 16S rRNA gene has determined the assignation of isolates to a definite species, as well as the synonymies of the proposed new species already described within the streptococcal group, such as

Enterococcus seriolicida with *L. garvieae*, or *S. shiloi* and *S. difficilis* with *S. iniae* and *S. agalactiae*, respectively (Domenech *et al.*, 1993; Eldar *et al.*, 1995b; Teixeira *et al.*, 1996; Vandamme *et al.*, 1997; Berridge *et al.*, 2001; Kawamura *et al.*, 2005).

Molecular techniques were also applied to epidemiological studies of these fish pathogens in which the heterogeneity within the different species were studied (Eldar *et al.*, 1997a, 1999; Hawkesford *et al.*, 1997; Meads *et al.*, 1998; Shoemaker and Klesius, 1998; Romalde *et al.*, 1999a; Ravelo *et al.*, 2000, 2003; Vela *et al.*, 2000; Fuller *et al.*, 2001; Kvitt and Colorni, 2004). Ribotyping, Random amplified polymorphic DNA (RAPD), pulsed-filed gel electrophoresis (PFGE) or restriction fragment length polymorphisms (RFLP), among other methods, have been employed demonstrating within some species the existence of different geno groups with epidemiological relevance (i.e., association of geno groups with specific hosts or geographical origins, as well as with virulence). Using such techniques, it has been possible to demonstrate that a single clone of *S. iniae* is present in wild and cultured fish, which suggests the possible role of wild fish as reservoir of infection in the environment (Zlotkin *et al.*, 1998a).

AETIOLOGY

There has been an important controversy about the number and the nature of the bacterial species involved with streptococcosis (Austin and Austin, 2007). Numerous Gram-positive cocci have been linked with pathology in fishes, which include, *Streptococcus agalactiae, S. equi, S. pyogenes, S. milleri* and *S. mutans* (Robinson and Meyer, 1966; Kusuda and Komatsu, 1978; Austin and Robertson, 1993). In addition, *Enterococcus faecalis* subsp. *liquefaciens, E. faecium*, or *Lactococcus lactis* have at various times been implicated with similar diseases in Atlantic salmon and rainbow trout (Boomker *et al.*, 1979; Ghittino and Prearo, 1992).

In early reports on streptococcosis in fish it was not always possible to assign isolates to a particular species; however, some attempt was made to group fish pathogenic strains on the basis of phenotypic traits such as haemolysis and correlate this characteristic with a range of pathologies (Miyazaki, 1982). Thus, α-haemolytic isolates were responsible for granulomatous inflammation and infected lesions, β-haemolytic isolates, causing systemic infection with septicaemia and suppurative eye inflammation, and non-haemolytic associated with meningoencephalitis episodes (Robinson and Meyer, 1966; Plumb et al., 1974; Kusuda et al., 1976; Minami et al., 1979; Kitao et al., 1981; Iida et al., 1986; Al-Harbi, 1994; Eldar et al., 1995a; Figueiredo et al., 2007).

With the development of taxonomic techniques and the application of molecular procedures to bacterial identification, it was possible to more accurately determine the precise taxonomic status of many isolates. There were a considerable number of new species descriptions and taxonomic re-appraisals (Pier and Madin, 1976; Wallbanks et al., 1990; Williams et al., 1990; Kusuda et al., 1991; Eldar et al., 1994, 1996; Domenech et al., 1996), which were helpful in clarifying the aetiology of streptococcosis.

Today, there is general acceptance for the division of streptococcosis into two forms according to the virulence of the agents involved at high or low temperatures (Ghittino, 1999). 'Warm water' streptococcosis, causing mortalities at temperatures higher than 15°C, typically involves *Lactococcus garvieae* (synonym *Enterococcus seriolicida*), *Streptococcus iniae* (synonym *S. shiloi*), *S. agalactiae* (synonym *S. difficilis*), *S. parauberis*, or *S. phocae*. On the other hand, 'cold water' streptococcosis is caused by *Vagococcus salmoninarum* and *L. piscium* and occurs at temperatures below 15°C.

Lactococcus garvieae

The first description of *Lactococcus garvieae* (formerly *Streptococcus garvieae*) came from an investigation of bovine mastitis in Great Britain (Collins et al., 1984). Later, *L. garvieae* was isolated from a variety of diseased freshwater and marine fish, and also from humans (Elliot et al., 1991), indicating the increasing importance of this bacterium both as a pathogen of fish and potential zoonotic agent and its ubiquitous distribution.

The identification criteria for *L. garvieae* based on biochemical and antigenic characteristics are very similar to *L. lactis* subsp. *lactis*, which has also been reported as a human pathogen (Collins et al., 1984; Mannion and Rothburn, 1990; Elliot et al., 1991; Domenech et al., 1993), and from *Enterococcus*-like strains isolated from diseased fish (Toranzo et al., 1994; Nieto et al., 1995). Gram-positive cocci which are capable of growth between 10 and 42°C, at pH 9.6, in the presence of 6.5% NaCl and on 0.3% methylene blue-milk agar can be identified as *L. garvieae*.

Further works of Eldar *et al.* (1999) and Vela *et al.* (1999) reveal the phenotypic heterogeneity of *L. garvieae*. These workers both proposed biotyping schemes that recognized three biotypes of *L. garvieae*. While based on the same phenotypic traits (acidification of tagatose, ribose and sucrose), there are some inconsistencies between the typing schemes described by the above authors. A possible explanation for this is the use of the API-20Strep and/or API-32Strep miniaturized systems for biochemical characterization of the strains. Ravelo *et al.* (2001) demonstrated that these systems may yield different results depending on the medium used for obtaining the bacterial inocula. In addition, the results achieved for some tests (i.e., acid production from: lactose, maltose, sucrose, tagatose and cyclodextrin) did not always correlate with results obtained with traditional plate and tube procedures. Moreover, although the strains studied by these authors showed variability for some characters, no biotypes with epidemiological value could be established. More recently, Vela *et al.* (2000) proposed a new intraspecies classification of *L. garvieae* with 13 biotypes, on the basis of acidification of sucrose, tagatose, mannitol, and cyclodextrin and the presence of the enzymes pyroglutamic acid arylamidase and N-acetyl-β-glucosaminidase, although only six of these biotypes were isolated from fish.

In 1991, Kusuda *et al.* proposed a new species, *Enterococcus seriolicida*, in order to bring together a number of Gram-positive isolates recovered from Japanese yellowtail over the preceding 20 years (Kusuda *et al.*, 1976, 1991). Subsequent phenotypic and molecular characterization of *E. seriolicida* demonstrated that this species should be reclassified as a junior synonym of *L. garvieae* (Domenech *et al.*, 1993; Eldar *et al.*, 1996; Pot *et al.*, 1996; Teixeira *et al.*, 1996). An interesting feature of these Japanese isolates is the existence of two serotypes, which could not be distinguished from one another biochemically. These two serotypes were associated with the presence (serotype KG⁻) or absence (serotype KG⁺) of a capsule. This capsule was reported to confer various properties on isolates, including a hydrophilic character, capacity of resistance to phagocytosis and higher pathogenicity (Kitao, 1982; Yoshida *et al.*, 1996, 1997). Freshly isolated cultures of this bacterium consist almost entirely of the KG⁻ serotype. Recently, five different genes were identified from KG⁻ but not from KG⁺ isolates of *L. garvieae*, coding for protease, dihydropteroate synthase, trigger factor and N-acetylglucosamine-6-phosphate deacetylase proteins which were the main immunogenic antigens in rabbit (Hirono *et al.*, 1999). On the other hand, Barnes and Ellis (2004), using trout sera, demonstrated capsular variation, and therefore serological differences, among *L. garvieae* strains related with the origin of the isolates. Two serovariants were defined by these authors, one comprising Japanese strains isolated from marine fish and the second one compiling European isolates from freshwater species. In addition, recent preliminary results in our laboratory indicate serological variability among strains isolated from rainbow trout in Spain on the basis of dot-blot and microagglutination assays.

Streptococcus iniae

Streptococcus iniae was first isolated from skin lesions on an Amazon freshwater dolphin (*Inia geoffrensis*) (Pier and Madin, 1976). Further, it has been described as the aetiological agent of septicaemia and meningoencephalitis in several cultured fish species such as rainbow trout, yellowtail, and hybrid tilapia among others (Eldar *et al.*, 1994; Perera *et al.*, 1994; Stoffregen *et al.*, 1996; Sugita, 1996). More recently, *S. iniae* has been implicated with cellulitis in humans with a history of injury while handling/cleaning fresh fish in different countries (Weinstein *et al.*, 1997; Berridge *et al.*, 1998; Facklam *et al.*, 2005; Lau *et al.*, 2006), with at least 25 cases confirmed to date. Therefore, this pathogen must be considered a zoonotic agent.

Although *S. iniae* is well characterized phenotypically, identification is complicated by the high degree of similarity with other pathogenic streptococci. Misidentification as *S. uberis* has been reported using miniaturized identification systems (Weinstein *et al.*, 1997). In addition, a further difficulty arises with the detection of *S. iniae*, as it grows relatively slowly it may be overgrown if primary cultures are grossly contaminated.

In 1994, Eldar and co-workers described a new species within the genus *Streptococcus* on the basis of the differential characteristics of a group of strains isolated from diseased rainbow trout in Israel. This species was named *Streptococcus shiloi* and was validated in 1995 (Ad Hoc Committee of the ICSB, 1995). The disease spread rapidly, and was responsible for significant economic losses in the Israeli fish farming industry (Eldar *et al.*, 1994). A wider taxonomic study of a large number of similar isolates from Israel and the USA, employing both biochemical and genetic traits, demonstrated that *S. shiloi* must be considered a junior synonym of *S. iniae* (Eldar *et al.*, 1995a).

In recent years, the first cases of infections by *S. iniae* in cultured seabass and gilthead seabream have been detected in Spain, which indicate the increasing importance of streptococcosis in these economically important fish species (Zarza and Padrós, 2007).

Streptococcus agalactiae

Streptococcus agalactiae, or group B streptococci, has been isolated predominantly from human and bovine sources but have been recovered occasionally from several homeothermic animals, such as cats or dogs, and also from some poikilothermic animals including frogs and fish (Kummeneje *et al.*, 1975; Kornblatt *et al.*, 1983; Dow *et al.*, 1987; Evans *et al.*, 2002). Most strains of the species show β-haemolysis, although a number of non-haemolytic, type Ib variants have been isolated from humans, cows and fish (Wilkinson *et al.*, 1973; Amborski *et al.*, 1983). The characterization of these variants by biochemical analysis (Wilkinson *et al.*, 1973) and whole-cell protein analysis (Elliott *et al.*, 1990) showed that although fish isolates presented several biochemical differences with isolates from human or cows, they were indistinguishable in whole-protein patterns. Today, it is well recognized that *S. agalactiae* is a major pathogen that causes serious economic losses in many species of freshwater, marine and estuarine fish worldwide (Pasnik *et al.*, 2005a, b).

Streptococcus difficilis (*S. difficile* [sic], the species epithet was corrected by Euzéby [1998]) was described to accommodate some isolates of a non-haemolytic, mannitol-negative Gram-positive coccus, that were perceived as constituting a new species, causing meningo-encephalitis in tilapia and rainbow trout cultured in Israel (Eldar *et al.*, 1994, 1995a). Some years later, Vandamme *et al.* (1997) reported that *S. difficilis* was a group B, serotype Ib streptococcus with whole-cell protein characteristics indistinguishable from those of *S. agalactiae*. Furthermore, Berridge *et al.* (2001) and Kawamura *et al.* (2005) determined a high genetic similarity between these two species, by analysis of the 16-23S intergenic rRNA gene sequence and comparison of five gene sequences (16S rRNA, *gyr*B, *sod*A, *gyr*A, and *par*C) respectively. On the basis of these findings, it was proposed that *S. difficilis* is a later synonym of *S. agalactiae*.

Streptococcus parauberis

Between 1993 and 1996, a streptococcal disease caused important economic losses to the turbot industry in the north of Spain (Toranzo *et al.*,

1994; Domenech *et al.*, 1996; Romalde and Toranzo, 1999), since the affected fish were unmarketable due to their poor external appearance (Fig. 4.2) (Nieto *et al.*, 1995). All the isolates from turbot showed a high phenotypic and serological homogeneity and were presumptively classified as *Enterococcus* sp. closely related to *Enterococcus seriolicida* (Toranzo *et al.*, 1994, 1995a). Further studies on sequencing of the 16S rRNA gene indicated that they should be classified within the *Streptococcus* group, as belonging to the species *Streptococcus parauberis* (Domenech *et al.*, 1996; Romalde *et al.*, 1999a, b).

Genetic characterization of the isolates employing the random amplified polymorphic DNA (RAPD) technique showed some variability among strains which could be related with the farm of isolation, indicating certain endemism within each farm. The high survival of the pathogen in the environment (up to 6 months) adopting a viable but non-culturable (VBNC) state (Currás *et al.*, 2002) could explain such endemicity.

Streptococcus phocae

From 1999, disease outbreaks occurred repeatedly during the summer months (temperatures higher than 15°C) in Atlantic salmon (*Salmo salar*) farmed in Chile affecting both smolts and adult fish cultured in estuary and marine waters (Romalde *et al.*, 2008; Valdés *et al.*, 2009). Cumulative mortality reached up to 20% of the affected population in some occasions. Diseased fish showed exophthalmia with accumulation of purulent and haemorrhagic fluid around eyes, and ventral petechial haemorrhages (Fig. 4.3). At necropsy, haemorrhage in the abdominal fat, pericarditis, and enlarged liver (showing a yellowish colour), spleen and kidney are

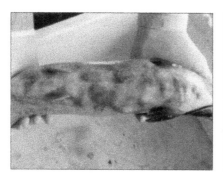

Fig. 4.3 Gross external symptoms of streptococcal infections: skin abscesses and ulceras with muscle liquefaction in diseased Atlantic salmon infected by *S. phocae*.

common pathological changes. Gram-stained smears revealed the presence of Gram-positive cocci, β-haemolytic, negative for oxidase and catalase test. Although biochemical characterization of the isolates using the miniaturized system rapid ID 32 Strep suggested their assignation to genus *Gemella*, sequencing and RFLP analysis of the 16S rRNA revealed that bacteria associated with the mortalities belong to *Streptococcus phocae*. Serological studies demonstrated that all the salmon isolates are antigenically homogeneous, which can facilitate the development of preventive measures and, although sharing some antigenical determinants, they belong to a different Lancefield group than the type strain isolated from seals. On the basis of these facts, we conclude that the species *S. phocae* is an emerging pathogen for salmonid culture in Chile, and it should be included as a new member of the warm water streptococcosis. Until these reports, *S. phocae* had only been involved in seal outbreaks causing pneumonia or respiratory infection (Henton *et al.*, 1999; Raverty and Fiessel, 2001; Skaar *et al.*, 2003; Raverty *et al.*, 2004; Vossen *et al.*, 2004).

Molecular typing of the fish isolates by different methods, such as pulsed-field gel electrophoresis (PFGE), RAPD, enterobacterial repetitive intergenic consensus sequence PCR (ERIC-PCR), repetitive extragenic palindromic PCR (REP-PCR) and restriction of 16S-23S rDNA intergenic spacer regions, demonstrated genetic homogeneity within the salmon isolates of *S. phocae*, suggesting the existence of a clonal lineage diverse from that of the type strain isolated from seal.

Vagococcus salmoninarum

During 1968, a bacterium similar to the lactobacilli was recovered from diseased adult rainbow trout in Oregon, USA. The isolate was further subjected to classical phenotypic and molecular taxonomic characterization, including the study of its 16S rRNA gene sequence. A 96.3% homology with *Vagococcus fluvialis* was recorded. Lower homology values were obtained with other related species such as *E. durans* (94.5%), *Carnobacterium divergens* (94.1%), *Enterococcus avium* (94.0%), *C. piscicola* (93.8%) and *C. movile* (93.7%). Despite the high similarities observed, this isolate became the type strain of a new species, *Vagococcus salmoninarum* (Wallbanks *et al.*, 1990).

Some years later, Schmidtke and Carson (1994) characterized new isolates of *V. salmoninarum* recovered from salmonid fishes in Australia, including Atlantic salmon (*Salmo salar*), rainbow trout, and brown trout

(*Salmo trutta*). These strains showed a high level of phenotypic similarity with the type strain. Interestingly, two strains isolated from brown trout in Norway were included in this study, constituting the first report of this pathogen in Europe.

Between 1989 and 1995, a Gram-positive chain forming diplococcus, identified as *V. salmoninarum* by DNA/DNA hybridization, caused significant losses in French rainbow trout farmed at low water temperatures. Mortality rates ranging up to 50% per year were reported (Nougayrède *et al.*, 1995; Michel *et al.*, 1997). The organism was isolated from two geographically distant locations, a trout farm in the southwest and Brest, in the northwest. These findings confirmed that the bacterium is widespread and much more common in Europe than firstly thought.

It is noteworthy that Michel *et al.* (1997) observed variability in some biochemical characteristics (i.e. carbohydrate reactions), among the French isolates. It was suggested that this variability may provide the basis for some useful epidemiological markers for this pathogen. However, much work is still needed to determine the overall level of phenotypic and/or genetic variability within the taxon before determining the significance or usefulness of this variability.

In the last years, outbreaks of streptococcosis caused by *V. salmoninarum* were described in Spain, affecting rainbow trout broodstocks (Ruiz-Zarzuela *et al.*, 2005). Mortality rates between 11 and 36% originated great economic losses in the farms.

Other Streptococci

Other species from the genus *Streptococcus* have been occasionally associated with fish pathologies, reinforcing the idea of the complicate aetiology of the fish streptococcosis. Thus, the Lancefield group C *S. dysgalactiae* was recovered in Japan from amberjack and yellowtail displaying necrotic lesions of the caudal peduncle (Nomoto *et al.*, 2004, 2006). Interestingly, those fish had been previously vaccinated against *L. garvieae*, another agent of the fish streptococcosis. On the other hand, *S. milleri* has been related with some pathological problems in Koi carp (Austin and Robertson, 1993).

From 2002 to 2004, four cases of suspected streptococcosis were recorded in Channel catfish (*Ictalurus punctatus*) farms at the Mississippi delta. Conventional biochemical characterization, 16S rRNA gene sequence analysis and DNA-DNA hybridization studies distinguished these isolates from previously described *Streptococcus* species, although they were phylogenetically related to *S. iniae*, *S. uberis* and *S. parauberis*. The name *S. ictaluri* was proposed for this new species (Shewmaker *et al.*, 2007). The potential significance of this emerging pathogen for the Channel catfish industry is still unknown.

CONTROL MEASURES

Effective control measures for Gram-positive infections in fish are important, not only because of the severe economic losses that these diseases can cause in aquaculture, but also because of the potential for some species such as, *Lactococcus garvieae*, *Streptococcus agalactiae*, and *S. iniae*, to infect humans (Elliot *et al.*, 1991; Wenstein *et al.*, 1997; Berridge *et al.*, 1998; Meads *et al.*, 1998; Sun *et al.*, 2007).

Several early works reported the effectiveness of antibiotics in treating streptococcal infections in fish (Robinson and Meyer, 1966; Katao, 1982), although this effectiveness is dependent on the fish species. Thus, increased survival was observed in *S. iniae* infected fish including hybrid striped bass treated with enrofloxacin (Stoffregen *et al.*, 1996), tilapia treated with amoxycillin (Darwish and Ismaiel, 2003; Darwish and Hobbs, 2005), or barramundi treated with erythromycin (Creeper and Buller, 2006). However, in the case of *L. garvieae*, although some drugs like erythromycin, oxytetracycline or enrofloxacin have proved to be active *in vitro*, they were ineffective in the field, probably due to the anorectic condition of diseased fish (Bercovier *et al.*, 1997; Romalde *et al.*, 2006).

Unfortunately, the indiscriminate use of these drugs has lead to the appearance of widespread antibiotic resistance. Experience in the field suggests that chemotherapy is now usually ineffective (Aoki *et al.*, 1990; Bercovier *et al.*, 1997; Romalde and Toranzo, 1999).

In the last years, an increasing interest in the use of probiotics as an alternative approach to control fish diseases, including streptococcosis, has been noticed (Irianto and Austin, 2002, 2003). Thus, Li *et al.* (2004) employed a strain of *Saccharomyces cereviseae* to stimulate the immune response against *S. iniae* in hybrid striped bass. On the other hand, Brunt and Austin (2005) and Brunt *et al.* (2007), respectively, isolated from the digestive tract of rainbow trout and ghost carp, strains of *Bacillus* sp. and *Aeromonas sobria* which were effective at preventing clinical streptococcal disease, caused by both *S. iniae* or *L. garvieae*, when used as feed additive.

More recently, Prado (2006) characterized an isolate of *Phaeobacter gallaeciensis* with activity against several gram-positive and gram-negative fish pathogens, including *S. parauberis*. Unfortunately, the efficacy of such strains was not tested under field conditions; hence all these results are based on *in vitro* experiments or controlled fish challenges.

Therefore, vaccination, together with proper management procedures, including the reduction of overfeeding, overcrowding, handling and transportation, has become essential in the control of fish streptococcosis.

VACCINATION

In the last 25 years, much research has been done to develop appropriate vaccination programmes against streptococcosis for several fish species (Iida *et al.*, 1981; Akhlaghi *et al.*, 1996; Bercovier *et al.*, 1997; Romalde *et al.*, 1999b, 2006; Agnew and Barnes, 2007), including active and passive immunization protocols. However, considerable variability in the protection achieved was observed depending on the fish and bacterial species, the vaccine formulation, the route of administration, the fish age, as well as the use of immunostimulants. We shall now summarize the progress on vaccination against the main aetiological agents of fish streptococcosis.

Lactococcus garvieae

As in the case of other gram-positive cocci pathogens for fish such as *S. iniae*, *S. agalactiae* or *S. parauberis* (Bercovier *et al.*, 1997; Romalde *et al.*, 1999b; Evans *et al.*, 2004), good levels of protection are only achieved when vaccines are intraperitoneally (i.p.) administered. Thus, in the case of bacterins composed by formalin-killed cells (FKC), immersion administration procedures always rendered a relative percentage of survival (RPS) values lower than 15 in both rainbow trout and yellowtail, while protection, in terms of RPS, reported when vaccines were i.p. administered ranged between 21 and 90 or around 100 for these two fish species, respectively (Table 4.1). Salati *et al.* (2005) proposed the use of preparations of a cellular component, namely Protein M, as a sub-unit vaccine on the basis of the results obtained in assays of antibody induction and phagocytosis index. However, more recent studies (Volpatti *et al.*, 2007) supported the higher protection conferred against *L. garvieae* with bacterial whole-cell preparations (RPS=95%) in comparison with

Table 4.1 Protection obtained by the different vaccines against *L.garvieae* four weeks after vaccination

Year	Reference	Fish	Type of vaccine	Administration method	Challenge method	Adjuvant/ encapsulation	RPS (%)
Injectable Vaccines							
1982	Iida et al.	Yellowtail	FKC [a]	i.p. [b]	i.p.	none	70
1995	Ghittino et al.	Rainbow trout	FKC	i.p.	i.p.	none	80-90
1996	Akhlaghi et al.	Rainbow trout	FKC	i.p.	i.p.	FCA	88.8
		Rainbow trout	FKC	inmersión	i.p.	none	11.1
1997	Bercovier et al.	Rainbow trout	FKC	i.p.	i.p.	none	90.0
1998	Ceschia et al.	Rainbow trout	FKC	i.p.	natural	none	21.0
1999	Ooyama et al.	Yellowtail	FKC	i.p.	i.p.	none	100.0 [c]
1999	Ghittino	Rainbow trout	FKC	i.p.	i.p.	none	64.7 [d]
		Rainbow trout	FKC	i.p.	i.p.	mineral oil	82.3 [d]
2002	Ooyama et al.	Yellowtail	FKC	i.p.	i.p.	none	100.0
2006	Ravelo et al.	Rainbow trout	FKC	i.p.	i.p.	none	82.6-100
		Rainbow trout	FKC	i.p.	i.p.	non-mineral oil	86.9-94
2007	Lee et al.	Grey mullet	FKC + ECP	i.p.	i.p.	none	100
Oral Vaccines							
1997	Sano et al.	Yellowtail	FKC	oral	i.p.	none	70.0
2004	Romalde et al.	Rainbow trout	FKC	oral	i.p.	none	7.0
		Rainbow trout	FKC	oral	i.p.	alginate	50.0
		Rainbow trout	FKC	i.p. + oral [e]	i.p.	none	87.5 [f]

[a] FKC, formalin killed cells; ECP, extracellular products; [b] i.p., intraperitoneal injection; [c] RPS at fourteen days after vaccination; [d] RPS at four months after vaccination; [e] First immunization by i.p. injection of aqueous bacterin and booster with oral vaccine three months later; [f] RPS value one month after booster.

formulations consisting in cellular fractions such as extracellular products (ECP) (35%) or membrane antigens (33%).

Passive immunization of rainbow trout against *L. garvieae* was also evaluated (Akhlaghi *et al.*, 1996) employing antibodies raised in sheep, rabbit or fish. The results obtained were comparable to those of active immunization in both, protective effect and duration. These observations indicate that passive immunization could have significant potential in the prevention of fish streptococcosis.

Regarding the duration of protection, most works have been done with active immunization. Thus, the aqueous bacterins (FKC) conferred long-term protection in yellowtail (Ooyama *et al.*, 1999). However, in the rainbow trout culture, the short duration of the immunity (3-4 months) constituted the main inconvenience for the success of these vaccines, since this period is not enough to cover the warm season (water temperature higher than 16°C) when most of the *L. garvieae* outbreaks occur. To overcome these problems, several approaches were considered in order to lengthen the duration of protection including the use of adjuvants in the vaccine formulation (Anderson, 1997; Schijns and Tangerås, 2005), and the use of booster immunization.

Ravelo *et al.* (2006) evaluated the effect of the inclusion of different mineral and non-mineral adjuvants in the vaccine formulation against *L. garvieae*, comparing the results with those of the aqueous bacterin. The aqueous and non-mineral adjuvant (Montanide-ISA-763-A and Aquamun) vaccines yielded good protection four weeks after immunization (Table 4.1). The protective rates obtained for the mineral oil adjuvant (Montanide-IMS-2212) and Carbomer (high molecular weight polymer organic of polyacrylic acid) were lower with RPS of 45.7 and 56.5, respectively. In addition, the non-mineral adjuvants in the fish did not induce the appearance of undesired side effects such as organ adhesions. On the other hand, the duration of protection conferred by the non-mineral adjuvanted vaccines was greatly increased in comparison with the aqueous bacterin with RPS values of 92 and 40%, respectively. Moreover, long-term protection was achieved with the adjuvanted vaccine (Fig. 4.4), obtaining high RPS values in the challenges performed at 6 (RPS of 90) and 8 (RPS of 83) months post-vaccination.

Another approach evaluated to lengthen the duration of protection in rainbow trout was a combined strategy consisting of a primary

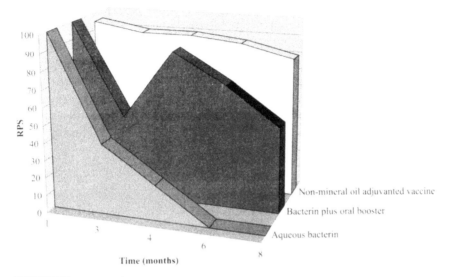

Fig. 4.4 Comparison of the level and duration of protection against *Lactococcus garvieae* achieved with an aqueous bacterin, an adjuvanted vaccine, and the combination of bacterin and oral booster. All the vaccines were developed at the University of Santiago de Compostela (Spain). The adjuvanted vaccine was the result of a collaboration with Hipra Laboratories (Spain). Data from Romalde *et al.* (2004) and Ravelo *et al.* (2006).

immunization with an aqueous bacterin and a booster immunization with an oral alginate-encapsulated vaccine (Romalde *et al.*, 2004). A previous study (Sano *et al.*, 1997) reported the efficacy of a non-encapsulated oral vaccine in the protection of yellowtail against *L. garvieae*, reaching considerable values of protection (RPS = 70). In the case of rainbow trout, and as a first step, several oral formulations were tested, including the non-encapsulated bacterial cells and bacteria encapsulated in alginate, pluronic F-68/alginate, and poly-L-lysine/alginate microparticles. The best protective rates by oral immunisation alone were obtained with the alginate-encapsulated vaccine (RPS 50%). However, values were low enough for not recommending the use of this formulation as primary immunization method. Therefore, the efficacy of this oral vaccine as booster immunization was evaluated (Fig. 4.4). Thus, fish were primary i.p. vaccinated with the aqueous-based vaccine and three months later were boostered with the oral vaccine (7 days, dose 1×10^9 cells/daily ration). Four weeks after revaccination, protection reached RPS values of 87.5%, which indicated the value of this encapsulated vaccine to increase the duration of protection of rainbow trout against *L. garvieae*. In addition, evidences for the effectiveness of this vaccination strategy were also obtained in a field experiment, since orally revaccinated trouts were protected during a natural outbreak of streptococcosis caused by *L. garvieae* occurred one month after the booster (Romalde *et al.*, 2004, 2006).

Recently, a vaccine was developed to protect grey mullet (*Mugil cephalus*) against *L. garvieae* in Taiwan (Lee *et al.*, 2007). The different formulations assayed, including FKC bacterin, FKC supplemented with ECP, and lysate of *L. garvieae* cells, rendered good levels of protection under laboratory conditions when i.p. administered, with RPS values ranging from 84 to 100% one month after vaccination.

Finally, it is interesting to point out that today vaccination against *L. garvieae*, mainly with adjuvanted vaccines, is a common and effective practice in the majority of trout farms in Spain, Portugal, and other European countries. A variety of commercial vaccines are available in Europe, marketed by major fish health companies including Novartis, Shering-Plough or Hipra Laboratories, among others. In addition, the majority of these companies, as well as public institutes or universities also produce autovaccines. Some preparations are specific for *L. garvieae*, while others include a mixture of *L. garvieae* and *S. iniae* in their formulations. Oral and injectable vaccines against *L. garvieae* infection in yellowtail have also been developed and commercialized in Japan (Hirokawa and Yoshida, 2003).

Recent evidence has identified several failures in both licenced and autogenous rainbow trout lactococcosis vaccines (which caused heavy losses in the farms) (Romalde *et al.*, 2005, 2006). The antigenic composition of these bacterins corresponded to avirulent non-capsulated strains of *L. garvieae* which gives little protection against a natural infection with virulent capsulated strains. This finding supported the results obtained previously regarding the necessity of inclusion of capsulated strains (serotype KG^-) in the vaccine formulation due to the limited level of protection conferred by the non-capsulated strains (serotype KG^+) (Ooyama *et al.*, 1999, 2002; Alim *et al.*, 2001; Ravelo, 2004; Shin *et al.*, 2007a). Alim *et al.* (2001) found a protective antigen of glycoproteic nature, only present in capsulated strains, which could explain these failures.

Streptococcus iniae

Apart from some experiments on dietary supplements including vitamins and nucleotides from yeast RNA among others as immunostimulants of the specific response of fish against *S. iniae* (Sealey and Gatlin, 2002; Li *et al.*, 2004), most research on the control of this pathological agent has focused on vaccination. Vaccines were developed for different fish species seriously affected by the pathogen including rainbow trout (Bercovier *et al.*, 1997; Eldar et al., 1997b), barramundi (Delamare-Deboutteville *et al.*, 2006), tilapia (Klesius *et al.*, 2000; Shelby *et al.*, 2002; Shoemaker *et al.*, 2006), hybrid striped bass (Buchanan *et al.*, 2005), and Japanese flounder (Shin *et al.*, 2007b; Shutou *et al.*, 2007) (Table 4.2).

A first vaccination programme was implemented in Israel from 1995 to 1997 using autovaccines consisting of whole-cell formalin inactivated *S. iniae* and administered by i.p. injection (Bercovier *et al.*, 1997; Eldar *et al.*, 1997b). Fish vaccinated at 50 g were protected for more than four months under laboratory or field conditions; time enough to cover the short trout production cycle in that country (Bercovier *et al.*, 1997). Such protection was related with the increased level of specific antibodies, generated in response to heat-labile protein-based antigenic determinants (Bercovier *et al.*, 1997). Annual mortalities due to *S. iniae* were reduced in the Upper Galilee area from 50 to less than 5% using this vaccine routinely. However, in 1997, massive outbreaks of streptococcosis occurred due to a new variant of the bacterium, with a different capsular composition and, therefore, antigenically diverse, which was designed as serotype II (Bachrach *et al.*, 2001). It was hypothetized that as a consequence of the selective pressure induced by the vaccination, a second serotype was able to colonize the environment surrounding the trout farms. This fact could be related to the demonstrated high survival capacity of the streptococcus group in the aquatic environment (Kitao *et al.*, 1979; Currás *et al.*, 2002; Nguyen *et al.*, 2002). Thus, a minoritary variant could remain in the water or mud around farms and become dominant under favourable circumstances (i.e., selective pressure by vaccination).

Autogenous vaccines have been also used to prevent *S. iniae* infections in barramundi in Australia (Creeper and Buller, 2006; Delamare-Deboutteville *et al.*, 2006). Again, the best results were obtained when the vaccine was i.p. administered. However, when applied under field conditions, vaccination has met with limited success (Agnew and Barnes,

Table 4.2   Protection obtained by different vaccines against S. iniae, S. agalactiae and S. parauberis, four weeks after vaccination.

Year	Reference	Fish	Type of vaccine	Administration method	Challenge method	Adjuvant	RPS (%)
S. iniae							
1997	Eldar et al.	Rainbow trout	FKC [a]	i.p. [b]	i.p.	none	80
1997	Bercovier et al.	Tilapia	FKC	i.p.	i.p.	none	80-90
2000	Klesius et al.	Tilapia	FKC + ECP	i.p.	i.p.	none	45-97
				i.m. [c]			17-59
2006	Shoemaker et al.	Tilapia	FKC + ECP	i.p.	i.p.	none	100
2006	Creeper and Buller	Barramundi	FKC	Oral	i.p.	Oraliject	34-63
2006	Delamare-Deboutteville et al.	Barramundi	FKC	i.p.	i.p.	none	Failure
2007	Shutou et al.	Japanese flounder	FKC	i.p.	i.p.	none	ND [d]
S. agalactiae							
1995	Eldar et al.	Tilapia	FKC	i.p.	i.p.	none	80
			Cell extract	i.p.	i.p.	alum	80
2004	Evans et al.	Tilapia	FKC + ECP	i.p.	i.p.	none	80
2005	Pasnik et al.	Tilapia	FKC + ECP	i.p.	i.p.	none	50 [e]
S. parauberis							
1995	Toranzo et al.	Turbot	FKC	i.p.	i.p.	none	100
1996	Romalde et al.	Turbot	FKC	i.p.	i.p.	none	70-80 [f]

[a] FKC, formalin killed cells; ECP, extracellular products; [b] i.p., intraperitoneal injection; [c] intramuscular injection; [d] Response determined by antibody titres; [e] RPS at six months after vaccination; [f] RPS at two years after vaccination.

2007; Tumbol *et al.*, 2007), with the re-emergence of infection within weeks after immunization. Some explanations for this lack of efficacy can be the serological diversity of the pathogenic strains (Tumbol *et al.*, 2007), or the rapid immune kinetics at high water temperatures where antibody titres subside no longer than 40 days (Agnew and Barnes, 2007).

Contrary to these results, positive protection against *S. iniae* was reported in tilapia using bacterins supplemented with ECP, administered either intramuscularly or i.p. routes (Klesius *et al.*, 2000). Taking into account the antigenic diversity within this fish pathogen for the vaccine formulation, the authors also developed bivalent vaccines including two serologically distinct isolates. Contrary to the monovalent vaccines, the bivalent formulation was able to protect against both serovariants (Klesius *et al.*, 2000). Further works (Shoemaker *et al.*, 2006; Klesius *et al.*, 2007) demonstrated the effectiveness of the ECP-enriched vaccine when delivered orally using a commercial adjuvant (Oralject), as well as when administered by immersion to newly hatched tilapia followed by sex reveal and immersion booster.

Other approaches to get improved vaccines with potential use in the future have been tried under laboratory conditions. Thus, a phosphoglucomutase mutant of a *S. iniae* isolate, showing greater size and decrease in capsule thickness, was employed as an experimental live vaccine in hybrid striped bass (Buchanan *et al.*, 2005). The attenuated mutant is able to disseminate to the blood, brain, and spleen but is eliminated by 24 h without any organ damage. In addition, it stimulates a protective immune response showing RPS values between 90 and 100%, being more effective than the FKC vaccines tested against *S. iniae* in different species. However, some problems have to be solved prior to its use as live attenuated vaccine, including the protective effect against the different serological variants of the pathogen or the question of reversion to a virulent form. In fact, it was described that reintroduction of an intact copy of the gene restored its virulence (Buchanan *et al.*, 2005). The reversion to virulence of attenuated pathogens under selective pressure in the environment has not been adequately studied, and the possibility of gene transfer under such conditions cannot be ruled out, specially when *S. iniae* is capable of survive in the environment (Nguyen *et al.*, 2002) and is present in wild fish (Colorni *et al.*, 2002).

Today, commercial vaccines are available in some parts of Asia to protect against *S. iniae* infection in different fish species such as barramundi and Asian sea bass (Intervet) or tilapia (Schering-Plough).

While the former one is a monovalent inactivated vaccine available in Indonesia that can be used injectable or by immersion, the latter is a bivalent vaccine against *S. iniae* and *L. garvieae* to be administered by immersion or orally in feed.

Streptococcus agalactiae

From the mid 1990s, efforts have been made to develop an effective vaccine to protect tilapia against streptococcosis caused by *S. agalactiae*, mainly in the USA and Israel (Eldar *et al.*, 1995c, 2004; Pasnik *et al.*, 2005a, b, 2006).

The first attempt to develop such vaccine was performed by Eldar and co-workers (1995c) in Israel. These authors formulated two vaccines, an aqueous bacterin containing formalin-killed cells and a vaccine based on an *S. agalactiae* extract containing 50% protein conjugated to alum. When i.p. administered, both formulations were able to protect tilapia against a challenge of 100 LD_{50}. In addition, a good correlation was observed between the level of protection and the development of specific agglutinins, and the Western blot analysis performed indicated that only a few proteins act as protective antigens in both whole-cell vaccine and streptococcal extract.

Almost ten years later, Evans *et al.* (2004) developed another vaccine formulated with formalin killed-cells and supplemented with concentrated ECP of *S. agalactiae*. The vaccine was effective in protecting 30 g tilapia against infection by *S. agalactiae* (RPS = 80), one month after vaccination), when i.p. administered, but no protection was achieved in 5 g tilapines or when the vaccine was administered by bath to both fish sizes. In addition, these authors observed a lack of cross-protection against *S. agalactiae* employing a vaccine against *S. iniae*, demonstrating the need of a specific vaccine. Further works (Pasnik *et al.*, 2005a, b) indicated the correlation of protection and the production of specific antibodies against ECP components, specially important being a fraction of about 55 kDa, as well as the duration of fish immunization for at least 180 days post-vaccination with RPS values around 50%. Unfortunately, the shelf-life of the vaccine, when stored at 4°C was limited, probably due to the degradation of the 55 kDa protective antigen (Pasnik *et al.*, 2005a), indicating that freshly prepared ECP is needed to be included in the vaccine formulation.

Passive immunization against *S. agalactiae* was also assayed in Nile tilapia (Pasnik *et al.*, 2006), employing serum obtained from *S. agalactiae* vaccinated fish. A significant higher survival was observed in the passively immunized fish (90%) in comparison with the non-immunized tilapia (27.3-36.7%) 72 h after challenge with a virulent isolate of *S. agalactiae*. As in the active immunization experiments, a correlation was observed between specific antibodies and protection.

Streptococcus parauberis

Due to the endemic condition of the *S. parauberis* infection, efforts to develop an effective vaccine against this streptococcal agent were only made in Spain and directed to the turbot industry. As mentioned earlier, biochemical and serological characterization of the isolates indicated a high homogeneity within this bacterial pathogen (Toranzo *et al.*, 1994, 1995a), although some genetic variability was observed (Romalde *et al.*, 1999a). On the basis of these characterization studies, two strains were selected for inclusion in the vaccine formulation. A toxoid-enriched whole-cell bacterin was prepared and tested in two turbot farms in the Northwest of Spain (Toranzo *et al.*, 1995b; Romalde *et al.*, 1996, 1999b), analysing not only the vaccine potency but also the influence of other variables such as the addition of immunostimulants, the inclusion of adjuvants in the vaccine formulation, the route of administration and the age of the fish. Moreover, the specific and non-specific immune responses were also analysed by means of level of antibodies and the phagocytosis rates.

The RPS obtained with the enriched bacterin administered by i.p. injection was 100% four weeks after vaccination (Toranzo *et al.*, 1995b), and remained higher than 80% at 6 and 12 months after immunization (Romalde *et al.*, 1996). A small decrease in RPS (70%) was only observed in the challenges performed 24 months post-immunization (Romalde *et al.*, 1996). No correlation could be established here between protection and level of specific antibodies. However, the phagocytosis rate increased significantly after immunization, indicating that the non-specific response played an important role in the protective effects recorded (Toranzo *et al.*, 1995b).

On the other hand, no significant differences in protection were detected between the water- and the oil-based vaccine, although a lower growth rate was observed in fish immunized with the adjuvanted formulation. The use of immunostimulants (yeast β-glucans) did not increase the protective effects of the enriched bacterin. As for the other streptococcal fish pathogens, no protection was achieved when the vaccines were administered by bath.

The use of the enriched vaccine against S. *parauberis* allowed the control of turbot streptococcosis, and was of great importance for the mainteance of this industry in Spain. In the previous few years, some occasional outbreaks were recorded in Spain in Portugal in farms without vaccination programmes against this disease (Zarza and Padrós, 2007), demonstrating the great efficacy of the developed vaccine in the field.

Other Streptococci

Autogenous vaccines were tested against S. *phocae* and V. *salmoninarum* with different results (Michel *et al.*, 1997; Ruiz-Zarzuela *et al.*, 2005; Sommerset *et al.*, 2005). While routine vaccination of Atlantic salmon in Chile during the last 2-3 years was of great help in controlling the infection by S. *phocae*, attemps at vaccinating rainbow trout against V. *salmoninarum* did not provide encouraging results.

FINAL REMARKS

A great effort has been made in recent years to develop appropriate vaccination programmes as a preventive control for fish streptococcosis. The main limitation of the majority of the designed vaccines was the short duration of protection. Recent advances using different approaches, such as the use of adjuvanted vaccines or the combination of aqueous bacterins and oral micro-encapsulated vaccines, allowed the lenghtening of protection against L. *garvieae* and, therefore, infections by this microorganism can be currently effectively prevented. Similar studies are still needed in the case of other streptococcal fish pathogens.

The development of vaccines composed by purified antigens, including ECP, proteins or other cellular components, and the recombinant subunit vaccines are another target for future research (Clark and Cassidy-Hanley, 2005). Knowledge of the real protective antigens for these pathogens in the susceptible fish species is necessary to reach such an objective. The possibility of DNA vaccines—until now developed mainly for viral fish diseases—is also a field to consider (Kurath, 2005). However, DNA vaccines are viewed as a genetic modification of the host and, although they appear safe, will encounter more difficulties for licensing than classical formulations. Another source of improved vaccines also being considered for the future use is the utilisation of attenuated mutants as live vaccines. It has been shown that a *S. iniae* mutant strain stimulates immune response higher than the killed bacterins in different fish species. However, the question of reversion to virulence is not well addressed, as this aspect is specially important in agents with zoonotic potential, such as the *Streptococcus* group. This fact makes the approval and licensing of attenuated vaccines difficult in most cases.

Another limitation of the anti-streptococcosis vaccines developed till date is the route of administration, since most of them have to be delivered by i.p. injection. The design of new delivery methods, including oral administration, should be also encouraged. In this sense, a percutaneous administration by immersion with application of a multiple puncture instrument has been tested in rainbow trout (Nakanishi *et al.*, 2002), proving to be as effective as i.p. injection.

Maternal transfer of immunity to the offspring, which could be important for the early defence against pathogens, is another field for future research, since it can help in the development of appropriate vaccination regimens for both broodstocks and larvae (Mulero *et al.*, 2007).

Finally, the combination of vaccination and selection of genetically resistant fish can improve the control of all these streptococcal diseases.

Acknowledgements

The studies from the University of Santiago reviewed in this work were supported in part by Grants PETRI95-0471, PETRI95-0685, MAR95-1848, and MAR96-1875 from the Ministerio de Educación y Ciencia, Spain.

References

Ad Hoc Committee of the ICSB. 1995. Validation of the publication of new names and new combinations previously effectively published outside the IJSB. List no. 52. *International Journal of Systematic Bacteriology* 45: 197-198.

Agnew, W. and A.C. Bartnes. 2007. *Streptococcus iniae*: an aquatic pathogen of global veterinary significance and a challenging candidate for reliable vaccination. *Veterinary Microbiology* 122: 1-15.

Akhlaghi, M., B.L. Munday and R.J. Whittington. 1996. Comparison of passive and active immunization of fish against streptococcosis (enterococcosis). *Journal of Fish Diseases* 19: 251-258.

Al-Harbi, A.L. 1994. First isolation of *Streptococcus* sp. from hybrid tilapia (*Oreochromis niloticus* × *O. aureus*) in Saudi Arabia. *Aquaculture* 128: 195-201.

Alim, S.R., M. Anwar, E.K. Chowdhury and R. Kusuda. 2001. G1 antigen: a cell surface immunoprotective 96 kDa glycoprotein from the virulen fish pathogen *Enterococcus seriolicida*, its purification and characterization. *Letters in Applied Microbiology* 32: 357-361.

Amborski, R.L., T.G. 3rd Snider, R.L. Thune and D.D. Culley, Jr. 1983. A non-hemolytic, group B *Streptococcus* infection of cultured bullfrogs, *Rana catesbeiana*, in Brazil. *Journal of Wildlife Diseases* 19: 180-184.

Anderson, D.P. 1997. Adjuvants and immunostimulants for enhancing vaccine potency in fish. In: *Fish Vaccinology*, R. Gudding, A. Lillehaug, P.J. Midlyng and F. Brown, (eds.). S. Karger, Basel, pp. 257-265.

Aoki, T., K. Takami and T. Kitao. 1990. Drug resistance in a non-hemolytic *Streptococcus* sp. isolated from yellowtail *Seriola quinqueradiata*. *Diseases of Aquatic Organisms* 8: 171-177.

Aoki, T., C.I. Park, H. Yamashita and I. Hirono. 2000. Species-specific polymerase chain reaction primers for *Lactococcus garvieae*. *Journal of Fish Diseases* 23: 1-6.

Atsuta, S., J. Yoshimoto, M. Sakai and M. Kobayashi. 1990. Streptococcicosis occurred in pen-cultured coho salmon *Oncorhynchus kisutch*. *Suisanzoshoku* 38: 215-219.

Austin, B. and P.A.W. Robertson. 1993. Recovery of *Streptococcus milleri* from ulcerated koi carp (*Cyprinus carpio* L.) in the UK. *Bulletin of European Association of Fish Pathologists* 13: 207-209.

Austin, B. and D.A. Austin. 2007. Bacterial Fish Pathogens. Disease of Farmed and Wild Fish. Springer/Praxis Publishing, Chichester. 3rd Edition.

Bachrach, G., A. Zlotkin, A. Hurvitz, D.L. Evans and A. Eldar. 2001. Recovery of *Streptococcus iniae* from diseased fish previously vaccinated with a streptococcus vaccine. *Applied Environmental Microbiology* 67: 3756-3758.

Barnes, A.C. and A.E. Ellis. 2004. Role of capsule in serotypic differences and complement fixation by *Lactococcus garvieae*. *Fish and Shellfish Immunology* 16: 207-214.

Baya, A.M., B. Lupiani, F.M. Hetrick, B.S. Roberson, R. Lukacovic, E. May and C. Poukish. 1990. Association of *Streptococcus* sp. with fish mortalities in the Chesapeake Bay and its tributaries. *Journal of Fish Diseases* 13: 251-253.

Bercovier, H., C. Ghittino and A. Eldar. 1997a. Immunization with bacterial antigens: infections with streptococci and related organisms. In: *Fish Vaccinology*, R. Gudding, A. Lillehaug, P.J. Midlyng and F. Brown (eds.). S. Karger, Basel, pp. 153-160.

Berridge, B.R., H. Bercovier and P.F. Frelier. 2001. *Streptococcus agalactiae* and *Streptococcus difficile* 16S-23S intergenic rDNA: genetic homogeneity and species-specific PCR. *Veterinary Microbiology* 78: 165-173.

Berridge, B.R., J.D. Fuller, J. Azavedo, D.E. Low, H. Bercovier and P.F. Frelier. 1998. Development of specific nested oligonucleotide PCR primers for the *Streptococcus iniae* 16S-23S ribosomal DNA intergenic spacer. *Journal of Clinical Microbiology* 36: 2778-2781.

Boomker, J., G.D. Imes, C.M. Cameror., T.W. Naudé and H.J. Schoonbee. 1979. Trout mortalities as a result of *Streptococcus* infection. *Onderstepoort Journal of Veterinary Research* 46: 71-77.

Bragg, R.R. 1988. The indirect fluorescent antibody technique for the rapid identification of streptococcosis of rainbow trout (*Salmo gairdneri*). *Onderstepoort Journal of Veterinary Research* 55: 59-61.

Bragg, R.R. and J.S.E. Broere. 1986. Streptococcosis in rainbow trout in South Africa. *European Association oF Fish Pathologists* 6: 89-91.

Bragg, R.R., J.M. Todd, S.M. Lordan and M.E. Combrink. 1989. A selective procedure for the field isolation of pathogenic *Streptococcus* spp. of rainbow trout (*Salmo gairdneri*). *Onderstepoort Journal of Veterinary Research* 56: 179-184.

Bromage, E.S., A. Thomas and L. Owens. 1999. *Streptococcus iniae*, a bacterial infection in barramundi *Lates calcarifer*. *Diseases of Aquatic Organisms* 36: 177-181.

Brunt, J. and B. Austin. 2005. Use of a probiotic to control lactococcosis and streptococcosis in rainbow trout, *Oncorhynchus mykiss* (Walbaum). *Journal of Fish Diseases* 28: 693-701.

Brunt, J., A. Newaj-Fyzul and B. Austin. 2007. The development of probiotics for the control of multiple bacterial diseases of rainbow trout, *Oncorhynchus mykiss* (Walbaum). *Journal of Veterinary Research* 30: 573-579.

Buchanan, J.T., J.A. Stannard, X. lauth, V.E. Ostland, H.C. Powell, M.E. Westerman and V. Nizet. 2005. *Streptococcus iniae* phosphoglucomutase is a virulence factor and a target for vaccine development. *Infection Immunulogy* 73: 6935-6944.

Carson, J. and B. Munday. 1990. Streptococcosis—an emerging disease in aquaculture. *Australian Aquaculture* 5: 32-33.

Carson, J., N. Gudkovs and B. Austin. 1993. Characteristics of an *Enterococcus*-like bacterium from Australia and South Africa, pathogenic for rainbow trout, *Oncorhynchus mykiss* (Walbaum). *Journal of Fish Diseases* 16: 381-388.

Clark, T.G. and D. Cassidy-Hanley. 2005. Recombinant subunit vaccines: potentials and constraints. In: *Progress in Fish Vaccinology*, P.J. Midtlyng, (ed.). S. Karger, Basel. pp. 153-163.

Collins, M.D., F.A.E. Farrow, B.A. Philips and O. Kandler. 1984. *Streptococcus garvieae* sp. nov. and *Streptococcus plantarum* sp. nov. *Journal of General Microbiology* 129: 3427-3431.

Colorni, A., A. Diamant, A. Eldar, H. Kvitt and A. Zlotkin. 2002. *Streptococcus iniae* infections in Red Sea cage-cultured and wild fishes. *Diseases of Aquatic Organisms* 49: 165-170.

Colorni, A., C. Ravelo, K.L. Romalde, A.E. Toranzo and A. Diamant. 2003. *Lactococcus garvieae* in wild Red Sea wrasse *Coris aygula* (Labridae). *Diseases of Aquatic Organisms* 56: 275-278.

Creeper, J. and N. Buller. 2006. An outbreak of *Streptococcus iniae* in barramundi (*Lates calcarifera*) in freshwater cage culture. *Australian Veterinary Journal* 84: 408-411.

Currás, M., B. Magariños, A.E. Toranzo and J.L. Romalde. 2002. Dormancy as a survival strategy of *Streptococcus parauberis* in the marine environment. *Diseases of Aquatic Organisms* 52: 129-136.

Darwish, A.M. and A.A. Ismaiel. 2003. Laboratory efficacy of amoxicillin for the control of *Streptococcus iniae* infection in sunshine bass. *Journal of Aquatic Animal Health* 15: 209-214.

Darwish, A.M. and M.S. Hobbs. 2005. Laboratory efficacy of amoxicillin for the control of *Streptococcus iniae* infection in tilapia. *Journal of Aquatic Animal Health* 17: 197-202.

Delamare-Deboutteville, J., D. Wood and A.C. Barnes. 2006. Response and function of cutaneous mucosal and serum antibodies in barramundi (*Lates calcarifer*) acclimated in seawater and freshwater. *Fish and Shellfish Immunology* 21: 92-101.

Domenech, A., J. Prieta, J. Fernández-Garayzábal, M.D. Collins, D. Jones and L. Domínguez. 1993. Phenotypic and phylogenetic evidence for a close relationship between *Lactococcus garvieae* and *Enterococcus seriolicida*. *Microbiología (SEM)* 9: 63-68.

Domenech, A., J.F. Fernández-Garayzábal, C. Pascual, J.A. García, M.T. Cutúli, M.A. Moreno, M.D. Collins and L. Domínguez. 1996. Streptococcosis in cultured turbot, *Scophthalmus maximus* (L.), associated with *Streptococcus parauberis*. *Journal of Fish Diseases* 19: 33-38.

Dow, S.W., R.L. Jones, T.N. Thomas, K.A. Linn and H.B. Hamilton. 1987. Group B streptococcal infection in two cats. *Journal of Veterinary Medical Association* 190: 71-72.

Eldar, A. and C. Ghittino. 1999. *Lactococcus garvieae* and *Streptococcus iniae* infections in rainbow trout *Oncorhynchus mykiss*: similar, but different diseases. *Diseases of Aquatic Organisms* 36: 227-231.

Eldar, A., Y. Bejerano and H. Bercovier. 1994. *Streptococcus shiloi* and *Streptococcus difficile*: Two new streptococcal species causing meningoencephalitis in fish. *Current Microbiology* 28: 139-143.

Eldar, A., Y. Bejarano, A. Livoff, A. Horovitcz and H. Bercovier. 1995a. Experimental streptococcal meningoencephalitis in cultured fish. *Veterinary Microbiology* 43: 33-40.

Eldar, A., P.F. Frelier, L. Assenta, P.W. Varner, S. Lawhon and H. Bercovier. 1995b. *Streptococcus shiloi*, the name for an agent causing septicemic infection in fish, is a junior synonym of *Streptococcus iniae*. *International Journal of Systematic Bacteriology* 45: 840-842.

Eldar, A., O. Shapiro, Y. Bejerano and H. Bercovier. 1995c. Vaccination with whole-cell vaccine and bacterial protein extract protects tilapia against *Streptococcus difficile* meningoencephalitis. *Vaccine* 13: 867-870.

Eldar, A., M. Goria, C. Ghittino, A. Zlotkin and H. Bercovier. 1999. Biodiversity of *Lactococcus garvieae* isolated from fish in Europe, Asia, and Australia. *Applied Environmental Microbiology* 65: 1005-1008.

Eldar, A., C. Ghittino, L. Asanta, E. Bozzetta, M. Goria, M. Prearo and H. Bercovier. 1996. *Enterococcus seriolicida* is a junior synonym of *Lactococcus garvieae*, a causative agent of septicemia and meningoencephalitis in fish. *Current Microbiology* 32: 85-88.

Eldar, A., S. Lawhon, P.F. Frelier, L. Assenta, B.R. Simpson, P.W. Varner and H. Bercovier. 1997a. Restriction fragment length polymorphisms of 16S rDNA and of whole rRNA genes (ribotyping) of *Streptococcus iniae* strains from the United States and Israel. *FEMS Microbiology Letters* 151: 155-162.

Eldar, A., A. Horovitcz and H. Bercovier. 1997b. Development and efficacy of a vaccine against *Streptococcus iniae* infection in farmed rainbow trout. *Veterinary Immunology and Immunopathology* 56: 175-183.

Elliot, J.A. and R.R. Facklam. 1996. Antimicrobial susceptibilities of *Lactococcus lactis* and *Lactococcus garvieae* and a proposed method to discriminate between them. *Journal of Clinical Microbiology* 34: 1296-1298.

Elliot, J.A., R.R. Facklam and C.B. Richter. 1990. Whole-cell prptein patterns of nonhemolytic group B, type Ib, streptococci isolated from humans, mice, cattle, frogs, and fish. *Journal of Clinical Microbiology* 28: 628-630.

Elliot, J.A., M.D. Collins, N.E. Pigott and R.R. Facklam. 1991. Differentiation of *Lactococcus lactis* and *Lactococcus garvieae* from humans by comparison of whole-cell protein patterns. *Journal of Clinical Microbiology* 20: 2731-2734.

Endo, H., J. Nakayama, H. Ushio, T. Hayashi and E. Watanabe. 1998. Application of flow cytometry for rapid detection of *Lactococcus garvieae*. *Applied Biochemistry and Biotechnology* 75: 295-306.

Euzéby, J.P. 1998. Taxonomic note: necessary correction of specific and subspecific epithets according to rules 12c and 13b of the International Code of Nomenclature of Bacteria (1990 revision). *International Journal of Systemic Bacteriology* 48: 1073-1075.

Evans, J.J., C.A. Shoemaker and P.H. Klesius. 2004. Efficacy of *Streptococcus agalactiae* (group B) vaccine in tilapia (*Oreochromis niloticus*) by intraperitonal and bath immersion administration. *Vaccine* 22: 3769-3773.

Evans, J.J., P.H. Klesius, P.M. Glibert, C.A. Shoemaker, M.A. Al Sarawi, J. Landsberg, R. Duremdez, A. Al Marzouk and S. Al Zenki. 2002. Characterization of beta-haemolytic group B *Streptococcus agalactiae* in cultured seabream, *Sparus aurata* (L.) and wild mullet, *Liza klunzingeri* (Day), in Kuwait. *Journal of Fish Diseases* 25: 505-513.

Facklam, R., J. Elliot, L. Shewmaker and A. Reingold. 2005. Identification and characterization of sporadic isolates of *Streptococcus iniae* isolated from humans. *Journal of Clinical Microbiology* 43: 933-937.

Figueiredo, H.C.P., G.F. Mian, D.T. Godoy, C.A.G. Leal and G.M. Costa. 2007. Insights in the natural history and virulence of *Streptococcus agalactiae* infections in Nile tilapia. *European Association of Fish Pathologists Grado (Italy)*, 48 pp.

Frerichs, G.N. 1993. Isolation and identification of fish bacterial pathogens. In: *Bacterial Diseases of Fish*, V. Inglis, R.J. Roberts and N.R. Bromage (eds.). Blackwell Scientific Publications, Oxford, pp. 257-286.

Fuller, J.D., D.J. Bast, V. Nizet, D.E. Low and J.C. de Azevedo. 2001. *Streptococcus iniae* virulence is associated with a distinct genetic profile. *Infection Immunology* 69: 1994-2000.

Ghittino, C. 1999. La estreptococosis en los peces. *AquaTic* 2, art. 605. Available at URL http://aquatic.unizar.es/n2/art605/lact_rev.htm.

Ghittino, C. and M. Prearo. 1992. Segnalizaciones di streptococcosi nella trota iridea (*Oncorhynchus mykiss*) in Italia: nota preliminare. *Bolletino Societa Italiana di Patologia Ittica* 8: 4-9.

Ghittino, C., M. Prearo, E. Bozzetta and A. Eldar. 1995a. Pathological characterization of the causative agent of fish streptococcosis in Italy and vaccination trials in rainbow trout. *Bolletino Societa Italiana di Patologia Ittica* 16: 2-12.

Ghittino, C., A. Eldar, M. Prearo, E. Bozzetta, A. Livvof and H. Bercovier. 1995b. Comparative pathology and experimental vaccination in diseased rainbow trout infected by *Streptococcus iniae* and *Lactococcus garvieae*. *Proceedings of the VII International Conference on Diseases of Fish and Shellfish. European Association of Fish Pathologists*. Sept 10-15, 1995, Palma de Mallorca, Spain. p. 27.

Goh, S.H., S. Potter, J.O. Wood, S.M. Hemmingsen, R.P. Reynolds and A.W. Chow. 1996. HSP60 gene sequences as universal targets for microbial species identification: studies with coagulase-negative staphylococci. *Journal of Clinical Microbiology* 34: 818-823.

Goh, S.H., Z. Santucci, W.E. Kloos, M. Faltyn, C.G. George, D. Driedger and S.M. Hemmingsen. 1997. Identification of *Staphylococcus* species and subspecies by the chaperonin 60 gene identification method and reverse checkerboard hybridization. *Journal of Clinical Microbiology* 35: 3116-3121.

Goh, S.H., D. Driedger, S. Gillet, D.E. Low, S.M. Hemmingsen, M. Amos, D. Chan, M. Lovgren, B.M. Willey, C. Shaw and J.A. Smith. 1998. *Streptococcus iniae*, a human and animal pathogen: specific identification by the chaperonin 60 gene identification method. *Journal of Clinical Microbiology* 36: 2164-2166.

Goh, S.H., R.R. Facklam, M. Chang, J.E. Hill, G.J. Tyrrell, E.C. Burns, D. Chan, C. He, T. Rahim, C. Shaw and S.M. Hemmingsen. 2000. Identification of *Enterococcus* species and phenotypically similar *Lactococcus* and *Vagococcus* species by reverse checkerboard hybridization to chaperonin 60 gene sequences. *Journal of Clinical Microbiology* 38: 3953-3959.

Hawkesford, T., B. Munday, J. Carson, C. Burke and M. Churchill. 1997. Comparison of Tasmanian isolates of *Streptococcus* sp. biovar 1 with verified strains of *Lactococcus garvieae* and *Enterococcus seriolicida* using microbiological, molecular biological and *in vivo* techniques. *Proceedings of the International Symposium on Diseases in Marine Aquaculture. Japanese Society of Fish Pathology*. Oct. 3-6, 1997, Hiroshima, Japan. p. 92.

Henton, M.M., O Zapke and P.A. Basson. 1999. *Streptococcus phocae* infection associated with starvation in Cape fur seals. *Journal of South African Veterinary Association* 2: 98-99.

Hirokawa, Y. and T. Yoshida. 2003. Role of immune serum in protection against *Lactococcus garvieae* infection in the yellowtail. *3rd International Symposium on Fish Vaccinology*. Bergen, Norway, p. 105.

Hirono, I., H. Yamashita, C.I. Park, T. Yoshida and T. Aoki. 1999. Identification of genes in a KG⁻ phenotype of *Lactococcus garvieae*, a fish pathogenic bacterium, whose proteins react with anti-KG⁻ rabbit serum. *Microbiological Pathogen* 27: 407-417.

Hoshina, T., T. Sano and Y. Morimoto. 1958. A *Streptococcus* pathogenic to fish. *Journal of Tokyo University of Fisheries* 44: 57-58.

Iida, T., H. Wakabayashi and S. Egusa. 1981. Vaccination for control of streptococcal disease in cultured yellowtail. *Fish Pathology* 16: 201-206.

Iida, T., K. Furukawa, M. Sakai and H. Wakabayashi. 1986. Non-haemolytic *Streptococcus* isolated from the brain of the vertebral deformed yellowtail. *Fish Pathology* 21: 33-38.

Irianto, A. and B. Austin. 2002. Probiotics in aquaculture. *Journal of Fish Diseases* 25: 633-642.

Irianto, A. and B. Austin. 2003. Use of death probiotic cells to control furunculosis in rainbow trout, *Oncorhynchus mykiss* (Walbaum). *Journal of Fish Diseases* 26: 59-62.

Katao, H. 1982. Erythromycin—the application to streptococcal infections in yellowtails. *Fish Pathology* 17: 77-82.

Kawahara, E. and R. Kusuda. 1987. Direct fluorescent antibody technique for diagnosis of bacterial diseases in eel. *Nippon Suisan Gakkaishi* 53: 395-399.

Kawamura, Y., Y. Itoh, N. Mishima, K. Ohkusu, H. Kasai and T. Ezaki. 2005. High genetic similarity of *Streptococcus agalactiae* and *Streptococcus difficilis*: S. *difficilis* Eldar *et al.* 1995 is a later synonym of *S. agalactiae* Lehmann and Neumann 1986 (Approved Lists 1980). *International Journal of Systematic Evolutionary Microbiology* 55: 961-965.

Kitao, T. 1982. The methods for detection of *Streptococcus* sp. causative bacteria of streptococcal disease of cultured yellowtail (*Seriola quinqueradiata*). *Fish Pathology* 17: 17-26.

Kitao, T. 1993. Streptococcal infections. In: *Bacterial Diseases of Fish*, V. Inglis, R.J. Roberts and N.R. Bromage (eds.). Blackwell Scientific Publications, Oxford, pp. 196-210.

Kitao, T., T. Aoki and K. Iwata. 1979. Epidemiological study on streptococcicosis of cultured yellowtail (*Seriola quinqueradiata*)-I. Distributiuon of *Streptococcus* sp. in seawater and muds around yellowtail farms. *Bulletin of the Japanese Society of Scientific Fisheries* 45: 567-572.

Kitao, T., T. Aoki and R. Sakoh. 1981. Epizootics caused by β-haemolytic *Streptococcus* species in cultured freshwater fish. *Fish Pathology* 15: 301-307.

Klesius, P.H., C.A. Shoemaker and J.J. Evans. 2000. Efficacy of a single and combined *Streptococcus iniae* isolate vaccine administered by intraperitoneal and intramuscular routes in tilapia (*Oreochromis niloticus*). *Aquaculture* 188: 237-246.

Klesius, P.H., C.A. Shoemaker and J.J. Evans. 2007. Immersion vaccination of newly hatched and sex reversed Nile tilapia (*Oreochromis niloticus*) using a *Streptococcus iniae* vaccine. *XIIIth International Conference of the European Association of Fish Pathologists* (EAFP). Grado, Italy, p. 92.

Kornblatt, A.N., R.L. Adams, S.W. Barthold and G.A. Cameron. 1983. Canine neonetal deaths associated with group B streptococcal septicemia. *Journal of American Veterinary Medical Association* 183: 700-701.

Kummeneje, K., T. Nesbakken and T. Mikkelsen. 1975. *Streptococcus agalactiae* infection in a hamster. *Acta Veterinaria Scandinavica* 16: 554-556.

Kurath, G. 2005. Overview of recent DNA vaccine development for fish. In: *Progress in Fish Vaccinology*, P.J. Midtlyng (ed.). S. Karger, Basel, pp. 201-213.

Kusuda, R. and I. Komatsu. 1978. A comparative study of fish pathogenic *Streptococcus* isolated from saltwater and freshwater fishes. *Bulletin of the Japanese Society of Scientific Fisheries* 44: 1073-1078.

Kusuda, R., I. Komatsu and K. Kawai. 1978. *Streptococcus* sp. isolated from an epizootic of cultured eel. *Bulletin of the Japanese Society of Scientific Fisheries* 44: 295.

Kusuda, R., T. Kawi, T. Toyoshima and I. Komatsu. 1976. A new pathogenic bacterium belonging to the genus *Streptococcus*, isolated from an epizootic of cultured yellowtail. *Bulletin of the Japanese Society of Scientific Fisheries* 42: 1345-1352.

Kusuda, R., K. Kawai, F. Salati, C.R. Banner and J.L. Fryer. 1991. *Enterococcus seriolicida* sp. nov., a fish pathogen. *International Journal of Systematic Bacteriology* 41: 406-409.

Kvitt, H. and A. Colorni. 2004. Strain variation and geographic endemism in *Streptococcus iniae*. *Diseases of Aquatic Organisms* 61: 67-73.

Lau, S.K.P., P.C.Y. Woo, W.K. Luk, A.M.Y. Fung, W.T. Hui, A.H.C. Fong, C.W. Chow, S.S.Y. Womg and K.Y. Yuen. 2006. Clinical isolates of *Streptococcus iniae* from Asia are more mucoid and β-hemolytic than those from North America. *Diagnostic Microbial Infectious Diseases* 54: 177-181.

Lee, J.-L., S.Y. Chu, C.Y. Lee, C.C. Lai, C.S. Hung, J.H. Wang, S.C. Chenb and S.S.Tsai. 2007. Immunogenicity of *Lactococcus garvieae* in grey mullet, *Mugil cephalus* L. XIIIth *International Conference of the European Association of Fish Pathologists* (EAFP). Grado, Italy, p. 197.

Li, P., D.H. Lewis and D.M. Gatlin, III. 2004. Dietary oligonucleotides from yeast RNA influence immune responses and resistance of hybrid striped bass (*Morone chrysops* × *Morone saxatilis*) to *Streptococcus iniae* infection. *Fish and Shellfish Immunology* 16: 561-569.

Mannion, P.T. and M.M. Rothburn. 1990. Diagnosis of bacterial endocarditis caused by *Streptococcus lactis* and assisted by immunoblotting of serum antibodies. *Journal of Infection* 21: 317-318.

Meads, S.V., E.B. Shotts and J. Maurer. 1998. Biochemical and molecular typing of *Streptococcus iniae* isolates from fish and human cases. *Proceedings of the 23rd Annual Eastern Fish Health Workshop.* March 30-April 2, 1998. Plymouth, MA, USA, p. 31.

Michel, C., P. Nougayrède, A. Eldar, E. Sochon and P. de Kinkelin. 1997. *Vagococcus salmoninarum*, a bacterium of pathological significance in rainbow trout *Oncorhynchus mykiss* farming. *Diseases of Aquatic Organisms* 30: 199-208.

Minami, T., M. Nakamura, Y. Ikeda and H. Ozaki. 1979. A beta-hemolytic *Streptococcus* isolated from cultured yellowtail. *Fish Pathology* 14: 33-38.

Miyazaki, T. 1982. Pathological study on streptococcicosis histopathology of infected fishes. *Fish Pathology* 17: 39-47.

Mulero, I., A. García-Ayala, J. Meseguer and V. Mulero. 2007. Maternal transfer of immunity and ontogeny of autologous immunecompetence of fish: A mini review. *Aquaculture* 268: 244-250.

Múzquiz, J.L., F.M. Royo, C. Ortega, I. de Blas, I. Ruiz and J.L. Alonso. 1999. Pathogenicity of streptococcosis in rainbow trout (*Oncorhynchus mykiss*): dependence of age of diseased fish. *Bulletin of the European Association of Fish Pathologists* 19: 114-119.

Nakanishi, T., I. Kiryu and M. Ototake. 2002. Development of a new vaccine delivery method for fish: percutaneous administration by immersion with application of multiple puncture instrument. *Vaccine* 20: 3764-3769.

Nakatsugawa, T. 1983. A streptococcal disease of cultured flounder. *Fish Pathology* 17: 281-285.

Nguyen, H.T. and K. Kanai. 1999. Selective agars for the isolation of *Streptococcus iniae* from Japanese flounder, *Paralichthys olivaceus*, and its cultural environment. *Journal of Applied Microbiology* 86: 769-776.

Nguyen, H.T., K. Kanai and K. Yoshikoshi. 2002. Ecological investigation *Streptococcus iniae* from Japanese flounder (*Paralichthys olivaceus*) using selective isolation procedures. *Aquaculture* 205: 7-17.

Nieto, J.M., S. Devesa, I. Quiroga and A.E. Toranzo. 1995. Pathology of *Enterococcus* sp. infection in farmed turbot, *Scophthalmus maximus* L. *Journal of Fish Diseases* 18: 21-30.

Nomoto, R., L.I. Munasinghe, D.H. Jin, Y. Shimahara, H. Yasuda, A. Nakamura, N. Misawa, T. Itami and T. Yoshida. 2004. Lancefield group C *Streptococcus dysgalactiae* infection responsible for fish mortalities in Japan. *Journal of Fish Diseases* 27: 679-686.

Nomoto, R., N. Unose, Y. Shimahara, A. Nakamura, T. Hirae, K. Maebuchi, S. Harada, N. Misawa, T. Itami, H. Kagawa and T. Yoshida. 2006. Characterization of Lancefield group C *Streptococcus dysgalactiae* isolated from farmed fish. *Journal of Fish Diseases* 29: 673-682.

Nougayrède, P., C. Michel, E. Sochon, A. Vuillaume and P. de Kinkelin. 1995. Vagococcosis, an emerging threat for farmed salmonid populations. *Proceedings of the VII International Conference on Diseases of Fish and Shellfish. European Association of Fish Pathologists*. Sept 10-15, 1995, Palma de Mallorca, Spain. p. 26.

Ooyama, T., A. Kera, T. Ookada, V. Inglis and T. Yoshida. 1999. The protective inmune response of yellowtail *Seriola quinqueradiata* to the bacterial fish pathogen *Lactococcus garvieae*. *Diseases of Aquatic Organisms* 37: 121-126.

Ooyama, T., Y. Hirokawa, T. Minami, H. Yasuda, T. Nakai, M. Endo, L. Ruangpan and T. Yoshida. 2002. Cell-surface properties of *Lactococcus garvieae* strains and their immunogenicity in the yellowtail *Seriola quinqueradiata*. *Diseases of Aquatic Organisms* 51: 169-177.

Pasnik, D.J., J.J. Evans and P.H. Klesius. 2006. Passive immunization of Nile tilapia (*Oreochromis niloticus*) provides significant protection against *Streptococcus agalactiae*. *Fish and Shellfish Immunology* 21: 365-371.

Pasnik, D.J., J.J. Evans, V.S. Panangala, P.H. Klesius, R.A. Shelby and C.A. Shoemaker. 2005a. Antigenicity of *Streptococcus agalactiae* extracellular products and vaccine efficacy. *Journal of Fish Diseases* 28: 205-212.

Pasnik, D.J., J.J. Evans and P.H. Klesius. 2005b. Duration of protective antigens and correlation with survival in Nile tilapia *Oreochromis niloticus* following *Streptococcus agalactiae* vaccination. *Diseases of Aquatic Organisms* 66: 129-134.

Perera, R.P., S.K. Johnson, M.D. Collins and D.H. Lewis. 1994. *Streptococcus iniae* associated with mortality of *Tilapia nilotica* × *T. aurea* hybrids. *Journal of Aquatic Animal Health* 6: 335-340.

Pier, G.B. and S.H. Madin. 1976. *Streptococcus iniae* sp. nov., a beta hemolytic streptococcus isolated from an Amazon freshwater dolphin, *Inia geoffrensis*. *International Journal of Systematic Bacteriology* 26: 545-553.

Plumb, J.A., J.H. Schachte, J.L. Gaines, W. Peltier and B. Carrol. 1974. *Streptococcus* sp. from marine fishes along the Alabama and northwest Florida coast of the Gulf of Mexico. *Transactions of American Fisheries Society* 103: 358-361.

Pot, B., L.A. Devriese, D. Ursi, P. Vandamme, F. Haesebrouck and K. Kersters. 1996. Phenotypic identification and differentiation of *Lactococcus* strains isolated from animals. *Systematic Applied Microbiology* 19: 213-222.

Prado, S. 2006. *Microbiota asociada a criaderos de moluscos. Patogénesis y probiosis*. Ph.D. Thesis. University of Santiago de Compostela, Spain.

Ravelo, C. 2004. *Caracterización del patógeno emergente de peces Lactococcus garvieae. desarrollo de un programa eficaz de vacunación*. Ph.D. Thesis. University of Santiago de Compostela, Spain.

Ravelo, C., B. Magariños, A.E. Toranzo and J.L. Romalde. 2001. Conventional versus miniaturized systems for the phenotypic characterization of Lactococcus garvieae strains. *Bulletin of the European Association of Fish Pathologists* 21: 136-144.

Ravelo, C., B. Magariños, S. Núñez, J.L. Romalde and A.E. Toranzo. 2000. Caracterización fenotípica, serológica y molecular de cepas de Lactococcus garvieae. *Proceedings of the III Reunión Científica del Grupo de Microbiología del Medio Acuático de la Sociedad Española de Microbiología*, May 14-16, 2000, Santiago de Compostela, Spain, pp. 129-130.

Ravelo, C., B. Magariños, S. López-Romalde, A.E. Toranzo and J.L. Romalde. 2003. Molecular fingerprinting of fish pathogenic Lactococcus garvieae strains by RAPD analysis. *Journal of Clinical Microbiology* 41: 751-756.

Ravelo, C., B. Magariños, M.C. Herrero, L.I. Costa, A.E. Toranzo and J.L. Romalde. 2006. Use of adjuvanted vaccines to lengthen the protection against lactococcosis in rainbow trout (*Oncorhynchus mykiss*). *Aquaculture* 251: 153-158.

Raverty, S. and W. Fiessel. 2001. Pneumonia in neonatal and juvenile harbor seals (*Phoca vitulina*) due to Streptococcus phocae. Animal Health Center. *Newsletter Diagnostic Diary* 11: 11-12.

Raverty, S., J.K. Gaydos, O. Nielsen and P. Ross. 2004. Pathologic and clinical implications of Streptococcus phocace isolated from pinnipeds along coastal Washington state, British Columbia, and Artic Canada. *35th Annual Conference of the International Association of Aquatic Animal Medicine*, Galveston, TX.

Ringø, E. and F.J. Gatesoup. 1998. Lactic acid bacteria in fish: A review. *Aquaculture* 160: 177-203.

Robinson, J.A. and F.P. Meyer. 1966. Streptococcal fish pathogen. *Journal of Bacteriology* 92: 512.

Romalde, J.L. and A.E. Toranzo. 1999. Streptococcosis of marine fish. In: *ICES Identification Leaflets for Diseases and Parasites of Fish and Shellfish. N°. 56.* G. Olivier (ed.). International Council for the Exploration of the Sea. Copenhagen, Denmark.

Romalde, J.L. and A.E. Toranzo. 2002. Molecular approaches for the study and diagnosis of salmonid streptococcosis. In: *Molecular Diagnosis of Salmonid Diseases*, C.O. Cunningham(ed.). Kluwer Academic Publishers, Dordrecht, pp. 211-233.

Romalde, J.L., C. Ravelo and A.E. Toranzo. 2006. Control of fish lactococcosis: efficacy of vaccination procedures. *CAB Reviews: Perspectives in Agriculture, Veterinary Science, Nutrition and Natural Resources* 1. http://www.cababstractplus.org/cabreviews

Romalde, J.L., R. Silva, A. Riaza and A.E. Toranzo. 1996. Long-lasting protection against turbot streptococcosis obtained with a toxoid-enriched bacterin. *Bulletin of the European Association of Fish Pathologists* 16: 169-171.

Romalde, J.L., B. Magariños, C. Villar, J.L. Barja and A.E. Toranzo. 1999a. Genetic analysis of turbot pathogenic Streptococcus parauberis strains by ribotyping and random amplified polymorphic DNA. *FEMS Microbiological Letters* 179: 297-304.

Romalde, J.L., B. Magariños and A.E. Toranzo. 1999b. Prevention of streptococcosis in turbot by intraperitoneal vaccination: a review. *Journal of Applied Ichthyology* 15: 153-158.

Romalde, J.L., A. Luzardo-Alvárez, C. Ravelo, A.E. Toranzo and J. Blanco-Méndez. 2004. Oral immunisation using alginate microparticles as a useful strategy for booster vaccination against fish lacotococcosis. *Aquaculture* 236: 119-129.

Romalde, J.L., C. Ravelo, S. López-Romalde, R. Avendaño-Herrera, B. Magariños and A.E. Toranzo. 2005. Vaccination strategies to prevent emerging diseases for Spanish aquaculture. In: *Progress in Fish Vaccinology*, P.J. Midtlyng (ed.). S. Karger, Basel, pp. 85-95.

Romalde, J.L., C. Ravelo, I. Valdés, B. Magariños, E. de la Fuente, C. San Martín, R. Avendaño-Herrera and A.E. Toranzo. 2008. *Streptococcus phocae*, an emerging pathogen for salmonid culture in Chile. *Veterinary Microbiology* 130: 198-207.

Ruiz-Zarzuela, I., I. de Blas, O. Gironés, C. Ghittino and J.L. Múzquiz. 2005. Isolation of *Vagococcus salmoninarum* in rainbow trout, *Oncorhynchus mykiss* (Walbaum), broodstocks: characterization of the pathogen. *Veterinary Research Communication* 29: 553-562.

Sakai, M., S. Atsuta and M. Kobayashi. 1986. A streptococcal disease of cultured Jacopever, *Sebastes schlegeli*. *Suisanzoshuku Aquiculture* 34: 171-177.

Sakai, M., R. Kubota, S. Atsuta and M. Kobayashi. 1987. Vaccination of rainbow trout (*Salmo gairdneri*) against β-haemolytic streptococcal disease. *Bulletin of the Japanese Society of Scientific Fisheries* 53: 1373-1376.

Salati, F., P. Tassi and P. Bronzi. 1996. Isolation of an *Enterococcus*-like bacterium from diseased Adriatic sturgeon, *Acipenser naccarii*, farmed in Italy. *Bulletin of the European Association of Fish Pathologists* 16: 96-100.

Salati, F., G. Angelucci, I. Viale and R. Kusuda. 2005. Immune response of gilthead sea bream, *Sparus aurata* L., to *Lactococcus garvieae* antigens. *Bulletin of the European Association of Fish Pathologists* 25: 40-448.

Sano, T., K. Kawano, I. Komatsu and Y. Hariya. 1997. Oral vaccination against enterococcosis in cultured yellowtail (*Seriola quinqueradiata*). International Workshop on Aquaculture Application of Controlled Drug and Vaccine Delivery. Universita di Udine, Villamanin di Passariano, Italy. p. 47.

Schijns, V.E.J.C. and A. Tangerås. 2005. vaccine adjuvant technology: from theoretical mechanisms to practical approaches. In: *Progress in Fish Vaccinology*, P.J. Midtlyng (ed.). S. Karger, Basel, pp. 127-134.

Schmidtke, L.M. and J. Carson. 1994. Characteristics of *Vagococcus salmoninarum* isolated from diseased salmonid fish. *Journal of Applied Bacteriology* 77: 229-236.

Sealey, W.M. and D.M. Gatlin III. 2002. Dietary vitamin C and vitamin E interact to influence growth and tissue composition of juvenile hybrid striped bass (*Morone chrysops* (female) × M. *saxatilis* (male)) but have limited effects on immune responses. *Journal of Nutrition* 132: 748-755.

Shelby, R.A., P.H. Klesius, C.A. Shoemaker and J.J. Evans. 2002. Passive immunization of tilapia, *Oreochromis niloticus* (L.), with anti-*Streptococcus iniae* whole sera. *Journal of Fish Diseases* 25: 1-6.

Shewmaker, P.L., A.C. Camus, T. Bailiff, A.G. Steigerwalt, R.E. Morey and M.G.S. Carvalho. 2007. *Streptococcus ictaluri* sp. nov., isolated from Channel catfish *Ictalurus punctatus* broodstock. *International Journal of Systematic Evolutionary Microbiology* 57: 1603-1606.

Shin, G.W., K.J. Palaksha, Y.R. Kim, S.W. Nho, J.H. Cho, N.E. Heo, G.J. Heo, S.C. Park and T.S. Jung. 2007a. Immunoproteomic analysis of capsulate and non-capsulate strains of *Lactococcus garvieae*. *Veterinary Microbiology* 119: 205-212.

Shin, G.W., K.J. Palaksha, Y.R. Kim, S.W. Nho, S. Kim, G.J. Heo, S.C. Park and T.S. Jung. 2007b. Application of immunoproteomics in developing a *Streptococcus iniae* vaccine for olive flounder (*Paralichthys olivaceus*). *Journal of Chromatography* B 849: 315-322.

Shoemaker, C. and P. Klesius. 1998. Pulsed-field gel electrophoresis of *Streptococcus iniae* from cultured fish. *Proceedings of the 23rd Annual Eastern Fish Health Workshop.* March 30-April 2, 1998. Plymouth, MA, USA, p. 33.

Shoemaker, C.A., P.H. Klesius and J.J. Evans. 2001. Prevalence of *Streptococcus iniae* in tilapia, hybrid striped bass, and channel catfish on commercial fish farms in the United States. *American Journal of Veterinary Research* 62: 174-177.

Shoemaker, C.A., G.W. Vanderberg, A. Désormeaux, P.H. Klesius and J.J. Evans. 2006. Efficacy of a *Streptococcus iniae* modified bacterin delivered using Oraljet technology in Nile tilapia (*Oreochromis niloticus*). *Aquaculture* 255: 151-156.

Shutou, K., K. Kanai and K. Yoshikoshi. 2007. Role of capsule in immunogenicity of *Streptococcus iniae* to Japanese flounder *Paralichthys olivaceus*. *Fish Pathology* 42: 101-106.

Sommerset, I., B. Krøssøy, E. Biering and P. Frost. 2005. Vaccines for fish in aquaculture. *Expert Review Vaccines* 4: 89-101.

Stoffregen, D.A., S.C. Backman, R.E. Perham, P.R. Bowser and J.G. Babish. 1996. Initial disease report of *Streptococcus iniae* infection in hybrid striped (sunshine) bass and successful therapeutic intervention with the fluoroquinolone antibacterial enrofloxacin. *Journal of World Aquaculture Society* 27: 420-434.

Sugita, A. 1996. A case of streptococcicosis in dusky spinefoot. *Fish Pathology* 31: 47-48.

Sun, J.R., J.C. Yan, C.Y. Yeh, S.Y. Lee and J.J. Lu. 2007. Invasive infection with *Streptococcus iniae* in Taiwan. *Journal of Medical Microbiology* 56: 1246-1249.

Teixeira, L.M., V.L.C. Merquior, M.C.E. Vianni, M.G.S. Carvalho, S.E.L. Fracalanzza, A.G. Steigerwalt, D.J. Brenner and R.R. Facklam. 1996. Phenotypic and genotypic characterization of atypical *Lactococcus garvieae* strains isolated from water buffalos with subclinical mastitis and confirmation of *L. garvieae* as a senior subjective synonym *of Enterococcus seriolicida. International Journal of Systematic Bacteriology* 46: 664-668.

Toranzo, A.E., S. Devesa, P. Heinen, A. Riaza, S. Núñez and J.L. Barja. 1994. Streptococcosis in cultured turbot caused by an *Enterococcus*-like bacterium. *Bulletin of the European Association of Fish Pathologists* 14: 19-23.

Toranzo, A.E., J.M. Cutrín, S. Nuñez, J.L. Romalde and J.L. Barja. 1995b. Antigenic characterization of *Enterococcus* strains pathogenic for turbot and their relationship with other Gram-positive bacteria. *Diseases of Aquatic Organisms* 21: 187-191.

Toranzo, A.E., S. Devesa, J.L. Romalde, J. Lamas, A. Riaza, J. Leiro and J.L. Barja. 1995b. Efficacy of intraperitoneal and immersion vaccination against *Enterococcus* sp. infection in turbot. *Aquaculture* 134: 17-27.

Tumbol, R.A., J. Baiano and A.C. Barnes. 2007. Opsonisation, pahogocytosis and killing of *Streptococcus iniae* by peritoneal leucocytes from barramundi, *Lates calcarifer*. XIIIth *International Conference of the European Association of Fish Pathologists* (EAFP). Grado, Italy. p. 45.

Valdés, I., B. Jaureguiberry, J.L. Romalde, A.E. Toranzo, B. Magariños and R. Avendaño-Herrera. 2009. Genetic characterization of *Streptococcus phocae* strains isolated from Atlantic salmon in Chile. *Journal of Fish Diseases* doi: 10.1111/j.1365-2761.2008.01014x.

Vandamme, P., L.A. Devriese, B. Pot, H. Kersters and P. Melin. 1997. *Streptococcus difficile* is a nonhemolytic group B type Ib *Streptococcus*. *International Journal of Systematic Bacteriology* 47: 81-85.

Vela, A.I., J. Vázquez, A. Gibello, M.M. Blanco, M.A. Moreno, P. Liebana, C. Albendea, B. Alcalα, A. Méndez, L. Domínguez and J.F. Fernández-Garayzábal. 2000. Phenotypic and genetic characterization of *Lactococcus garvieae* isolated in Spain from lactococcosis outbreaks in comparison with isolates of other countries and sources. *Journal of Clinical Microbiology* 38: 3791-3795.

Vela, I., P. Liebana, C. Albendea, A. Gibello, M. Blanco, M. Moreno, M. Cutúli, J.A. García, A. Domenech, L. Domínguez, J. Vαzquez and J. Fernández-Garayzábal. 1999. Caracterización fenotípica y molecular de *Lactococcus garvieae*. *Proceedings of the XIII Congreso Nacional de Microbiología, Sociedad Española de Microbiología*, Sept. 19-24, 1999, Granada, Spain. p. 47.

Volpatti, D., B. Contessi, E. Buonasera and M. Galeotti. 2007. Protective role and *in vitro* activity of fractions extracted from *Lactococcus garvieae*, the lactococcosis agent in rainbow trout (*Oreochromis mykiss*). *XIIIth International Conference of the European Association of Fish Pathologists* (EAFP). Grado, Italy. p. 215.

Vossen, A., A. Abdulmawjood, C. Lämmler, R. Weiß and U. Siebert. 2004. Identification and molecular characterization of beta-hemolytic streptococci isolated from harbor seals (*Phoca vitulina*) and grey seals (*Halichoerus grypus*) of the German north and Baltic seas. *Journal of Clinical Microbiology* 42: 469-473.

Wallbanks, S., A.J. Martínez-Murcia, J.L. Fryer, B.A. Phillips and M.D. Collins. 1990. 16S rRNA sequence determination for members of genus *Carnobacterium* and related lactic acid bacteria and description of *Vagococcus salmoninarum* sp. nov. *International Journal of Systematic Bacteriology* 40: 224-230.

Weinstein, M.R., M. Litt, D.A. Kertesz, P. Wyper, D. Rose, M. Coulter, A. McGeer, R.R. Facklam, C. Ostach, B.M. Willey, A. Borczyk and D.E. Low. 1997. Invasive infections due to a fish pathogen, *Streptococcus iniae*. *New England Journal of Medicine* 337: 589-594.

Wilkinson, H.W., L.G. Thacker and R.R. Facklam. 1973. Nonhemolytic group B streptococci of human, bovine, and ichthyc origin. *Infection Immunology* 7: 496-498.

Williams, A.M., J.L. Fryer and M.D. Collins. 1990. *Lactococcus piscium* sp. nov., a new *Lactococcus* species from salmonid fishes. *FEMS Microbiological Letters* 68: 109-114.

Yoshida, T., M. Endo, M. Sakai and V. Inglis. 1997. A cell capsule with possible involvement in resistance to opsonophagocytosis in *Enterococcus seriolicida* isolated from yellowtail *Seriola quinqueradiata*. *Diseases of Aquatic Organisms* 29: 233-235.

Yoshida, T., T. Eshima, Y. Wada, Y. Yamada, E. Kakizaki, M. Sakai, T. Kitao and V. Inglis. 1996. Phenotypic variation associated with an antiphagocytic factor in the bacterial fish pathogen *Enterococcus seriolicida*. *Diseases of Aquatic Organisms* 25: 81-86.

Zarza. C. and F. Padrós. 2007. Enfermedades emergentes en la piscicultura marina española. *Skretting Informa* 12: 23-31.

Zlotkin, A., H. Hershko and A. Eldar. 1998a. Possible transmission of *Streptococcus iniae* from wild fish to cultured marine fish. *Applied Environmental Microbiology* 64: 4065-4067.

Zlotkin, A., A. Eldar, C. Ghittino and H. Bercovier. 1998b. Identification of *Lactococcus garvieae* by PCR. *Journal of Clinical Microbiology* 36: 983-985.

5

Behavioral Defenses against Parasites and Pathogens

Brian D. Wisenden[1, *], Cameron P. Goater[2] and Clayton T. James[2]

INTRODUCTION

Parasites exert profound and pervasive costs on their hosts through mounting an immunity based defense, causing reduced growth and reproduction, and immunopathology (Sheldon and Verhulst, 1996; Zuk and Stoehr, 2003). Several chapters in this volume attest to the central importance of the host's immune system—and its effectiveness—in addressing these costs. Yet, natural selection should favor hosts that develop and maintain diverse anti-parasite behavioral strategies independent of host immunity and typical tissue reactions that either limit their exposure to parasites or that counter their negative effects (reviewed by Goater and Holmes, 1997).

Authors' addresses: [1]Department of Biosciences, Minnesota State University Moorhead, Moorhead, MN, USA.
[2]Department of Biological Sciences, University of Lethbridge, Lethbridge, AB, Canada.
E-mail: cam.goater@uleth.ca, clayton.james@uleth.ca
Corresponding author: E-mail: wisenden@mnstate.edu

Hart (1994) was among the first to emphasise the importance of parasite-mediated selection for parasite avoidance behaviors, especially in the face of expensive immunity. Combes (2001) also emphasized avoidance behaviors in the context of 'exposure filters' that may limit infection rates. Such behaviors range in complexity from simple adjustments to host movement, posture and habitat choice, for example, to avoid biting arthropods, to sophisticated avoidance behaviors linked to fine-tuned parasite-detection strategies. Following intensive interest by both parasitologists and behavioral ecologists over the past decade, there is now a much better understanding of the extent of parasite avoidance behaviors across a broad range of both parasite and host taxa (recent reviews by Combes, 2001; Moore, 2002). However, most empirical tests regarding the effectiveness and extent of host avoidance behaviors involve the visible ectoparasites of ungulates, the avoidance by grazing mammals of fecal patches containing larvae of gastrointestinal nematodes, and the avoidance of parasitoids by insects. Much less attention has been devoted to studies aimed to evaluate parasite avoidance behaviors in aquatic systems, especially those involving fish (but see also Barber et al. 2000). The aim of this chapter is to review parasite avoidance behaviors in fish.

Hart (1994) and Moore (2002) stipulate two requirements that must be met before designating certain behaviors as 'avoidance'. First, parasites should be demonstrated to have a negative effect on host fitness, and second, anti-parasite behaviors should be demonstrated to decrease parasite intensity, or to ameliorate their negative effects. Implicit in these conditions is a third requirement: infective stages must be detectable. In Figure 5.1, we diagrammatically represent our view of potential host defense strategies that include host-avoidance behaviors. This framework provides the conceptual outline for our chapter. While very few studies support many of the links in Figure 5.1, there is abundant evidence from predator-prey interactions that fish can and do develop well-tuned behavioral responses to risk (Lima and Dill, 1990). Indeed, one underlying theme of our chapter is that the large number of studies on mechanisms of detection and avoidance of aquatic predators (and other aquatic stressors) provides a solid theoretical and empirical foundation for future studies involving parasite avoidance.

For our purposes, we define parasites to include typical ectoparasites of fish (copepods, brachyurans, and monogenean trematodes), helminthes (digenean trematodes, cestodes, acanthocephalans, and nematodes) and certain single-celled types (myxozoans and microsporidians). While our

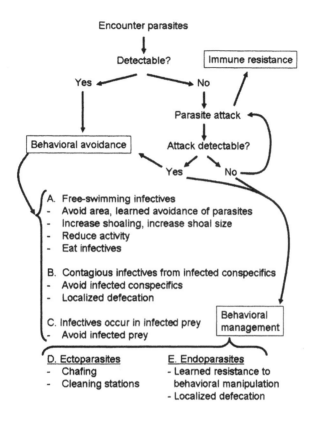

Encounter parasites

Detectable?   Immune resistance

Yes   No

Behavioral avoidance   Parasite attack

Attack detectable?

Yes   No

A. Free-swimming infectives
- Avoid area, learned avoidance of parasites
- Increase shoaling, increase shoal size
- Reduce activity
- Eat infectives

B. Contagious infectives from infected conspecifics
- Avoid infected conspecifics
- Localized defecation

C. Infectives occur in infected prey
- Avoid infected prey

Behavioral management

D. Ectoparasites
- Chafing
- Cleaning stations

E. Endoparasites
- Learned resistance to behavioral manipulation
- Localized defecation

Fig. 5.1 The first step in parasite behavioral resistance is detection of the parasite. Parasite detection may occur either before the parasite contacts the host or after initial attack. Where behavioral avoidance fails, the immune system offers a second line of defense. When parasites are detected, behavioral avoidance of infection depends on the nature of the parasite threat. Generally, these fall into three broad categories: (A) avoidance of free-swimming infective stages, (B) avoidance of conspecifics infected with contagious parasites, or (C) avoidance of prey that are infected with early or intermediate forms of a parasite. Behavioral management of existing parasite load includes different strategies for ectoparasites (D) and endoparasites (E).

focus is on fish/parasite interactions, we also consider other aquatic host/parasite interactions where appropriate. We further place our discussion of behavioral avoidance into the context of the notoriously variable transmission strategies and life cycles of aquatic parasites. Generally, parasitic arthropods and nematodes actively seek their hosts, while parasitic viruses, bacteria, microsporidia, and fungi tend to rely on host contact. However, parasites defy tidy taxon-based categorization in the sense that many aquatic parasites such as aquatic trematodes (and some cestodes) possess a combination of active and passive stages within their complex life cycles (e.g., Haas, 1994). The larval stages have two obligate free-living stages (the miricidia and cercariae) that vary in the extent to which they rely on passive versus active contact with hosts. In this chapter, our focus is on avoidance behaviors of parasites that actively infect their host, primarily because these stages tend to be relatively large and perhaps more easily detectable.

EVIDENCE FOR BEHAVIORAL AVOIDANCE IN FISHES

Behavioral Avoidance of Motile Infective Stages

Regardless of the active versus passive nature of parasite transmission strategies, high rates of encounter between hosts and parasites should lead to high rates of infection. One way for motile aquatic hosts to reduce exposure to motile infective stages is to reduce the overall activity. In a laboratory experiment, larval green frogs, *Rana clamitans*, and wood frogs, *Rana sylvatica*, reduced their activity by 25-33% when exposed to cercariae of the trematode *Echinostoma* sp. (Theimann and Wassersug, 2000). The authors concluded that reduced host activity in the presence of cercariae is an adaptive response to reduce their risk of exposure. Unfortunately, there are few comparable studies, and none involves fish hosts. Moreover, we do not know if similar host responses are present under natural conditions of exposure.

Experiments with ectoparasites provide indirect evidence that changes in host activity can affect the outcome of parasite/fish interactions. The time that brook trout (*Salvelinus fontinalis*) spent active in laboratory aquaria was positively associated with exposure to the copepod parasite *Salminicola edwardsii* (Poulin et al., 1991). Thus, inactive fish acquired fewer parasites. When trout were re-exposed to copepodids of *S. edwardsii*, individuals with high infections from an initial exposure were found to be more active and they further increased their rates of exposure. Thus, innately active individuals were infected with more parasites on each exposure. In this example, it is unknown whether fish could detect infective stages of *S. edwardsii* and thereby reduce their risk of exposure via reduction in swimming activity.

Many species of fish are known to shoal in response to risk imposed by a wide range of aquatic stressors (Krause and Ruxton, 2002). Although several studies have evaluated the shoaling behavior of infected fish (e.g., Ward et al., 2005), only one has evaluated shoaling as a direct response to parasite exposure. Poulin and Fitzgerald (1989) showed that three-spined stickleback (*Gasterosteus aculeatus*) and blackspot stickleback (G. *wheatlandi*) increased shoal attendance and increased the number of individuals per shoal (in large shoals of G. *aculeatus*) in response to the presence of the brachyuran *Argulus*. Increased group size and increased group cohesion is similar to the response of mammal herds to the threat of biting flies (Duncan and Vigne, 1979; Coté and Poulin, 1995; Moore, 2002). Although individual parasites achieved higher attack success in dense shoals than in small or sparse shoals, individual host fish reduced probability of parasite attack in large shoals—the greater the size of the shoal, the greater the benefit. Thus, in the case of this visually detectable parasite, risk of parasite infection evoked a shoaling response in hosts. It would be worthwhile to test for similar types of shoaling response in appropriate fish species that are exposed to trematode cercariae or penetrating myxozoan larvae.

Use of parasite-free refugia has also been shown to effectively reduce exposure. Karvonen *et al.* (2004) showed that rainbow trout avoided shelters when cercariae of *Diplostomum spathaceum* were released into it. The longer the trout waited to leave the shelter, the heavier was the rate of infection. In this case, the mechanism of parasite detection remains unknown. The trout were not parasite-naïve and, thus, learned recognition of chemical or other cues of the parasite is a possibility (see below). Alternatively, trout may have responded to tactile detection of penetrating cercariae without learning. The use of cercariae-conditioned water would help distinguish these alternatives. More studies of this type (Karvonen *et al.*, 2004) are needed to evaluate the degree to which behavioral responses are finely tuned, for example, to distinguish between cercariae that penetrate that particular host, from the many in a given habitat that do not.

Another form of behavioral avoidance can occur at the level of host habitat. One possibility is for hosts to avoid habitats in which direct evaluation of infection risk is possible. Evidence for this type of direct assessment is most common for the avoidance of biting insects and nematode larvae by grazing animals (reviewed in Moore, 2002). An empirical test of this idea involves sticklebacks (*Gasterosteus* spp.) exposed to the brachyuran, *Argulus canadensis* (Poulin and Fitzgerald, 1989). In parasite-free aquaria, sticklebacks were active near the bottom, adjacent to vegetation. In tanks containing free-swimming stages of A. *canadensis*, the fish swam along the surface away from vegetation. For this large, visually detectable parasite, shift in habitat use to avoid infection is probably an effective means of avoiding infection.

A second possibility is for hosts to avoid patches in which detection occurs via indirect indicators of infective stages. A crude way to evaluate this possibility is to determine whether variation in parasite avoidance behavior is associated with spatial or seasonal variation in the presence of infective stages. In north-temperate ecosystems, fishes are most at risk of cercariae infection during a narrow window of transmission in late summer and autumn (review by Chubb, 1979). Fathead minnows, *Pimephales promelas*, in lakes in northern Alberta, Canada are exposed to a maximum of approximately 30 cercariae of a brain-encysting trematode in their first fall (Sandland *et al.*, 2001) and up to approximately 500 in their second fall (C.P. Goater, unpubl. obs.). Similar examples of seasonally and spatially restrictive pulses of transmission are common in many different types of aquatic parasites (Chubb, 1979). Unfortunately, the extent to which variation in the expression of avoidance behaviors is associated with variation in crude indicators of transmission, such as site and season, has not been tested.

Evidence for more finely tuned assessment of infection status in aquatic systems is also rare, but enticing. Female grey tree frogs *Hyla versicolor* avoided ovipositing in experimental ponds in which they detected the presence of cercariae-releasing snails (Kiesecker and Skelly, 2000). Similarly, female mosquitoes avoided ovipositing in sites that contained infected snails (Lowenberger and Rau, 1994). In both cases, avoidance of infected snails would lead to reduction in exposure of the host's offspring to free-swimming cercariae released from snails. Kiesecker and Skelly (2000) showed that, in addition to detecting cercariae-releasing snails, female frogs could distinguish ponds with infected versus uninfected snails. These results indicate that behavioral avoidance of parasite infective stages can be finely tuned. Currently, the mechanisms underlying the detection of cercariae and/or the presence or absence of appropriate snails are unknown, but presumably it occurs via chemical cues.

There are several important components of parasite avoidance behavior that have not been studied in aquatic systems. One involves the avoidance of feces. Thus, for parasites with direct life cycles that spread to new hosts by infective stages contained in fecal matter, one might predict selection on hosts to: (1) avoid feces; and (2) to defecate in areas separate from foraging areas. Feces avoidance and localized defecation are well documented in terrestrial vertebrates as a strategy to minimize exposure to infective stages of parasites (e.g., Haufsater and Meade, 1982). Fecal avoidance is especially important for species with limited home ranges or permanent stations where they reside much of the time. Certainly, nesting male fish would qualify as candidates for localized defecation but, to our knowledge, such data have never been collected. Sit-and-wait ambush predators face a similar problem in that they remain in place for long periods of time. Moreover, many prey species can detect conspecific chemical alarm cues released from ingested and digested prey contained in the feces of predators (Chivers and Mirza, 2001). Northern pike (*Esox lucius*), a sit-and-wait ambush predator, designates a specific area to defecate, and does so away from the area where it forages (Brown *et al.*, 1995). The authors argued that localized defecation could be explained by selection to avoid chemical labeling by their prey. An interesting alternative is that localized defecation could also limit the risk of infection with parasites originating from the host's own feces. Future studies may reveal similar attention to fecal management in other species of fish.

The role of host learning in parasite and/or habitat avoidance is also poorly understood. This is an important shortcoming. Predator-naïve fish do not recognize predators as dangerous until after they have had an opportunity to associate an olfactory (Chivers and Smith, 1994a), visual (Chivers and Smith, 1994b) or auditory (Wisenden *et al.*, 2007) stimulus with a predation event. Commonly, the releasing stimulus for this form of learning are chemical alarm cues released from injured epidermal tissue that occurs as a natural consequence of predatory attack. The same classes of chemical cues are reportedly released following exposure of juvenile rainbow trout to cercariae of *Diplostomum spathaceum*, even when the odor of the cercariae themselves invoked no response (Poulin *et al.*, 1999). Prey may learn to avoid predators directly by their odors or images. Additional cues for parasite avoidance can form with many other correlates of infection risk. For example, minnows can rapidly acquire recognition and avoidance of a specific habitat type associated with predation (Chivers and Smith, 1995). Alternatively, minnows can learn to associate novel odors with risk after watching a shoal mate exhibit alarm behavior to an odorant (Mathis *et al.*, 1996). Both conspecifics and heterospecifics can serve as models to impart acquired recognition of novel indicators of predation risk. The sophisticated learning mechanisms that arose to mediate risk of predation can be applied equally to risk of infection.

Likewise, it is conceivable that fish may learn to associate cercariae infection with certain species of snails (at least at certain times of the year) or with the types of habitats that contain infected snails. Fish might also learn to associate certain habitat types with the presence of highly pathogenic spores of microsporidians (e.g., *Glugea* in sticklebacks) or the habitats favored by the oligichaetes that serve as primary hosts of myxozoans.

Behavioral Avoidance of Infected Conspecifics

For parasites that are directly transmitted, selection should favor avoidance of hosts harboring infective stages. These would include some of the single-celled microparasites (e.g., the ciliate protozoan, *Ichthyophthirius* and other protist or fungal parasites) and certain macroparasites such as the monogenean trematodes. Consistent with this prediction, Milinski and Bakker (1990) showed that female sticklebacks avoided mating with males whose nuptial coloration had been dulled by the ectoparasitic ciliate *Ichthyophthirius*. In this case, choosy females accrued direct fitness benefits from reducing their risk of exposure to infective stages, and indirect fitness benefits by avoiding genes linked to parasite susceptibility. Outside of the mate choice context, bullfrog tadpoles (*Rana catesbeiana*) spent more time adjacent to uninfected tadpoles than tadpoles infected with the yeast *Candida humicola*, a pathogen that reduces host growth and survival (Kiesecker *et al.*, 1999). Further, tadpoles could express this preference based only upon chemical cues released from infected tadpoles, but could not do so when limited to visual cues alone. Kiesecker *et al.* (1999) elegantly demonstrated that spatial avoidance of infected conspecifics reduced an individual's risk of infection. Here again, the reliance of aquatic animals on semiochemicals for information management is remarkably similar to the mechanisms used for managing predator-prey interactions and reproductive decision making (Wisenden, 2003; Wisenden and Stacey, 2005).

Avoidance of conspecifics has also been evaluated in the context of shoaling behavior. Three-spined sticklebacks preferred to join shoals of conspecifics that were not infected with tumor-like growths caused by the microsporidian, *Glugea anomala*, perhaps to reduce the risk of direct transmission (Ward *et al.*, 2005). Likewise, sticklebacks preferred to shoal with uninfected conspecifics over those infected with *Argulus* (Dugatkin *et al.*, 1994). The simplest interpretation of these results is that individuals reduced their risk of exposure by avoiding hosts with transmissible parasites. However, the explanation may be more complex in the light of results indicating that killifish (*Fundulus diaphanus*) detect shoals of conspecifics infected with cysts of the trematode *Crassiphiala bulboglossa* (these are one of the causative agents of 'black spot' in fishes) and discriminate against them (Krause *et al.*, 1999). In this case, direct transmission from fish to fish is impossible. Perhaps individuals simply discriminate against 'sick' conspecifics, especially if they are detected at high density. Wedekind (1992) showed that female roach, *Rutilus rutilis*, could discriminate among males on the basis of the species of parasite with which they were infected. Thus, it is conceivable that individuals can assess the health status of conspecifics and assort themselves accordingly. Alternatively, hosts may avoid all parasites that infect epidermal sites of conspecifics to avoid potentially pathogenic secondary infections involving fungi (e.g., *Saprolegnia*) or bacteria. The adaptive significance of parasite-assortative shoaling needs further study.

The cues used to evaluate infection status in conspecifics are poorly known. In the case of *Glugea*-infected sticklebacks and black-spot infected killifish, detection is likely via visual cues. However, sticklebacks did not seem to avoid infective stages of *Argulus* when they were presented alone, even when chemical and visual stimuli from the parasite were made available to focal fish (Dugatkin *et al.*, 1994). In this case, the recognition and avoidance of the infected fish was cued either by altered behavior of infected hosts or perhaps by a chemical cue released by infected fish. It is also important to note that in each of the examples described above, fish were collected from the wild and thus presumably had opportunity to learn to recognize the parasite from previous experiences.

Behavioral Avoidance of Infected Prey

Many aquatic parasites have complex life cycles, with many involving the ingestion of resting stages. Thus, any discussion of avoidance of infective stages must also include avoidance of infected intermediate hosts. This feature has been covered in several reviews (Barber *et al.*, 2000; Moore, 2002). We will not duplicate that coverage here, other than to emphasize two points. First, despite the fact that some of the best empirical tests of this idea involve fish as hosts, no evidence for avoidance of infected intermediate hosts exists. Contrary to predictions, two empirical tests showed that fish prefer infected hosts to uninfected hosts. In one case, sticklebacks selected amphipods infected with larval stages of an acanthocephalan worm over uninfected ones (Bakker *et al.*, 1997), largely due to their ease of visual detection (the larvae are bright orange, and presumably easily detected through the exoskeleton). In the second case, three-spined sticklebacks strongly preferred to eat copepods infected with procercoids of the cestode *Schistocephalus solidus* over uninfected ones (Wedekind and Miliniski, 1996). Both cases involved parasites that have a negative effect on stickleback reproduction (e.g., Barber and Svensson, 2003). Therefore, they are precisely the types of systems where selection should be strong for avoidance, yet the outcome was prey attraction—*not* avoidance. Why should this be so? Lafferty (1992) explored this conundrum in a theoretical context, concluding that in such cases, the costs of infection must be balanced by the benefits of foraging on conspicuous (in the case of infected amphipods) or unhealthy prey that are easier to catch or to handle. When cost of infection to the final host is relatively small or delayed, a final host may potentially benefit from eating infected prey by using mature parasites in its gut to infect and compromise the antipredator competence of its prey (Lafferty, 1992). The costs and benefits of feeding on infected vs. uninfected prey is a ripe area for future study.

The second point we wish to emphasize is that learned avoidance of infected prey has not yet been evaluated. In the experiments described above and others reviewed by Moore (2002), discriminating hosts were exposed only once to infected prey. In the case of infected amphipods, the adult worms take approximately 30 days to reach maturity (longer at cooler temperatures). Thus, the pathogenic consequences of infection will almost certainly lag behind the point of ingestion. To what extent is it possible for fish to associate parasite-induced pathology to a prior ingestion event? We do not know the answer to this question, but we can predict that learned avoidance should be most likely to occur for parasites that are conspicuous within their intermediate hosts (e.g., *Plagiorhynchus* in *Gammarus*) and for parasites for which the lag between ingestion and pathology is shortest (e.g., packages of bivalve glochidia larvae within prey mimics).

Behavioral Management of Ectoparasites

Fish hosts have some options for behavioral management of parasite intensities after parasites have successfully contacted the host.

Ectoparasites may be dislodged and removed by chafing behavior, whereby a fish scrapes its body against a firm surface to remove a parasite (reviewed by Wyman and Walters-Wyman, 1985). Chafing is most likely to occur at the moment of parasite-host contact because dislodging a parasite is most likely to succeed if it is done before the parasite can firmly attach itself to the host fish. Chafing can be accompanied by body shakes, coughing motions or rapid starts (e.g., Thieman and Wassersug, 2000) or 'wiggling' (Baker and Smith, 1997) that serve to interrupt attachment of ectoparasites. Larval damselflies groom themselves in response to exposure to parasitic larval mites by rubbing a leg against an antenna, head, abdomen or another leg (Forbes and Baker, 1990). They also attempt to flee by rapid swimming.

While it may seem intuitive that chafing behavior should reduce infection, explicit evidence of such is difficult to find in the literature. Larval damselflies groom themselves in response to contact with the parasitic mite *Arrenurus* and successfully dislodge the parasite (Baker and Smith, 1997). Wyman and Walters-Wyman (1985) experimentally induced significant increases in chafing behavior in two species of fish by carefully loosening a scale or inserting a small particle of charcoal under a scale. Fish naturally infected with an external fungal infection also showed heightened levels of chafing.

If chafing and shaking are analogous to autogrooming by terrestrial vertebrates, cleaning stations by coral reef fishes are analogous to allogrooming, or perhaps 'anting' behavior of birds (Clark *et al.*, 1990). Client fishes visit cleaning stations (the territory of an individual of a cleaner species, often a member of the wrasse family) to be rid of their ectoparasites. Unlike other types of behavioral resistance to parasite attack, there is an impressive amount of literature documenting the benefits and interrelationships between the individual cleaner fish and their client fishes (Rhode, 1993; Losey *et al.*, 1999). Cleaners occasionally 'cheat' by nipping healthy mucus and scales from clients rather than searching diligently for ectoparasites and dead and infected tissue (Bshary and Grutter, 2002). However, because the majority of non-predatory reef fishes continue to actively visit cleaning stations, the benefits from doing so must outweigh the risk of encountering a cheating cleaner. Indeed evidence clearly shows a net benefit of cleaner fish in reducing parasite load of client fish (Grutter, 1999). The cleaner wrasse *Labroides dimidiatus* consumes 1200 ectoparasitic gnathiid isopods per day from the client species *Hemigymnus melapterus*. Individuals of *H. melapterus* that visit a cleaner show a 4.5-fold reduction in the number of isopods compared to individuals that were prevented from visiting a cleaner. This benefit occurred within a 12-h time span, strongly suggesting that behavioral management of parasite load occurs daily through visits to a cleaner station.

Behavioral Management of Endoparasites

The opportunities for management of endoparasites in fishes and other aquatic animals are probably limited in scope. Behavioral thermoregulation is one possibility, but has only rarely been assessed in fish. In a laboratory test, sunfish (*Lepomis macrochirus*) and largemouth bass (*Micropterus salmoides*) injected with *Aeromonas hydrophila* showed a 2.6° C increase in mean preferred temperatures compared to unexposed controls (Reynolds *et al.*, 1975). Follow-up experiments involving goldfish (*Carassius auratus*) as hosts indicated that short-term 'behavioral fever' decreased host mortality relative to controls, presumably through temperature-induced enhancement of the immune response (Covert and Reynolds, 1977). Evaluation of behavioral fever in fish exposed to other types of parasites would be a useful addition.

Although some terrestrial vertebrates have been documented to consume the leaves of specific plants to reduce infection by endoparasites (reviewed by Lozano, 1998), there are no comparable examples in fish. Perhaps herbivory pressure that selected for noxious secondary plant compounds in terrestrial plants does not occur to the same degree in aquatic plants, thus pharmacological opportunities may be more limited for aquatic animals. A more likely behavior to arise in fishes is the consumption of roughage to dislodge intestinal parasites from the gastrointestinal tract as is known to occur in some primates (Wrangham, 1995; Huffman *et al.*, 1996).

CONSTRAINTS ON THE EVOLUTION AND EXPRESSION OF ANTI-PARASITE BEHAVIOR

In this chapter, we have emphasized the shortage of supportive evidence for parasite avoidance behaviors in fish. The lack of devoted attention to this topic must certainly be a contributing factor. Yet it is also possible that selection for avoidance behaviors is opposed by various constraints and trade-offs that may make them too costly or unlikely to evolve.

The first constraint is that infective stages of many fish parasites may not be detectable. Although chemical detection thresholds have been evaluated for fish in the context of mate selection and predator avoidance (Wisenden and Stacey, 2005), no such data exist for parasite infective stages. Thus, it may be no coincidence that the best examples of avoidance behavior come from fish exposed to large and visible ectoparasites. Although it is not possible to generalise on the relative costs of ecto- versus endoparasites of fishes, many of the parasitic brachyuran, copepod, and isopod arthropods, and also the monogenean trematodes, certainly have strong negative effects on fish growth and reproduction (review by Rhode, 1993; Barber et al., 2000). Indeed, for many aquatic parasites, selection is likely to favor cryptic infective stages that restrict detection by visual, chemical, or tactile cues. Further, some species of aquatic trematodes have infective stages that are shaped and colored to encourage attraction, not avoidance, by potential fish hosts (e.g., Dronen, 1973; Beuret and Pearson, 1994). We should not expect parasite avoidance behaviors in those systems where strong counter-selection of this sort is common.

Avoidance behaviors may also be costly, both energetically and in the form of trade-offs with conflicting demands. Direct energetic costs associated with grooming are well documented in mammalian and avian host-parasite interactions (Hart, 1994), but have not been evaluated in aquatic systems. A second energetic cost is reductions in foraging opportunities. Predator-induced reductions in fish activity have strong negative effects on the foraging behavior of individuals. Similar costs are likely to exist for anti-parasite behaviors. Thus, avoidance behaviors that alter host activity (Poulin and FitzGerald, 1989; Thiemann and Wassersug, 2000; Karvonen et al., 2004) should also be expected to reduce foraging opportunities. This hypothesis is untested.

A third potential constraint is that anti-parasite behaviors may conflict with anti-predator behaviors. Thus, alteration in habitat choice to open, non-vegetated regions of a pond by sticklebacks to avoid an ectoparasite may come at a cost of increased predation (Poulin and FitzGerald, 1989). However, there are very few tests of potential trade-offs between parasite and predator avoidance strategies. In a laboratory test, exposure to predator kairomones and cercariae of Echinostoma sp. reduced the swimming activity of Rana clamitans tadpoles by 48% and 30%, respectively (Thiemann and Wassersug, 2000). Predator-induced reduction in host swimming activity led to a 16% increase in the numbers of encysted *Echinostoma* sp. found in the kidneys. The authors speculate that the presence of fish predators masks the typical bursts of activity (Taylor *et al.*, 2004) that tadpoles elicit when they detect penetrating cercariae. Thus, a reduction of activity is beneficial in the presence of cercariae only, but it promotes attack when predator cues are present. Likewise, *Daphnia magna* avoid surface water during the day to avoid visually hunting predators. However, increased time at the bottom exposes *Daphnia* to the spores of *Pasteuria ramos*, a bacterial endoparasite (Decaestecker *et al.*, 2002). We need more studies that assess parasite avoidance behaviors in the context of predation and other aquatic stressors. Larval damselflies also suffered increased predation when they engaged in anti-mite behaviors (Baker and Smith, 1997).

Lastly, it is also conceivable that changes in host activity and other avoidance behaviors to one parasite may come at a cost of increased exposure to others. Thus, inactivity induced by motile cercariae may lead to increased exposure to parasites that require direct contact with substrate. Thus, selection may not exist for specific anti-parasite behaviors directed to one species, but for a low-level, generalized response to parasite risk.

CONCLUSION

We conclude that the requirements for parasite avoidance behaviors (Hart, 1997) are met in fish/parasite interactions. Fish are exposed to an enormous diversity of types and numbers of parasites, possibly on a daily or even hourly basis. This diversity is probably paralleled by a diversity of behavioral responses involving detection and then avoidance of infective stages. The evidence for avoidance behaviors is strongest for pathogenic ectoparasites that tend to have large, visible infective stages. For other parasites, the evidence is enticing that fish possess sophisticated detection capabilities that lead to avoidance behaviors that reduce infection risk. However, for this latter group, the evidence is scant, being restricted to only a handful of empirical studies. Thus, for the five anti-parasite behaviors that we have identified in this chapter, there are only one or two solid examples of each that involve fish as hosts. Not surprisingly, our understanding of parasite avoidance strategies lags far behind that for predator avoidance strategies. Can hosts associate risk of infection with seasonal or microhabitat cues and then engage avoidance behaviors to minimise that risk? What role does past exposure experience and learning play in the development and expression of subsequent avoidance behaviors? Do the risk avoidance behaviors that fish employ in their aquatic habitats include parasites at all, and if so, are they traded off with behaviors associated with features such as predation and foraging? In answering these and other questions, we should recognise that chemical ecology is at the forefront of ecological interactions in aquatic environments. We predict that parasite-host interactions will prove to be no exception.

References

Baker, R.L. and B.P. Smith. 1997. Conflict between antipredator and antiparasite behaviour in larval damselflies. *Oecologia* 109: 622-628.

Bakker, T.C.M., D. Mazzi and S. Zala. 1997. Parasite-induced changes in behavior and color make *Gammarus pulex* more prone to fish predation. *Ecology* 78: 1098-1104.

Barber, I. and P.A. Svensson. 2003. Effects of experimental *Schistocephalus solidus* infections on growth, morphology and sexual development of female three-spined sticklebacks, *Gasterosteus aculeatus*. *Parasitology* 126: 359-367.

Barber, I., D. Hoare and J. Krause. 2000. Effects of parasites on fish behaviour: a review and evolutionary perspective. *Reviews in Fish Biology and Fisheries* 10: 131-165.

Beuret, J. and J.C. Pearson. 1994. Description of a new zygocercous cercariae (Opisthorchioidea: Heterophyidae) from prosobranch gastropods collected at Heron Island (Great Barrier Reef, Australia) and a review of zygocercariae. *Systematic Parasitology* 27: 105-125.

Brown, G.E., D.P. Chivers and R.J.F. Smith, 1995. Localized defecation by pike: a response to labeling by cyprinid alarm pheromone? *Behavioral Ecology and Sociobiology* 36: 105-110.

Bshary, R. and A.S. Grutter. 2002. Asymmetric cheating opportunities and partner control in a cleaner fish mutualism. *Animal Behaviour* 63: 547-555.

Chivers, D.P. and R.J.F. Smith. 1994a. The role of experience and chemical alarm signaling in predator recognition by fathead minnows, *Pimephales promelas*. *Journal of Fish Biology* 44: 273-285.

Chivers, D.P. and R.J.F. Smith. 1994b. Fathead minnows, *Pimephales promelas*, acquire predator recognition when alarm substance is associated with the sight of unfamiliar fish. *Animal Behaviour* 48: 597-605.

Chivers, D.P. and R.J.F. Smith. 1995. Fathead minnows (*Pimephales promelas*) learn to recognize chemical stimuli from high risk habitats by the presence of alarm substance. *Behavioural Ecology* 6: 155-158.

Chivers, D.P. and R.J.F. Smith. 1998. Chemical alarm signaling in aquatic predator-prey systems: a review and prospectus. *Écoscience* 5: 338-352.

Chivers, D.P. and R.S. Mirza. 2001. Predator diet cues and the assessment of predation risk by aquatic vertebrates: a review and prospectus. In: *Chemical Signals in Vertebrates*, A. Marchlewska-Koj, J.J. Lepri and D. Muller-Schwarze (eds.). Plenum Press, New York, pp. 277-284.

Chubb, J.C. 1979. Seasonal occurrence of helminths in freshwater fishes. Part II. Trematoda. *Advances in Parasitology* 17: 142-313.

Clark, C.C., L. Clark and L. Clark. 1990. Anting behavior by common grackles and European starlings. *Wilson Bulletin* 102: 167-169.

Combes, C. 2001. *Parasitism the Ecology and Evolution of Intimate Interactions.* The University of Chicago Press, Chicago.

Coté, I.M. and R. Poulin. 1995. Parasitism and group size in social animals: a meta-analysis. *Behavioural Ecology* 6: 159-165.

Covert, J.B. and W.W. Reynolds. 1977. Survival value of fever in fish. *Nature (London)* 267: 43-45.

Decaestecker, E., L.D. Meester and D. Ebert. 2002. In deep trouble: habitat selection constrained by multiple enemies in zooplankton. *Proceedings of the National Academy of Sciences of the United States of America* 99: 5481-5485.

Dronen, N.O. 1973. Studies on the macrocercous cercariae of the Douglas Lake, Michigan area. *Transactions of American Microscopical Society* 92: 641-648.

Dugatkin, L.A., G.J. Fitzgerald and J. Lavoie. 1994. Juvenile three-spined sticklebacks avoid parasitized conspecifics. *Environmental Biology of Fishes* 39: 215-218.

Duncan, P. and N. Vigne. 1978. The effect of group size in horses on the rate of attacks by blood-sucking flies. *Animal Behaviour* 27: 623-625.

Forbes, M.R.L. and R.L. Baker. 1990. Susceptibility to parasitism: experiments with the damselfly *Enallagma ebrium* (Odonata: Coenegrionidae) and larval water mites, *Arrenurus* spp. (Acari: Arrenuridae). *Oikos* 58: 61-66.

Goater, C.P. and J.C. Holmes. 1997. Parasite-mediated natural selection. In: *Host-Parasite Evolution: General Principles and Avian Models*, D. Clayton and J. Moore (eds.). Oxford University Press, Oxford, pp. 9-29.

Grutter, A.S. 1999. Cleaner fish really do clean. *Nature (London)* 398: 672-673.

Haas, W. 1994. Physiological analyses of host-finding behaviour in trematode cercariae: adaptations for transmission success. *Parasitology* 109: S15-S29.

Hart, B.L. 1994. Behavioural defense against parasites: interaction with parasite invasiveness. *Parasitology* 109: S139-S151.

Hart, B.L. 1997. Behavioural defence. In: *Host-Parasite Evolution, General Principles and Avian Models*, D.H. Clayton and J. Moore (eds.). Oxford University Press, Oxford, pp. 59-77.

Hausfater, G. and Meade B.J. 1982. Alternation of sleeping groves by yellow baboons (*Papio cynochephalus*) as a strategy for parasite avoidance. *Primates* 23: 287-297.

Huffman, M. A., J.E. Page, M.V.K. Sukhdeo, S. Gotoh, M.S. Kalunde, T. Chandrasiri and G.H.N. Towers. 1996. Leaf swallowing by chimpanzees: A behavioral adaptation for the control of strongyle nematode infections. *International Journal of Primatology* 17: 475-503.

Karvonen, A., O. Seppälä and E.T. Valtonen. 2004. Parasite resistance and avoidance behaviour in preventing eye fluke infections in fish. *Parasitology* 129: 159-164.

Kiesecker, J.M. and D.K. Skelly. 2000. Choice of oviposition site by gray treefrogs: The role of potential parasitic infection. *Ecology* 81: 2939-2943.

Kiesecker, J.M., D.K. Skelly, K.H. Beard and E. Preisser. 1999. Behavioral reduction of infection risk. *Proceedings of the National Academy of Sciences of the United States of America* 96: 9165-9168.

Krause, J., G.D. Ruxton and J.-G. J. Godin. 1999. Distribution of *Crassiphiala bulboglossa*, a parasitic worm, in shoaling fish. *Journal of Animal Ecology* 68: 27-33.

Lafferty, K.D. 1992. Foraging on prey that are modified by parasites. *American Naturalist* 140: 854-867.

Lima, S.L. and L.M. Dill. 1990. Behavioral decisions made under the risk of predation: a review and prospectus. *Canadian Journal of Zoology* 68: 619-640.

Losey, G.C., A.S. Grutter, G. Rosenqvist, J.L. Mahon and J.P. Zamzow. 1999. Cleaning symbiosis: a review. In: *Behaviour and Conservation of Littoral Fishes*, V.C. Almada, R.F. Oliveira and E.J. Gonvales (eds.). Instituto Superior de Psichologia Aplicada, Lisbon, pp. 379-395.

Lowenberger, C.A. and M.E. Rau. 1994. Selective oviposition by *Aedes aegypti* (Diptera: Culicidae) in response to a larval parasite, *Plagiorchis elegans* (Trematoda: Plagiorchiidae). *Environmental Entomology* 23: 1269-1276.

Lozano, G.A. 1991. Optimal foraging theory: A possible role for parasites. *Oikos* 60: 391-395.

Mathis, A., D.P. Chivers and R.J.F. Smith. 1996. Cultural transmission of predator recognition in fishes: Intraspecific and interspecific learning. *Animal Behaviour* 51: 185-201.

Milinski, M. and T.C.M. Bakker. 1990. Female sticklebacks use male coloration in mate choice and hence avoid parasitized males. *Nature (London)* 344: 330-333.

Moore, J. 2002. *Parasites and the Behavior of Animals*. Oxford University Press, New York.

Poulin, R. and G.J. FitzGerald. 1989a. Risk of parasitism and microhabitat selection in juvenile sticklebacks. *Canadian Journal of Zoology* 67: 14-18.

Poulin, R. and G.J. FitzGerald. 1989b. Shoaling as an anti-ectoparasite mechanism in juvenile sticklebacks (*Gasterosteus* spp.). *Behavioural Ecology and Sociobiology* 24: 251-255.

Poulin, R., M.E. Rau and M.A. Curtis. 1991. Infection of brook trout fry, *Salvelinus fontinalis*, by ectoparasitic copepods: The role of host behaviour and initial parasite load. *Animal Behaviour* 41: 467-476.

Poulin, R., D.J. Marcogliese and J.D. McLaughlin. 1999. Skin-penetrating parasites and the release of alarm substances in juvenile rainbow trout. *Journal of Fish Biology* 55: 47-53.

Reynolds, W.W., M.E. Casterlin and J.B. Covert. 1976. Behavioural fever in teleost fishes. *Nature (London)* 259: 41.

Rhode, K. 1993. *Ecology of Marine Parasites*. CAB International, Wallingford.

Sandland, G.J., C.P. Goater and A.J. Danylchuk. 2001. Population dynamics of *Ornithodiplostomum ptychocheilus* metacercariae in fathead minnows (*Pimephales promelas*) from four northern-Alberta lakes. *Journal of Parasitology* 87: 744-748.

Sheldon, B.C. and S. Verhulst. 1996. Ecological immunology: Costly parasite defenses and trade-offs in evolutionary ecology. *Trends in Ecology and Evolution* 11: 317-321.

Taylor, C.N., K.L. Oseen and R.J. Wassersug. 2004. On the behavioural response of *Rana* and *Bufo* tadpoles to echinostomatoid cercariae: Implications to synergistic factors influencing trematode infection in anurans. *Canadian Journal of Zoology* 82: 701-706.

Thiemann, G.W. and R.J. Wassersug. 2000. Patterns and consequences of behavioural responses to predators and parasites in *Rana* tadpoles. *Biological Journal of Linnean Society* 71: 513-528.

Ward, A.J.W., A.J. Duff, J. Krause and I. Barber. 2005. Shoaling behaviour of sticklebacks infected with the microsporidian parasite, *Glugea anomala*. *Environmental Biology of Fishes* 72: 155-160.

Wedekind, C. 1992. Detailed information about parasites revealed by sexual ornamentation. *Proceedings of the Royal Society of London* 247: 169-174.

Wedekind, C. and M. Milinski. 1996. Do three-spined sticklebacks avoid consuming copepods, the first intermediate host of *Schistocephalus solidus*? — An experimental analysis of behavioural resistance. *Parasitology* 112: 371-383.

Wisenden, B.D. 2003. Chemically-mediated strategies to counter predation. In: *Sensory Processing in the Aquatic Environment*, S.P. Collin and N.J. Marshall (eds.). Springer-Verlag, New York, pp. 236-251.

Wisenden, B.D. and N.E. Stacey. 2005. Fish semiochemicals and the network concept. In: *Animal Communication Networks*, P. K. McGregor (ed.). Cambridge University Press, Cambridge, pp. 540-567.

Wisenden, B.D. and D.P. Chivers. 2006. The role of public chemical information in antipredator behaviour. In: *Communication in Fishes*, F. Ladich, S.P. Collins, P. Moller and B.G. Kapoor (eds.). Science Publisher, Enfield, N.H., USA, pp. 259-278.

Wisenden, B.D., J. Pogatshnick, D. Gibson, L. Bonacci, A. Schumacher and A. Willet. 2007. Sound the alarm: learned association of predation risk with novel auditory stimuli by fathead minnows (*Pimephales promelas*) and glowlight tetras (*Hemigrammus erythrozonus*) after single simultaneous pairings with conspecific chemical alarm cues. *Environmental Biology of Fishes* 81: 141-147.

Wrangham, R.W. and M.F. Walters-Wyman. 1985. Relationship of chimpanzee leaf swallowing to a tapeworm infection. *American Journal of Primatology* 37: 297-303.

Wyman, R.L. and M.F. Walters-Wyman. 1985. Chafing in fishes: Occurrence, ontogeny, function and evolution. *Environmental Biology of Fishes* 12: 281-289.

Zuk, M. and A.M. Stoehr. 2002. Immune defense and host life history. *The American Naturalist* 160: S9-S22.

Pharmacology of Surfactants in Skin Secretions of Marine Fish

Eliahu Kalmanzon[#] and Eliahu Zlotkin

INTRODUCTION

Our studies deal with the chemistry and action of natural offensive and defensive substances (allomonal systems) from the chemo-ecological and pharmacological points of view. The marine environment, especially the biologically diverse and densely populated tropical reefs (such as those found in the Bay of Eilat at the Red Sea and the coastal regions of the Sinai peninsula), is extremely attractive from a zoo-ecological point of view due to the abundance of chemical interactions and their chemical and functional diversity. The present chapter deals with one aspect of the chemical ecology of the marine environment, namely the defensive role of polypeptides and surfactants in fish skin secretions.

Authors' addresses: Department of Cell and Animal Biology, Institute of Life Sciences, Hebrew University of Jerusalem, Jerusalem 91904, Israel.

Present address: Nitzana – Educational Community, Doar Na, Halutza, 84901, Nitzana, Israel, Jerusalem 91904, Israel.

Corresponding author: E-mail: elizon@gmail.com

#I wish to dedicate the final version of this paper to the memory of Prof. E.Z. who through his boundless curiosity and enthusiasm, his determination and perseverence, introduced me to the exciting field of chemical ecology of marine fishes.

In the terrestrial environment, the compounds used for defense against predators are often small, volatile organic molecules. Polypeptides and proteins are usually not used in topical or medium applications since they cannot be efficiently distributed around the defending organism (are not volatile) and cannot penetrate into the attacking predator. In the marine environment, however, due to their solubility in the high ionic strength of seawater, polypeptides are readily delivered through a medium application and may fulfill more diverse biological functions, such as those carried out by volatile organic compounds in the terrestrial environment. Thus, they are often secreted onto the skin of fish and other marine organisms and can participate in the chemical defense of the producer.

This chapter deals with the phenomenon of defensive-toxic fish skin secretions from ecological and pharmacological points of view. The first section will provide an overview of the current knowledge on the production, delivery and action of toxic fish skin secretions. The next two sections, comprising the main part of this chapter, will focus on two examples of such secretions studied in our laboratory.

EPIDERMAL FISH SECRETIONS — AN OVERVIEW

Secretory Cells in the Fish Integument

Secretory epidermal cells of vertebrates are currently divided in two major categories (Quay 1972; Cameron and Endean 1973): the mucus secreting cells and those that produce proteinaceous material. In fish, the proteinaceous cells (composed of several morphological types) are readily distinguished from mucus producing cells by histological-chemical dyes or tests (Birkhead 1967; Halstead 1970; Randall et al., 1971). The mucus cells are, in essence, goblet cells—which open to the exterior—in contrast to the proteinaceous cells, which store their secretory products after these have been produced. However, in spite of the fact that the turbulence-reducing slime (Rosen and Cornford, 1971) is a major product of the mucus cells, it undoubtedly fulfills certain other defensive functions. The latter is expressed by the inclusion of antibodies in the mucous (Fletcher and Grant, 1969) and the mucous provides protection against pathogenic epibionts and parasites (Nigrelli et al., 1955) as well as microorganisms (Hildemann, 1962; Kitzan and Sweeney, 1968). The above effects are probably not due to the mucus itself (Cameron and Endean, 1973), although the antibacterial substance (see below) derived from the fish epidermal secretion is primarily attributed to mucus cells. A recent comprehensive survey on fish epidermis (Zaccone *et al.*, 2001) reveals that the mucus produced by mucous goblet cells is a carrier of various biologically active substances which are stored in the sacciform and club cells of fish skin and the glandular cells in the skin of amphibians (Zaccone *et al.*, 2001).

The proteinaceous cells can be roughly subdivided into two major categories. The first category, associated with a venom apparatus, are clustered in groups and form epidermal glands, located in the vicinity of some puncturing apparatus. In the elaborated teleost venom glands, such as those of the stonefish *Synanceja* and *Scorpaena*, the pungent fin spines provide such an apparatus. A second category of cells are those whose proteinaceous layer is not associated with a venom apparatus and whose content is secreted onto the surface of the fish. The secretion of these cells, described in greater detail below, has been termed ichthyocrinotoxic by Halstead (1970). Production of toxins by cells in the integument of the fish seems to be in correlation to a reduction of mobility, and its adoption of a stationary sluggish mode of life (Cameron and Endean, 1973).

Against this background, the fish skin secretions in this chapter are subdivided into three categories, namely, antibacterial substances, venom glands and ichthyocrinotoxins.

Antibacterial Substances

Antibacterial peptides are part of the host defense systems of plants, insects, fish, amphibians, birds and mammals (Gallo and Huttner, 1998). They have been isolated from the mucus of common and edible fish. Typical examples include the eel (*Anguilla*) and the rainbow trout, which produce antimicrobial polypeptides (Ebran *et al.*, 1999) and glycosilated pore-forming proteins. The pore-forming activity was well correlated with a strong antibacterial activity at the micromolar range (Ebran *et al.*, 2000). Additional examples of this phenomenon are provided by a more recent study which revealed that skin extracts of rainbow trout contain a potent 13.6 kDa antimicrobial protein, active against gram-positive bacteria at the submicromolar range (Fernandes *et al.*, 2002). A similar and complementary example is the antibacterial proteins with a molecular mass of 31 and 27 kDa from the skin mucus of carp (Lemaitre *et al.*, 1996). The skin epithelial mucus cells produce one such antimicrobial peptide, the 25 residue linear antimicrobial peptide, from the skin mucus of the winter flounder.

Epidermal Venom Glands

For specific information on fish venom glands, the reader is addressed to the article of Maretic (1988) on fish venoms. Certain groups of slow-moving, usually sessile fish (see below), possess a stinging apparatus composed of toxic spines. The spines contain grooves, containing clusters of venom glands cells. A membranal sheath covers the entire arrangement. Mechanical pressure on this apparatus—such as when a diver steps on these fish—results in penetration of the spine, followed by breaking of the sheath and the associated glandular venom cells and envenomation. Similar to a typical venom apparatus, the spine punctures the integument and delivers the venom into the body cavity and its circulation. However, unlike a typical venom apparatus (such as those of reptiles and arachnids), the toxic spine is not a site- and time-directed injection device; nor is it operated by a contractile venom reservoir and its collecting duct. Furthermore, as shown in the stone fish *Synanceja horrida* (Gopalakrishnakone and Gwee, 1993), the venom secretory cells do not reveal the features of a classic protein-secreting cell, such as a golgi apparatus or rough endoplasmic reticulum. Instead, the entire cell completely transforms into granules, suggesting a holocrine type of secretion. The above fundamental arrangement of the spine-mediated venom delivery is common to various groups of benthic—sessile fish such as Stingrays (Chondrichthyes, Batsidea) (Halstead, 1970), Weeverfish (Skeie, 1962), Stargazers (Uranoscopidae) and the most dangerous scorpion fish (Scorpaenidae).

The Scorpaenidae, represented by the dangerous genera of *Scorpaena* (Dragon head), *Pterois* (zebrafish) and *Synanceja* (stone fish), include *Synanceja horrida*, the most venomous fish known (Maretić, 1988). The lethal factor from the stonefish (*S. horrida*) venom was identified by Ghadessy *et al.* (1996) as a multifunctional lethal protein termed stonustoxin (Stn). Stn comprises two subunits termed (alfa and beta) with respective molecular masses of 71 and 73 kDa, which reveal 50% homology in their primary structures. Stn elicits an array of biological responses, particularly a potent hypotension that appears to be mediated by the nitric oxide pathway (Low *et al.*, 1994).

Neurotoxins

It is a well-known phenomenon that fish of the Tetraodontidae family are highly toxic upon ingestion due to the presence of the alkaloid neurotoxin

Tetrodotoxin (TTX) in the various tissues of the fish (Prince, 1988; Soong and Venkatesh, 2006). TTX blocks nerve transmission by binding to and blocking voltage-gated Na+ channels. While the most toxic tissues are the liver and ovaries, the skin of these fish is also highly toxic (Matsui *et al.*, 1981). In the skin, TTX is found in special secretory glands (Tanu *et al.*, 2002) and is secreted into the surrounding seawater when the fish is agitated (Kodama *et al.*, 1986). Very low concentrations of TTX in seawater (10^{-7} M) cause electrical responses from the palatine nerve innervating the palate in rainbow trout and arctic char, probably through interaction with a specific receptor on the palate. These results explain why predatory fish avoid food containing TTX (Yamamori *et al.*, 1987), and reveal how a neurotoxin—which by definition must interact with a neural target—can serve as a defensive allomone in the marine environment.

One interesting aspect of the defensive role of TTX in fish is that the toxin is not produced by the fish, but rather by symbiotic bacteria of the genera *Vibrio*, *Alteromonas* and *Pseudoalteromonas* (Prince, 1988). This explains how many different animals (fish, salamanders, sea stars, flatworms and octopi, to name just a few) can contain such a complex alkaloid, which necessitates a complex and specific biosynthetic pathway. Another interesting question is how do the fish protect themselves from the toxic effect of TTX? This is a general problem faced by chemically defended organisms, and will be treated below in more detail with regard to the ichthyocrinotoxin pardaxin. In the case of TTX, the sodium channels of the fish are resistant to the toxin, as a result of specific mutations (Venkatesh *et al.*, 2005).

Finally, TTX does not only serve as a defensive allomone in Tetraodontidae. Matsumura (1995) has shown that the TTX found in high concentrations in the egg mass serves as a pheromone, attracting the male to fertilize the eggs.

Ichthyocrinotoxins

It was always known among fishermen that certain fish are able to cause lethality of other fish when kept together. As far back as the nineteenth century, this phenomenon was attributed to the toxic effect of fish skin secretions, defined as Ichthyocrinotoxins (Halstead, 1967).

Ichthyocrinotoxins comprise a second category of epidermal fish secretions that are devoid of spines or teeth to deliver the venom into the target organism. The secretion is supposedly (Halstead, 1970) released to the surrounding water for defensive purposes. The fish crinotoxic secretions occupy a position similar to that of amphibians (such as salamandrae and toads), which possess in their skin venomous glands without any delivery device, intoxicating predators by contact (Maretić, 1988).

Ichthyocrinotoxins have been recorded in some 50 teleost species belonging to at least 14 families (Halstead, 1967, 1978). Two functions have been proposed: firstly, that Ichthyocrinotoxins provide protection against infection and fouling organisms (Cameron and Endean, 1973; Cameron, 1974). This possibility was supported by the reduced squamation and sedentary habits of many crinotoxic fish (Cameron and Endean, 1973). Secondly, that Ichthyocrinotoxins afford protection from predation (Winn and Bardach, 1959; Randall, 1967; Randall et al., 1971). The studies mentioned below were directed and motivated by the anti-predatory aspect of fish skin secretions or, more precisely, were aimed to characterize the pharmacological and ecological significance of their amphipathic (detergent-like) constituents.

DETERGENTS IN THE MARINE ENVIRONMENT

A large number of species of fish have been reported to be ichthyocrinotoxic (Halstead, 1967, 1978). However, the effect and the chemistry of skin secretions from only a few species belonging to four families (Ostraciidae, trunkfish or box fish; Serranidae, soap fish; Batrachoididae, toad fish and Soleidae, flat fish) have been studied in detail (Nair, 1989). These skin secretions share the following common features (Nair, 1989; Hashimoto, 1979):

1. Their toxic secretions are collected essentially in the same manner — by immersing the fish in distilled water, accompanied by various degrees of mechanical agitation.

2. The resulting viscous opaque secretions are foamy — indicating the inclusion of substances possessing surfactant activity.

3. The above secretions are lethal to fish (ichthyotoxic) when placed in their surrounding water, and are also cytolytic.

4. The active isolated components, in spite of their versatile chemistry, could all be defined as amphipathic (detergent-like) substances.

Surfactants in Fish Secretions

According to their composition and structure, the ichthyocrinotoxic substances described above can be subdivided into two categories: simple detergents (Fig. 6.1) and polypeptides, themselves comprised of two further groups (Fig. 6.2). The first group of polypeptides is represented by grammistins, associated with a tertiary or quaternary amine, with molecular masses of around 3-4 KD and which are responsible for the ichthyotoxicity and cytotoxicity of the skin secretion derived from the soap fish *Grammistes* and *Pogonoperca* (Hashimoto and Oshima, 1972). The second and more investigated group of polypeptides is the Pardaxins, derived from the flat fish *Pardachirus marmoratus* (Fig. 6.2) and *P. pavonicus*. These are amphipathic polypeptides of molecular weight around 3.5 KD that possess a hydrophobic N-terminal region with a short polar and acidic C-terminus (Thompson *et al.*, 1986, 1988). Pardaxins were claimed to exert their cytolytic effects either as solubilizers: detergents of cell membranes at high concentrations (10^{-7}-10^{-4} M) or as cationic pore formers at low concentrations (Lazarovici, *et al.*, 1986; Shai *et al.*, 1991).

Pahutoxin, a typical cationic detergent, can represent the non-peptide detergents of skin secretions. In Hawaiian box fish *Ostracion lentiginosus* (Boylan and Scheuer, 1967) (Fig. 6.1) and the Japanese box fish *Ostracion immaculatus* (Fusetani and Hashimoto, 1987), Pahutoxin is a choline chloride ester of 3-acetoxypalmitic acid, while in the Caribbean trunk fish (box fish) *Lactophrys triqueter* (Goldberg *et al.*, 1982) it is the choline chloride ester of palmitic acid. The skin secretion of the flat fish *Pardachirus pavonicus* was shown to possess 'Pavoninins'— steroid-N-acetylglucosaminides (Tachibana *et al.*, 1984; Tachibana and Gruber, 1988), claimed to be responsible for 40% of the ichthyotoxicity of the crude secretion. In parallel, the secretion of *Pardachirus marmoratus* was also shown to contain steroid monoglycosides (Mosesins, Fig. 6.1), where the sugar was galactose or its monoacetate, in contrast to N-acetylglucosamines of pavoninins (Tachibana *et al.*, 1984). It is noteworthy that the occurrence of amphipathic defensive substances is not limited only to fish but exists also in invertebrate marine organisms. The saponinic-holothurins of sea cucumbers may serve as a typical example (Fig. 6.1).

When dealing with the biological significance of the above skin secretions and their derived amphipathic toxic substances, there are two basic arguments suggesting their defensive (allomonal) function. The first

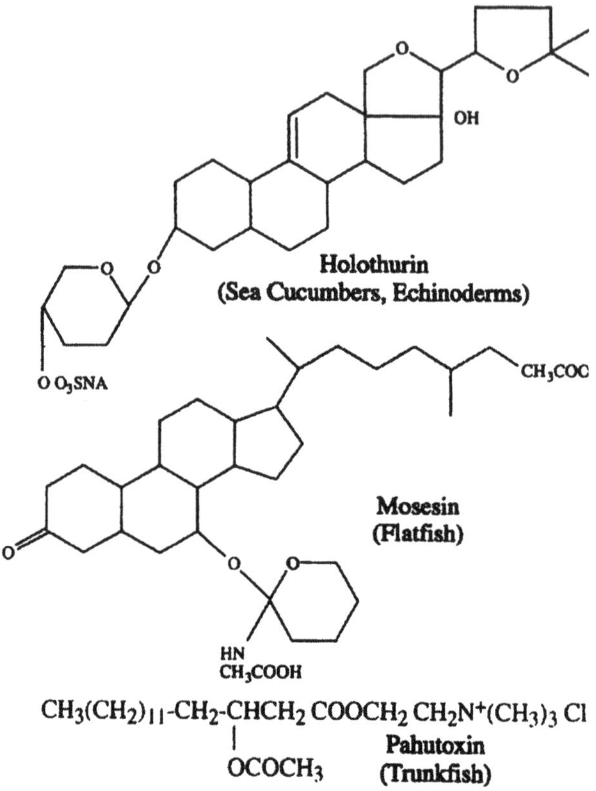

Holothurin
(Sea Cucumbers, Echinoderms)

Mosesin
(Flatfish)

$CH_3(CH_2)_{11}$-CH_2-$\underset{\underset{OCOCH_3}{|}}{CH}CH_2$ $COOCH_2$ $CH_2N^+(CH_3)_3$ Cl
Pahutoxin
(Trunkfish)

Fig. 6.1 Typical examples of natural surfactants from marine organisms. The two steroidic glycosides are derived from the secretions of a flatfish (Mosesin) and echinoderms (Holothurin). The polar hydrocarbon (Pahutoxin) is collected from reef trunkfish. Pahutoxin (choline chloride ester of 3-acetoxypalmitic acid) is a typical cationic detergent with a quaternary ammonium head group and a branched chain hydrocarbon as a hydrophobic portion.

Gly-Phe-Phe-Ala-Leu-Ile-Pro-Lys-Ile-Ile-Ser-Ser-Pro-Leu-Phe-Lys-Thr-Leu-Leu-
Ser-Ala-Val-Gly-Ser-Ala-Leu-Ser-Ser-Ser-Gly-Asp-Gln-Glu

Fig. 6.2 Pardaxin the amphipathic polypeptide derived from the defensive skin secretion of the Red Sea Moses sole *Pardachirus marmoratus*.

argument is based on simple eco-zoological considerations, claiming that the various ichthyocrinotoxic fish are generally slow and sluggish swimmers, subjected in their natural habitats to constant threat from predators. The secretions are thought to be a part of their defense mechanisms to repel the predators (Nair, 1989). This notion is supported by an additional assumption that a marine organism such as a predatory fish (shark, for example) may be vulnerable to an effect of an externally released surfactant since it exposes to the surrounding—an extremely large surface area of unprotected and accessible gill membranes. The second argument is supplied by a series of observations and field assays performed by Clark (1974, 1983), suggesting that the flat fish *Pardachirus marmoratus* is not eaten by sharks (unpalatable), because of its skin secretion.

Within 10 to 15 years following the above finding (Clark, 1974), it has been shown that the major constituents of the flat fish *Pardachirus* skin secretion are shark repellents. The latter included the polypeptide Pardaxin (Fig. 6.2) that was shown to act when delivered only via the external bathing medium and is specifically targeting the gills and/or the pharyngeal cavity (Primor, 1983). Furthermore, Tachibana and Gruber (1988) have revealed that shark repellency of the *Pardachirus* secretion is due not only to the polypeptide Pardaxin but also to a lipophilic ichthyotoxins which are steroidic glycosides (see Fig. 6.2). However, the most significant support to the notion of the defensive role of the marine amphipathic substances was supplied by our previous findings (Gruber *et al.*, 1984; Zlotkin and Gruber, 1984) that certain commercial detergents can be employed as shark repellents.

Synthetic Detergents as an Approach to Shark Repellents

Shark attacks may be deterred either by physical devices (Wallett, 1972) or by altering the stimulus qualities of a human so as to render the swimmer aversive (Johnson, 1963; Zahuranec, 1975). The latter forms the logical basis for the development of chemical shark repellents (Gruber, 1983). Our approach to shark repellency is a follow up from considerations of the chemistry and actions of the skin secretion of the Red Sea flatfish *Pardachirus marmoratus* (PMC) and its derived toxin Pardaxin (PXN) (Zlotkin and Barenholz, 1983). We suggested that the extremely diverse and versatile pharmacology of PMC and PXN expressed in lethal (Primor and Zlotkin, 1975), neurotoxic (Parnass and Zlotkin, 1976; Spira *et al.*, 1976), cytolytic (Helenius and Simons, 1975), histopathologic, enzyme blocking (Spira *et al.*, 1976; Primor *et al.*, 1980), permeability modifying (Hashimoto, 1979) and shark-repellent properties (Clark, 1974, 1983), could be attributed to amphipathic-surfactant activity. PXN was shown to cause foaming and drop volume reduction in aqueous solutions, to affect the integrity of artificial liposomes, to deform and increase permeability of the enveloped vesicular stomatitis virions and possess a strongly hydrophobic amino-terminal sequence (Zlotkin and Bernholz, 1983) followed by a polar and negatively charged carboxy-terminal segment (Fig. 6.2, Thompson *et al.*, 1988). The suggestion that PXN is involved in hydrophobic interactions with membranal phospholipids, thus disrupting membrane integrity and function (Zlotkin and Barenholz, 1983), led us to hypothesize that synthetic surfactants may repel sharks.

We chose the lemon shark, *Negaprion brevirostris*, as an experimental subject because it is a dangerous species known to attack humans (Gruber, 1983), is common and easily captured, and can be rapidly and reliably trained and subjected to experimental manipulations (Gruber and Myrberg, 1977; Gruber, 1980). The biological activity of PMC and various commercial detergents was examined through their ability to kill fish (killifish—*Floridichtyes carpio*), irritate immobilized sharks (the tonic immobility assay) and prevent the feeding of an aggressive hungry shark (the shark feeding test).

Shark Feeding Test: A group of 15 lemon sharks, placed in a separate pool, were deprived of food for 48 hours. Prior to assaying, the sharks were stimulated by dipping a defrosted fish into the water. This activated the sharks and caused them to swim close to the water surface at the site of experimentation. The sharks were then offered a whole baitfish with a syringe attached to it (Fig. 6.3A). They readily attacked the bait and could be induced to take the fish's head into their mouth (Fig. 6.3B). Thus, substances could be introduced into the shark's mouth as it attempted to feed. With effective repellents, the sharks immediately broke off the attack, quickly turned and rapidly left the feeding site, leaving the intact bait behind. With higher concentrations (>5 mg ml^{-1}), the sharks often sank to the bottom, strongly contracting and expanding their buccal cavity. In Table 6.1, the data for the shark feeding studies are presented in the form of a range of threshold concentration as determined on 10-20 sharks for each substance. These data indicate that above the higher concentration, all test animals are repelled and below the lower concentration, there is no repellency (Table 6.1).

The Tonic Immobility Assay: Tonic immobility is a quiescent state of inactivity induced by restraining an animal in an inverted position (Carli, 1977) (Fig. 6.3). It is also known as catalepsy or death feigning. Lemon

Fig. 6.3 Assays of shark repellence. A. Feeding bioassay: A 20 cm long blue runner (*Caranx crysos*) is prepared as bait by attaching a 25 ml syringe to the fish. The plastic tube extends out of the bait's mouth. B. Feeding bioassay: An 80 cm long lemon shark, *Negaprion brevirostris*, attacks the bait and grasps the head in its mouth. Simultaneously the experimenter releases 15 ml of the substance into the shark's mouth. C. Tonic immobility bioassay: An 85 cm lemon shark is inverted under tonic immobility. A shark will remain essentially immobile for at least 10 min except for respiratory movements of the mouth and gills. Experimenter releases a test substance into the immobilized shark's mouth. D. Tonic immobility bioassay; a shark terminated tonic immobility after a test substance has been released into its mouth. (Taken from Zlotkin and Gruber, 1984).

Table 6.1 Shark repellency and fish lethality of different surfactants and *Pardachirus* secretion (PMC)[a].

Substance	Formula Commercial names and sources[a]	Killifish lethality LD_{50} ($\mu g*ml^{-1}$)	Shark feeding assay range ($mg*ml^{-1}$)	Shark tonic immobility assay ED_{50} ($mg*ml^{-1}$)
1	Lauryl sulfate sodium salt (SDS, Sigma, USA)	3.0	0.2-2.0	0.45
2	Lauryl sulfate lithium salt (LDS, Sigma, USA)	6.0	0.2-2.0	0.62
3	Lyophilized *Pardachirus*[b] secretion (PMC, Laboratory prepared)	16.0	0.8-3.0	0.66
4	Polyethoxylated octylphenol (Triton-X-100, Packard, USA)	36.0	3.0-8.0	10.0
5	Lauryl trimethyl ammonium bromide (Sigma, USA)	60.0	3.0-8.0	8.0
6	Cholic acid-sodium salt (Sigma, USA)	100	8.0-10.0	8.1
7	Ethoxylated (20) sorbitan monolaurate (Tween 20, Casali Inst. Hebrew Univ.)	100	10.0-20.0	10

[a]The substances are listed in order of their fish lethality.
[b]Dissolved immediately prior to the experiment.

sharks do not habituate ('desensitize') to tonic immobility and are naturally resistant to its termination, thus enabling experimental manipulation and even minor surgery (Gruber and Watsky, unpublished). In our experiments, 3 ml of gradually increasing concentrations of a test substance were injected into the buccal cavity of a tonically immobilized shark (Fig. 6.3C) and the concentration resulting in the righting of the shark (Fig. 6.3D) was recorded. With higher concentrations of active substances (>3 $\mu g m l^{-1}$), termination of the tonic immobility occurred in less than 1 s and was often preceded by a violent jump accompanied by rapid gill contraction. In Table 6.1, the data on tonic immobility are presented in the form of the 50% effective dose (ED_{50}) as sampled and estimated according to Reed and Muench (1938). Each of the different concentrations of the test solutions was assayed on three sharks, each in three repetitions. The seven most potent of the 15 substances assayed are listed in Table 6.1, in order of toxicity to killifish. Excluding the toxic secretion PMC, the six others represent the principal types of surfactants. Substances 1 and 2 are anionic; substance 5 is cationic; substances 4 and 7 are nonionic and substance 6 represents a natural surfactant (bile acid). Three additional nonionic industrial detergents (Brij 35,10.G.1.o and Myrj 59 — obtained from the Casali Institute, Hebrew university) as well as saponin (a mixture of steroidic glycosides — Sigma Co) had weak activity. Killifish LD_{50} of these substances exceeded 100 $\mu g m l^{-1}$ and slight behavioral effects of shark feeding and tonic immobility occurred in the range of 50-100 $\mu g m l^{-1}$.

As a result of the relatively strong activity of substances 1 and 2 (Table 6.1), several additional derivatives of lauric acid were assayed, namely lauryl bromide, sodium laurate, ethyl ester of lauric acid, lauryl alcohol, and lauryl amine (Sigma Co). All of them proved to be ineffective. In contrast to the other test substances, the lauric acid derivatives are practically insoluble in seawater. We attribute their inactivity in all three assays to this fact. Against this background, the weak activity of the completely soluble, cationic, quaternary ammonium derivative of lauric acid (substances 1 and 2 — Table 6.1) is noteworthy, because it provides clues as to the mode of action of these detergents in repelling sharks, and suggests further experiments.

With regard to the shark repellency assays employed in the present study, it is noteworthy that the feeding tests are self-evident and can be easily interpreted. In this case, we are not dealing with the simple inhibition of feeding, but with active repulsion of the shark from a highly motivated and aggressive behavior which is induced by hunger and amplified by a preliminary sensory stimulation (the dipping of the dead fish). The feeding attack, a directional and highly oriented behavior of the shark, can be completely interrupted by the aversive effect of the repellent substance. The feeding assay, however, is limited because complete control over the position of the animal, the exact timing and direction of release of the repellent substance is not possible. Since feeding trials also depend on motivational states, the number of trials that could be run on a single day is limited. The tonic immobility assay corrected some of the drawbacks of the feeding assay, and may serve as a quantifiable behavioral system since it is stereotyped, resistant to habituation (at least in sharks) and has a definite starting point (the immobile position of the shark) and end point (its spontaneous recovery). The close relation between data concerning the feeding and the tonic immobility assays (Table 6.1) suggests that the latter may be employed for the rapid screening of potential shark repellents.

The shark repellent capacity of SDS was also revealed in open sea assays with blue sharks, *Prionace glauca*. The blue shark is one of the most common and the most wide ranging of all sharks. It is found in the epilagic zone of all tropical to cool–temperate seas (Gruber *et al.*, 1984). It is the most commonly encountered shark in the surface waters of Los Angeles, California (where the tests were carried out). So, in this sense, the blue shark chose us. Under ordinary conditions, blue sharks can be attracted to a boat with ground fish (chum). Once at the boat, they usually swim slowly, remain in the vicinity and will take a bait. Thus, they are excellent subjects for field tests. Nevertheless, blue sharks are dangerous and have bitten human beings. In the field studies, the following test methods were used (Gruber *et al.*, 1984):

1) Delivery to the oral cavity — either from a reservoir via a flexible plastic tube which terminates inside a bait fish or a measured amount of chemical in a latex 'balloon' packet tied to a bait fish.

2) Squirt application to the circulating shark using extended bulb-type syringe.

3) Delivery into an odor corridor formed by the delivery of chum into the seawater. A measured quantity of the test solution was added to the outgoing attractant, observing the time and site (down the odor corridor) of the sharks turning away.

In the above assays, SDS in concentrations of 15, 3 and 1% in seawater revealed obvious shark repellence. For example, about 100 ml of

15% SDS solution caused an immediate rejection of the bait, nictating membrane closure and rapid withdrawal with the mouth held wide open.

Our prediction that surfactants possess shark repellent properties was in principle verified. On the basis of weight, the lauryl sulfate salts were found to be superior in repellence to the *Pardachirus* secretion (Table 6.1), potent detergents and foaming agents, and could be distinguished from all the other compounds tested by being extremely hydrophilic, anionic and also by possessing sulfate as a functional group. These characteristics are further clues in experimentation for effective shark repellents.

To summarise, from the point of view of the chemical ecology of the defensive fish skin secretions, sharks provide the perfect classical model of a natural enemy against which such secretions are used. SDS is available, widely used and chemically as also structurally known as a synthetic detergent, which provides a model for the study of the mechanism of action of the amphipathic–surfactant constituents of the ichthyocrinotoxic secretions.

Concerning the shark repellent action of SDS, we have studied the possible relationship between the shark-repelling capacity of SDS and its physicochemical mode of interaction with lipid bilayers in the natural shark habitat, seawater. The reader is addressed to some research papers by Kalmanzon et al. (1989, 1992, 1996, 1997). Briefly, it has been shown that SDS was the detergent with the highest decrease in critical micelle concentration (CMC) in transition from distilled water to seawater (28.5 fold). Such a phenomenon was expected since SDS possesses a strong negative charge in seawater and the electrolyte content of seawater neutralizes part of the electrostatic antagonism among the negatively charged polar heads of SDS molecule. It has been proposed that the unique shark repellent potency of SDS is not simply a consequence of its detergent-solubilizing properties, but rather represents specific interactions with biological membranes at high ionic strength, presumably through a pore-forming process. We suggest that SDS forms negatively charged pores in the lipid bilayer that resemble inverted micelles. These pores can serve as cation channels and, thus, induce disturbances in externally exposed shark sensory-neuronal tissues ('pain production'). This hypothesis may explain the significantly superior effectiveness of SDS as a broad-spectrum shark repellent, as opposed to nonionic detergents such as Triton X-100, positively charged detergents such as Dodecyl Trimethyl Ammonium Bromide (DTAB), and negatively charged detergents such as cholic acid salts, which are uncharged in seawater.

Against the background of the above-mentioned hypothesis—that the repellent function of an anionic detergent (such as SDS) in seawater is mediated through pore formation—the most recent information concerning a marine cationic detergent reveals a different form of function and specificity.

COOPERATIVE COCKTAIL IN A CHEMICAL DEFENSE MECHANISM OF A TRUNKFISH SKIN SECRETION

The Double Paradox

The colorful trunkfish, *Ostracon cubicus*, is a classical example of a slow and 'lumbering' organism, which is chemically defended against its predators, and advertises this defense using colorful aposomatic markings (Fig. 6.4A). Previous studies (Boylan and Scheuer, 1967; Mann and Povich, 1969; Fusetani and Hashimoto, 1987; Goldberg *et al.*, 1988) have shown that the major active factor in the defensive skin secretion of trunkfish is a fish killing (ichthytoxic) compound designated as pahutoxin (PHN), which affects fish by medium application within the surrounding water. PHN is a choline chloride ester of 3-acetoxypalmitic acid (Boylan and Scheuer, 1967) and reveals an obvious structural resemblance to synthetic cationic long chain quaternary ammonium detergents. Thus, its ichthyotoxicity was attributed to its surfactant activity (Mann and Povich, 1969) (Fig. 6.4B).

The present chapter was motivated and directed by two considerations/problems ('the double paradox'):

1. PHN by itself is responsible for only 3% of the fish toxicity of the entire crude secretion of the Red Sea trunkfish (Table 6.5). How can the total toxicity be explained?
2. It is highly unlikely that a defensive role in a marine environment is carried out through a surfactant–detergent-like action due to problems of dilution and lack of biological specificity.

The Solution of the First Problem: Proteins

As will be presented here, PHN (or PHN-like lipophilic substances) are not the only active compounds in the crude secretion of the trunkfish. Proteins that function as PHN-chelators, ichthyotoxins and PHN regulators accompany PHN.

B. $CH_3(CH_2)_{12}$-CHCH$_2$COO-CH$_2$-CH$_2$-N(CH$_3$)$_3$-Cl$^-$

Fig. 6.4   A. The Red Sea trunkfish *Ostracion cubicus*, 20-30 cm long, aposematic, reef dweller. **B**. Pahutoxin (PHN) — choline chloride ester of 3-acetoxypalmitic acid.

Trunkfish, when placed for several minutes into distilled water, secrete an opaque foamy secretion, which when lyophilized, forms a whitish powder. As exhibited in Table 6.2, the supernatant obtained from an aqueous suspension of the above powder was shown to contain about 15% proteins by dry weight. The data presented in Figure 6.5 and Table 6.3 demonstrate the occurrence of protein–PHN complex in the crude trunkfish secretion.

The data presented in Table 6.3 indicates that about 60% of the hemolytic activity of the crude trunkfish secretion was recovered from the organic supernatant. The above data clearly specifies that the recovered activity is derived from a lipophilic factor, which is associated with proteins (Fig. 6.5). However, the remaining activity (40%) may be attributed to a protein factor which was denatured by the organic extraction. The latter assumption, the presence of toxic protein(s) in the trunkfish secretion, was

Table 6.2   Proteins in the Red Sea trunkfish crude secretion

Reaction	Protein content (%)
Bradford reagent (Bradford, 1976)	13 ($\pm$1.7, n=4)
Lowry reagent (Lowry *et al*., 1951)	18 ($\pm$1.4, n=4)
*TCA precipitate	12 (n=2)

*Solubilized in phosphate buffered saline (PBS) and determined by Lowry reaction.

Table 6.3 Protein content (μg, Lowry et al., 1951) and hemolytic activity (H.U.) of samples (and their organic extracts) derived from the trunkfish secretion.

Assay	Original sample	Organic extract	
		Sediment	Supernatant
Protein content	560	580	7
	(n=1)	(SD±110, n=3)	(SD±4, n=3)
Hemolysis	57	0	34
	(n=1)	(n=3)	(SD±11, n=3)

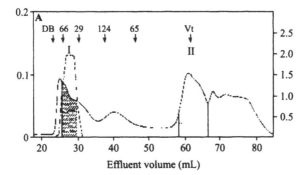

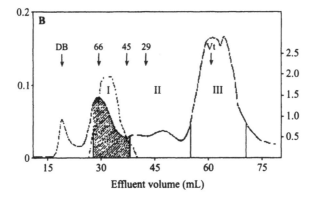

Fig. 6.5 Separation of the crude skin secretion of the Red Sea trunkfish by a molecular exclusion column chromatography: Sephadex G-50 (A) and Sephadex G-100 (B). Each of the two identical columns was equilibrated and eluted by the same buffer and flow rate and charged by 100 mg of crude skin secretion. DB indicates the void volume (Dextran blue) and arrows correspond to various molecular weight markers (kDa). Vt indicates the total volume of the column. The continuous line indicates hemolytic activity, which entirely coincides with the ichthyotoxicity (marked fraction). Two identical volumes of Sephadex G-100 fraction were sampled. The first sample was used for the determination of its protein content and hemolytic activity (Table 6.2, original sample). The second sample was used for protein determination, dried under nitrogen, extracted by hexane: isopropanol, and the supernatant were assayed for protein, and hemolytic activity (Table 6.3, organic extract).

proven by the data presented in Figure 6.6. The Sephadex G-50 fraction I (Fig. 6.5) was lyophilized and 10 mg of the dry material (composed substantially of proteins) was separated under conditions identical to those used to purify PHN. The data presented in Figure 6.6 indicate a clear distinction between two groups of ichthyotoxic substances namely PHN (Fig. 6.6A, C) and Proteins (Fig. 6.6C). The distinction between proteins and PHN was based on three criteria: (1) Spectrophotometry, in which protein is assayed by absorbance at 280 nm as well as 254 nm, and PHN, at 254 nm (Fig. 6.6A, C); (2) Fluorescamine and Folin phenol assays, which specifically detect proteins, and the Dragendorf assay (Boylan and Scheuer, 1967), which detects quaternary amines and therefore identifies PHN; and (3) qualitative assays of ichthyotoxicity, performed with the fractions obtained by the preparative run (Fig. 6.6C), which show that both substances are ichthyotoxic. The protein fraction (Prot., Fig. 6.6C) was lethal to juvenile *Sparus aurata* fish at concentrations of 5 to 10 μgml^{-1}, and pahutoxin (PHN Fig. 6.6C) was toxic from 2 to 4 μgml^{-1}.

The common way to attribute biological activity to a polypeptide factor is by subjecting it to heat treatment or proteolytic digestion. The protein fraction (Prot., Fig. 6.6C) resisted heat (60 min. at 95°C) and Trypsin (5% E/S, 37°C, 2 hours) treatments. However, it completely lost its ichthyotoxicity upon incubation with Pronase (5% E/S, 37°C, 2 hours). Thus, it may be concluded that the skin secretion of the Red Sea trunkfish contains, in addition to PHN, stable ichthyotoxic protein(s). Recently, a toxic protein designated as Boxin was isolated and purified from this secretion (Kalmanzon and Zlotkin, 2000). Boxin is a stable, heat and proteolysis resistant protein with a molecular mass of 18 kDa. Its protein nature was assessed by spectral analysis, strong proteolysis, amino acid analysis and amino acid sequence determination (data not shown; Kalmanzon and Zlotkin, 2000). Similar to PHN, boxin also affects the marine fish through medium application.

The above data shows that proteins exist in the trunkfish skin secretion and function either as ichthyotoxins or as PHN chelators. The pharmacological significance of the PHN-protein association is presently unclear. However, it is 'tempting' to assume that in such a PHN-protein complex, the protein may function as an 'affinity probe'-leading-navigating the PHN molecule to a critical site of action. The data presented in Figure 6.7 and Table 6.4 reveal that proteins in the trunkfish skin secretion fulfill an additional role as regulators-potentiators of PHN.

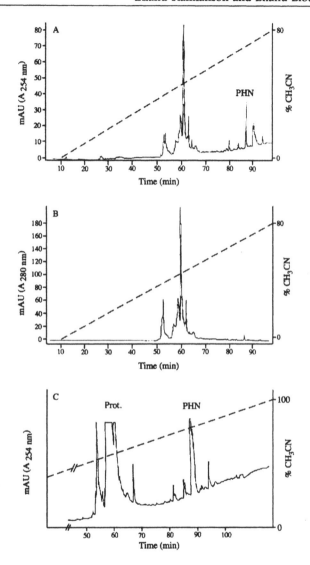

Fig. 6.6 Separation of the ichthyotoxic Sephadex G-50 fraction I (Fig. 6.5A) by analytical (A, B) and semi-preparative (C) RP-HPLC. (A) One milligram of lyophilized substance was separated on an analytical Vydac (Hesperia, CA) C-18 RP column (250×4.6 mm), 5 μm pore size. Buffer A: 0.1% Trifluoroacetic acid (TFA) in double distilled water (DDW), buffer B: 0.1% TFA in acetonitrile (CH$_3$CN). Flow rate was 0.5 ml min^{-1}, the gradient of CH$_3$CN is graphically presented (dashed line). Absorbance was monitored at 254 nm. (B) Same run as A but absorbance was monitored at 280 nm. (C) Ten milligrams of the lyophilized substance was separated on a semi-preparative Vydac C-18 RP column (250 × 22 mm), 10 μm pore size. Same buffer system as in A and B. Flow rate 5 ml min^{-1}, gradient of CH$_3$CN is graphically presented (dashed line). Absorbance was monitored at 254 nm. (PHN pahutoxin; Prot. Protein.)

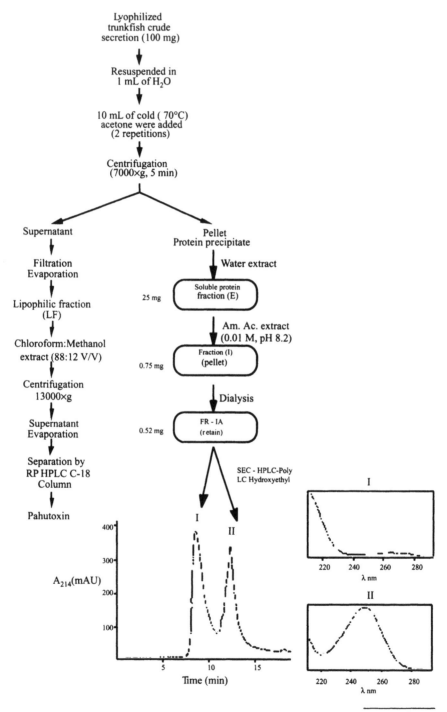

Fig. 6.7 Contd.

Table 6.4 Potentiation of ichthyotoxicity by protein factors

Substance	Concentration ($\mu g\ ml^{-1}$)	Effect
SPF	500	NA
LF+SPF	3+40	Lethal, within 10 min
PHN+SPF	0.9+50	Lethal, within 10 min
LF+BSA	3+40	NA
LF+SPF (p)	3+40	NA
LF+SPF (t)	3+40	NA
LF+SPF (b)	3+40	NA

LF — lipophilic fraction (Fig. 6.7); SPF — soluble protein fraction (Fig. 6.7); p,t,b — following incubation with Pronase (p), Trypsin (t) and boiling water (b); NA — not active.

As shown in Figure 6.7, the separation of proteins and lipophilic substances can be achieved not only by column chromatography (Figures 6.5 and 6.6) but also by extraction with acetone. The data presented in Table 6.4 shows, firstly, that toxicity possessed by the purified toxins PHN and boxin corresponds to only 3% of the total toxicity of the crude secretion for each. Furthermore, it was shown that the crude secretion (the entire mixture) reveals the highest toxicity; namely, it is more toxic than the isolated toxic constituents. This suggests that it acts as a cooperative cocktail of organic surfactants and stable proteins.

Fig. 6.7 Contd.

Fig. 6.7 Flow diagram of a solvent fractionation of the Red Sea trunkfish skin secretion. The lyophilized soluble protein fraction (SPF) did not possess any ichthyotoxicity, in contrast to the lipophilic fraction (LF) (LC_{50}, 3.5 µg ml^{-1}).
Table 6.4 summarizes a series of simple assays monitoring fish lethality and shows that (a) LF fraction is ichthyotoxic, (b) the SPF fraction is not toxic, (c) the SPF fraction is able to potentiate (synergize) the ichthyotoxic effect of LF and of PHN and (d) the synergic effect of SPF is lost upon heat treatment or proteolytic digestion. Thus, it may be concluded that the crude trunkfish secretion possesses a protein factor(s) that increases the ichthyotoxic potency of PHN. The protein precipitate was extracted with distilled water and centrifuged, yielding a soluble supernatant (E), which was then lyophilized. The lyophilizate was dissolved in ammonium acetate (0.01 M, pH 8.2) yielding a precipitate which was centrifuged at 13000×g for 3 min. The pellet (1) was resuspended in ammonium acetate buffer and dialyzed against distilled water. The dialyzate (Fr-IA) was lyophilized and then dissolved in buffer composed of 200 mM NHSO$_4$, 5 mM KH$_2$PO$_4$ (pH-3.0), and 25% (v/v) CH$_3$CN, separated by size exclusion chromatography using a poly LC Hydroxyethyl (5 µm, 4.6×200 mm, Poly LC, USA) column. The column was equilibrated with the above buffer and eluted at a flow rate of 0.5 ml/min. As shown, two fractions, I and II, were eluted. Automated spectral analysis of the fraction peaks is demonstrated in the bottom right.

A Solution to the Second Problem: Receptor Mediated Toxicity of Pahutoxin

The notion that PHNs fish killing capacity is linked to its surfactant activity (Mann and Povich, 1969) is supported, firstly, by the well-known occurrence of surfactants in fish defensive skin secretions (Nair, 1988), and the finding, described above, that commercial detergents can function as shark repellents (Zlotkin and Gruber, 1984).

However, a chemical defense mechanism in a marine environment based on surfactants is paradoxical. Firstly, a surfactant-detergent like substance affects biological membranes either as a solubiliser in its micellar association (CMC) or as a pore former in an oligomeric association. In the infinite volume of seawater, the surfactant is readily diluted to its monomeric form at which it is unlikely to act. Secondly, an effect, based on a detergent-surfactant action is devoid of the proper selectivity in order to distinguish between the self-defending organism and its predators. The experimental treatment of the dilution specificity paradox demanded the synthesis of a radioactive PHN and two derivatives (Table 6.6) and the determination of PHNs critical micelle concentration (CMC) in seawater (69 μM, 30 μg ml^{-1}, data not shown). As shown in Table 6.6, the various derivatives of PHN were assayed for their ability to kill a marine fish upon medium application, and to permeabilize fluorescein-loaded liposomes suspended in seawater. The data presented in Table 6.6 indicates that:

1. PHNs fish lethality is achieved at a concentration almost 30 times below the CMC.
2. Liposomal permeation of PHN is in the range of CMC values.
3. The desmethyl derivative, which is deprived of the positive charge, obviously loses its ability to kill fish in contrast to its ability to permeabilize the liposomes.
4. The removal of the branched acetoxy group does not modify the ichthyotoxic ability.
5. The crude secretion reveals the highest toxicity to fish but is devoid of the ability to affect the liposomal integrity.

The above data and considerations suggest that two different forms of PHN organization are responsible for its ichthyotoxic and its liposome-disrupting effects. Ichthyotoxicity is probably caused by the monomeric form and requires chemical specificity (see below), while the liposomal permeation is affected by the surfactant properties of PHN and requires

Table 6.5   Total and specific ichthyotoxicities of fractions and toxins derived from Red Sea trunkfish skin secretion*

Substance	Content in total secretion (% dry weight)	Specific toxicity LD_{50} value ($\mu g\ ml^{-1}$)	Recovery of toxicity (%)
Entire crude skin secretion	100	1.1	100
Acetonic extract[a]	52	1.12	51
Acetonic precipitate[b]	48	3.5	15
Pahutoxin	3.5	1.25	3
Boxin	4.4	1.57	3

*The fifty percent lethal concentration (LD_{50}) was determined by medium application on *Sparus aurata* fries,
[a]Following evaporation under nitrogen
[b]Resuspended in seawater

the presence of micelles. If ichthyotoxicity is not a consequence of the surfactant capacity of PHN, then a reasonable alternative is that it affects its targets via its binding to a specific receptor. The latter hypothesis is supported by certain conceptual as well as experimental considerations: (1) If PHN plays an allomonal role in the trunkfish secretion, then it should act through a mechanism which is able to distinguish between the trunkfish and its potential enemies. Receptors supply the most reasonable and most common solution for problems of biological specificity. (2) The fact that PHN affects the experimental fish only by application to the medium, and is absolutely ineffective when injected, suggests that PHN identifies externally located receptors exposed to the surrounding water but absent from tissues inside the fish body. The fish gill membranes, due to their large surface area and exposure to the surrounding seawater, are the natural candidates to possess such receptors. The possibility that gill membranes are specifically targeted by PHN was supported by two assays where *Sparus aurata* fries (100-150 mg) were incubated with the radio labeled PHN in seawater. The first assay revealed that the toxin bound to the fish membranes according to a Michaelis-Menten plot of binding saturation to toxin concentration increase. The second assay showed that the experimental fish head portion, which includes the gills, possessed significantly higher relative radioactivity (data not shown) compared to the entire body. Therefore, for the purpose of binding assays, a preparation of gill membranes was prepared according to Barbier (1976), and the radioactive toxin [^{14}C]PHN was employed in binding assays.

Here, Figure 6.8 presents a typical equilibrium saturation-binding assay, which reveals the occurrence of several receptor types at a wide range of [^{14}C]PHN concentrations, and a single type of receptor at lower and limited range of PHN concentrations close to and lower than its ichthyotoxic value. Thus, we can conclude that the latter higher affinity site is the pharmacologically functional site, which is responsible to pahutoxins ichthyotoxicity. However, from the point of view of the

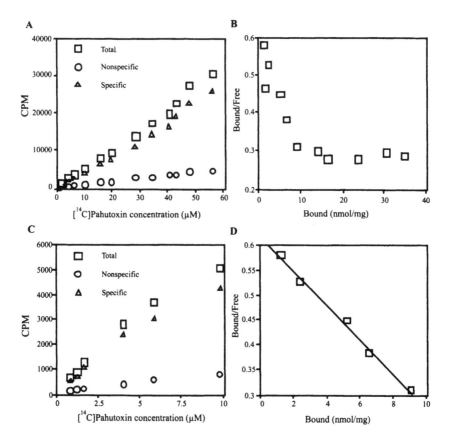

Fig. 6.8 Equilibrium saturation binding assay of [^{14}C]PHN to gilt-head sea bream fish gill preparation (*Sparus aurata*).

(A) The entire experiment with a wide range of concentrations. (B) Scatchard analysis of A. The analysis reveals at least two classes of binding sites. (C) A detailed presentation of a limited range of the low concentrations. A K_D of 0.3 µM was calculated. (D) Scatchard analysis of C, revealing a linear plot with a Bmax of 9 nmoles mg^{-1} membrane protein. These values correspond to the high-affinity binding sites shown in (B). Analysis of binding assays was performed using the iterative computer program LIGAND (P.J. Munson and D. Rodbard, modified by G.A. McPherson, 1985).

defensive role, it is suggested that lower-affinity, higher-capacity binding sites (Fig. 6.8A, B) may play an essential role. We assume that when an offensive fish approaches the trunkfish within a certain critical distance, its gills become loaded with the PHN secreted by the trunkfish. The relatively abundant PHN molecules, which are weakly attached to the lower-affinity sites, may easily dissociate and translocate to the functional high-affinity sites, thus prolonging and strengthening the effect. In other words, the lower-affinity, higher-capacity binding sites are synergistic to the higher-affinity site by serving as a reservoir of active PHN molecules, which may function at a more advanced stage of the encounter.

Thus, it may be concluded that the ichthyotoxic effect of PHN is mediated by specific gill membrane receptors. The question arises vis-à-vis issue specificity: How can PHN distinguish between the trunkfish and a threatening fish? The answer is presented in Figure 6.9, revealing that the gill membrane of the trunkfish is devoid of PHN receptors (Kalmanzon et al., 2003).

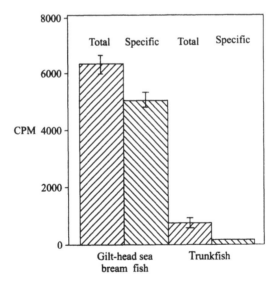

Fig. 6.9 Binding of [^{14}C]PHN to the gilt-head sea bream (*Sparus aurata*) and trunkfish gill membrane preparations. Gill membranes were prepared and incubated with radiolabeled PHN (1.9 nmole = 9050 cpm). Reaction mixtures (300 µl) contained Hanks' buffer (Wolf and Quimby, 1969), with 1 mg ml^{-1} bovine serum albumin (BSA), 250 µg of tissue protein, and 1.5 µg of [^{14}C]PHN. The values of total and specific binding are presented in the histograms. Briefly, as shown, the trunkfish is devoid of PHN receptors.

Endogenous Regulation of the Functional Duality of Pahutoxin

The previous data (Table 6.6) has revealed that PHN's ichthyotoxicity and its membrane disruption effect are provided by two separate mechanisms performed by two separate physicochemical domains or 'pharmacologic determinant' in the PHN molecule. A study (Kalmanzon *et al.*, 2004) has revealed the occurrence of a natural mechanism, which regulates PHN's functional duality.

Figure 6.7 reveals a process of fractionative solubilization coupled to column chromatography, which enabled the isolation of two fractions. The first fraction (I) is suspected due to its UV absorption spectrum (Fig. 6.7) to be a protein, unlike the second fraction (II). Figure 6.10 demonstrates that each of the above fractions specifically affects either the ichthyotoxicity of PHN or its liposomal permeabilization. However, the effects are in opposite directions: factor I enhances PHN's fish lethality (Fig. 6.10A), while factor II suppresses its liposomal permeabilization (Fig. 6.10B). This data suggests that each of the above two activities, the

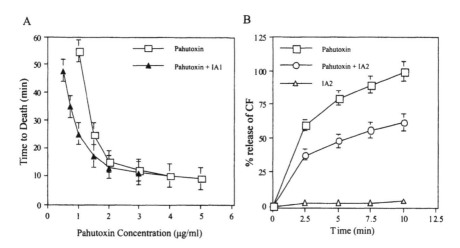

Fig. 6.10 Effects of fractions I and II from Figure 6.7, respectively, on enhancing fish lethality (A) and inhibiting liposomal permeation (B).

A. Three fries were used per experimental point. The concentration of factor I was twice as that of PHN by mass. The time to lethality was monitored. As seen, at the lower concentrations of PHN the two curves differ significantly.

B. PHN (10 µg/ml) in the presence of 100 µg/ml of factor II. As shown factor II suppresses the PHN-induced increase in the liposome permeability. Factor II by itself was not effective. Liposome permeability was assessed by monitoring fluorescence released from preloaded liposomes with carboxy flourescein (CF) according to the reported (Kalmanzon *et al.*, 2003) protocol.

Table 6.6   Effect of various substances on fish lethality and liposomal permeation in seawater.

Substance	Ichthyotoxicity (LC_{50} μg/ml)	Liposomal content release (ED_{50} μg/ml)
Crude secretion	0.73¶	not active (<200 μg/ml)
Lipophilic fraction (Fig. 6.1)	2.42	not tested
Natural Pahutoxin‡ (Mr. = 526)	1.25 (2.87 μm)	20.0

$CH_3(CH_2)_{12}$–CH–CH_2–COO–CH_2–CH_2–N^+–$(CH_3)_3$–Cl^-
      |
      $OCOCH_3$		

Synthetic Pahutoxin (Mr. = 618)	0.95 (1.8 μm)	10.0

$CH_3(CH_2)_{12}$–CH–CH_2–COO–CH_2–CH_2–N^+–$(CH_3)_3$–I^-
      |
      $OCOCH_3$		

Desmethyl Pahutoxin (Mr. = 385)	35.1 (91.1 μM)	20.0

$CH_3(CH_2)_{12}$–CH–CH_2–COO–CH_2–CH_2–N–$(CH_3)_2$
      |
      $OCOCH_3$		

Synthetic (smooth Trunkfish) toxin (Mr. = 467) $CH_3(CH_2)_{12}$–CH_2–CH_2–COO–CH_2–CH_2–N^+–$(CH_3)_3$–Cl^-	1.12 (2.38 μM)	not tested

(‡) CMC of pahutoxin in sea water is 69 μm (30 μg/ml)

(§) Fish lethality was assayed according to experimental procedures.

(¶) The crude secretion contains in addition to PHN other ichthyotoxic factors (data not shown). ED_{50} refers to the concentration of material (μg/ml) which induces 50% of maximal fluorescence intensity following 5 min incubation.

Fish lethality was assayed on *Sparus aurata* fries by medium application. The crude secretion contains, in addition to PHN, toxic and potentiating polypeptides. CMC of pahutoxin in sea water is 69 μM (30 μg/ml). ED_{50} refers to the concentration of material (μg/ml) which induces 50% of maximal fluorescence intensity following 5 min incubation, LC_{50} refers to medium concentration which induces 50% fish lethality following 60 min exposure.

ichthyotoxicity and membrane permeabilization (cytolysin), are affected by a separate pharmacology.

The distinction between PHN's ichthyotoxicity and its liposomal permeabilisation effect is revealed by active concentrations and a chemical modification of the PHN molecule. However, the most significant indication of the functional duality of PHN is provided by a pharmacologic distinction between its ichthyotoxicity and membrane permeation-cytolytic activity. The latter is revealed by the occurrence of the two endogenous regulatory factors, derived from the secretion of Red

Sea trunkfish. The first factor (I) strengthens ichthyotoxicity without affecting the cytolytic-liposomal effect (Fig. 6.10A). The second factor (Fig. 6.10B) exclusively suppresses the liposomal effect without affecting the ichthyotoxic effect.

The occurrence of the endogenous regulatory mechanism located in the trunkfish secretion suggests that PHN's dual functionality possesses an ecological relevance. As specified earlier, the trunkfish is fully protected from the receptor-mediated ichthyotoxicity. Therefore, the amplification of the ichthyotoxicity by the aid of factor I is a device to strengthen the defensive role of the trunkfish secretion. On the other hand, the non-selective surfactant lipid disruption effect of PHN, may risk the trunkfish itself, justifying its regulated suppression provided by factor II. Thus, the endogenous regulatory mechanism is aimed to balance the advantages and hazards of PHN to its producer, the trunkfish itself.

CONCLUDING REMARKS

Substances used for defensive repulsive purposes by terrestrial animals are airborne and should possess considerable volatility (Barbier, 1976). In the marine environment, however, proteins can replace low molecular weight organic allomones, typical for the terrestrial environment. The latter are ideal candidates to fulfill allomonal functions in the marine environment due to the high information content inherent in their structures and their high solubility in water (Fainzilber et al., 1994). In the skin secretions of trunkfish, polypeptides were shown to cooperatively interact with typical detergent-like surfactants such as Pahutoxin (PHN). We show that PHN performs its ichthyotoxicity by a monomeric form, which requires chemical specificity mediated by receptors located on the predator fish gills. Its liposomal permeation effect, however, is due to its surfactant properties and requires the presence of micelles.

The notion that a 'detergent-like' molecule can act as a defensive allomone via interaction with specific receptors may possess far-reaching implications for two aspects of marine biology. The first concerns the chemical ecology of defensive allomones of marine organisms. A receptor-mediated action of a surfactant implies that the substance can act in its monomeric form without affecting the allomone-producing organism, which is devoid of the specific receptor-binding sites. The second aspect concerns environmental implications related to the pollution of the marine environment by detergents. The possibility exists that polluting detergents in seawater, in addition to functioning as solubilizers and pore formers, may affect marine biology in their monomeric forms through a receptor-mediated action.

Acknowledgement

The studies presented were supported by Israel Science Foundation grants 464/92 and 494/96. The authors are grateful to Daniel Sher (Life Sciences, Hebrew University) and to Professor Giacomo Zaccone (Faculty of Science, University of Messina) for consultation and conceptual assistance.

References

Barbier, M. 1976. *Introduction to Chemical Ecology.* Plenum Press, New York.

Birkhead, W.S. 1967. The comparative toxicity of stings of the ictalurid catfish genera *Ictalurus* and *Schilbeodes. Comparative Biochemistry and Physiology* 22: 101-111.

Boylan, B.D. and J.P. Scheuer. 1967. Pahutoxin a fish poison. *Science* 155: 52-56.

Bradford, M.M. 1976. A rapid and sensitive method for the quantitation of microgram quantities of protein utilizing the principle of protein-dye binding. *Analytical Chemistry* 72: 248-252.

Cameron, A.M. 1974. Toxicity phenomena in coral reef waters. In: *Proceedings of the Second International Coral Reef Symposium. Great Barrier Reef Committee,* Brisbane, Vol. 1, p. 513.

Cameron, A.M. and R. Endean. 1973. Epidermal secretions and the evolution of venom glands in fishes. *Toxicon* 11: 401-410.

Carli, G. 1977. Animal hypnosis in the rabbit. *Psychological Research* 1: 123-143.

Clark, E. 1974. The Red Sea's shark proof fish. *National Geographical Magazine* 146: 719-727.

Clark, E. 1983. Shark repellent effect of the Red Sea Moses aole. In: *Shark Repellents from the Sea. AAAS Selected Symposium,* B. Zahuranec (ed.). Westview Press, Boulder, 83, pp. 735-750.

Ebran, N.S., N. Julien, N. Orange, C. Saglio, C. Lemaitre and G. Molle. 1999. Pore-forming properties and antibacterial activity of proteins extracted from epidermal mucus of fish. *Comparative Biochemistry and Physiology* 122(2): 181-189.

Ebran, N.S., N. Julien, N. Orange, B. Auprin and G. Molle. 2000. Isolation and characterization of novel glycoproteins from fish epidermal mucus: correlation between their pore-forming properties and their antibacterial activites. *Biochimique et Biophysique Acta* 1467(2): 271-280.

Fainzilber, M., I. Napchi, D. Gordon and E. Zlotkin. 1994. Marine warning via peptide toxin. *Nature (London)* 369: 192-193.

Fernandes, J.M.O., G.D. Kemp, M.G. Molle and V.J. Smith. 2002. Anti-microbial properties of histone H2A from skin secretions of rainbow trout, *Oncorhynchus mykiss. Biochemical Journal* 368: 611-620.

Fletcher, T.C. and P.T. Grant. 1969. Immunoglobulins in the serum and mucus of the plaice. *Biochemical Journal* 115: 65.

Fusetani, N. and K. Hashimoto. 1987. Occurrence of Pahutoxin and Homopahutoxin in the mucous secretion of the Japanese box fish. *Toxicon* 25: 459-461.

Gallo, R.L. and K.M. Huttner. 1998. Antimicrobial peptides: an emerging concept in cutaneous biology. *Journal of Investigative Dermatology* 111: 739-743.

Ghadessy, F.J., D. Chen, R.M. Kini, M.C. Chung, K. Jeyaseelan, H.E. Khoo and R. Yuen. 1996. Stonustoxin is a novel lethal factor from stonefish (*Synanceja horrida*) venom. cDNA cloning and characterization. *Journal of Biological Chemistry* 271: 25575-25581.

Goldberg, A.S., A.M. Duffield and K.D. Barrow. 1988. Distribution and chemical composition of the toxic skin secretions from trunkfish (family Ostraciidae). *Toxicon* 26: 651-663.

Goldberg, A.S., J. Wasylyk, S. Renna, H. Refsman and M.S.R. Nair. 1982. Isolation and structural elucidation of an ichthyocrinotoxin from the smooth trunkfish (*Lactophrys triqueter*, Linnaeus). *Toxicon* 20: 1069-1074.

Gopalakrishnakone, P. and M.C.E. Gwee. 1993. The structure of the venom gland of stonefish *Synanceja horrida*. *Toxicon* 31: 979-988.

Gruber, S.H. 1980. Keeping sharks in captivity. *Journal of Aquaculture* 1: 6-14.

Gruber, S.H. 1983. Shark repellents: Protocols for a behavioral assay. In: *Shark Repellents from the Sea. AAAS Selected Symposium*, T. Zahuranec (ed.). Westview Press, Boulder, 83. pp. 91-113.

Gruber, S.H. and A.A. Myrberg. 1977. Approaches to the study of the behavior of sharks. *American Zoologist* 17: 471-486.

Gruber, S.H., E. Zlotkin and D.R. Nelson. 1984. Shark repellents: behavioral bioassays in laboratory and field. In: *Toxins, Drugs and Pollutants in Marine Animals*, L. Bolis, J. Zadunaisky and R. Gilles (eds.). Springer-Verlag, Berlin, pp. 26-42.

Halstead, B.W. 1967. *Poisonous and Venomous Marine Animals of the World*, U.S. Government Printing Office, Washington, DC, Vol. 2: Vertebrates.

Halstead, B.W. 1970. Poisonous and Venomous Marine Animals of the World. Washington: U.S. Government Printing Office, Vol. 3: Vertebrates.

Halstead, B.W. 1978. Poisonous and Venomous Animals of the World. The Darwin Press, Princeton.

Hashimoto, Y. 1979. *Marine Toxins and Other Bioactive Marine Metabolites*. Japan Scientific Press, Tokyo.

Hashimoto, Y. and Y. Oshima. 1972. Separation of grammistin A, B and C from a soapfish *Pogonoperca punctata*. *Toxicon* 10: 279-284.

Helenius, A. and K. Simons. 1975. Solubilization of membranes by detergents. *Biochimique et Biophysique Acta* 15: 29-39.

Hildemann, W.H. 1962. Immunogenetic studies of poikilothermic animals. *American Naturalist* 96: 195.

Johnson, C.S. 1963. Anti-shark devices and testing methods at naval undersea center. In: *Sharks and Man*, W. Seaman Jr. (ed.). Sea grant Report No. 10, Gainsville, Florida.

Kalmanzon, E. and E. Zlotkin. 2000. An ichthyotoxic protein in the defensive skin secretion of trunkfish. *Marine Biology* 136: 471-476.

Kalmanzon, E., E. Zlotkin and Y. Barenholz. 1989. Detergents in sea water: Properties and effect of detergents in water of various salinity conditions: Possible relevance to shark repellency. *Tenside Surfactant Detergents* 5: 33-342.

Kalmanzon, E., E. Zlotkin and Y. Barenholz. 1996. Liposomes as a model system to study shark repellency. In: *Nonmedical Applications of Liposomes*, D.D. Lasic and Y. Barenholz (eds.). CRC Press, New York, pp. 183-198.

Kalmanzon, E., E. Zlotkin, R. Cohen and Y. Barenholz. 1992. Liposomes as a model for the study of the mechanism of fish toxicity of sodium dodecyl sulfate in sea water. *Biochimique et Biophysique Acta* 1103: 148-156.

Kalmanzon, E., Y. Barenholz, S. Carmeli, S.H. Gruber and E. Zlotkin. 1997. Detergent in the marine environment. Thirteenth School on Biophysics of Membrane Transport. School Proceedings, Poland.

Kalmanzon, E., Y. Rahamim, Y. Barenholz, S. Carmeli and E. Zlotkin. 2003. Receptor-mediated toxicity of Pahutoxin, a marine trunkfish surfactant. *Toxicon* 42: 63-71.

Kalmanzon, E., Y. Rahamim, Y. Barenholz, S. Carmeli and E. Zlotkin. 2004. Endogenous regulation of the functional duality of Pahutoxin, a marine trunkfish surfactant. *Toxicon* 44: 939-942.

Kitzan, S.M. and P.R. Sweeney. 1968. A light and electron microscope study of the structure of protopterus annectens epidermis. I. Mucus production. *Canadian Journal Zoology* 46: 767-772.

Kodama, M., S. Sato T. Ogata, Y. Suzuki, T. Kaneko and K. Aida. 1986. Tetrodotoxin secreting glands in the skin of puffer fishes. *Toxicon* 24: 819-829.

Lowry, O.H., N.A. Rosenbrough, A.L. Fair and R.J. Randall. 1951. Protein measurement with the Folin phenol reagent. *Journal of Biological Chemistry* 193: 265-275.

Lazarovici, P., N. Primor and L.M. Loew. 1986. Purification and pore-forming activity of two hydrophobic polypeptides from the secretion of the Red Sea Moses sole (*Pardachirus marmoratus*). *Journal of Biological Chemistry* 261: 16704-16713.

Lemaitre, C., N. Orange, P. Saglio, N. Saint, J. Gagnon and G. Molle. 1996. Characterization and ion channel activities of novel antibacterial proteins from the skin mucosa of carp (*Cyprinus carpio*). *European Journal of Biochemistry* 240: 143-149.

Low, K.S.Y., M.C.E. Gwee, R. Yuen, P. Gopalakrishnakone and H.E. Khoo. 1994. Stonustoxin: Effects on neuromuscular function *in vitro* and *in vivo*. *Toxicon* 32: 573-581.

Mann, J.A. and M.J. Povich. 1969. Correlation of toxicity with the air-solution adsorption properties of Pahutoxin. *Toxicology and Applied Pharmacology* 14: 584-589.

Maretić, Z. 1988. Fish venoms. In: *Marine Toxins and Venoms*, A. Tu. (ed.). Marcel Dekkar, New York, pp. 445-476.

Matsui, T., S. Hamada and S. Konosu. 1981. Differences in accumulation of puffer fish toxin and crystalline tetrodotoxin in the puffer fish, *Fugu rubripes*. *Bulletin of the Japanese Society of Scientific Fisheries* 47: 535-537.

Matsumura, K. 1995. Tetrodotoxin as a pheromone. *Nature (London)* 378: 563-564.

Nair, M.S.R. 1988. Fish skin toxins. In: *Marine Toxins and Venoms*, A.T. Tu (ed.). Marcel Dekker, New York, pp. 212-225.

Nigrelli, R.F., S. Jakowska and M. Padnos. 1955. Pathogenecity of epibionts in fishes. *Journal of Protozoology* 2: 7.

Parness, T. and E. Zlotkin. 1976. Action of the toxic secretion of the flat fish *Pardachirus marmoratus* on the guinea pigileum. *Toxicon* 14: 85-91.

Primor, N. 1983. Pardaxin produces sodium influx in the teleost gill-like opercular epithelia. *Journal of Experimental Biology* 105: 83-94.

Primor, N. and E. Zlotkin. 1975. On the ichthyotoxic and hemolytic action of the skin secretion of the flat fish *Pardachirus marmoratus* (Soleidae). *Toxicon* 13: 227-231.

Primor, N., I. Sabnay, V. Lavie and E. Zlotkin. 1980. Toxicity to fish, effect on gill ATPase and gill ultrastuctural changes induced by *Pardachirus* secretion and its derived toxin Pardaxin. *Journal of Experimental Biology* 211: 33-34.

Prince, R.C. 1988. Tetrodotoxin. *Trends in Biochemical Sciences* 13: 76-77.

Quay, W.B. 1972. Integument and the environment: glandular composition, function, and evolution. *American Zoologist* 12: 95-108.

Randall, J.E. 1967. Food habits of reef fishes of the West Indies. *Studies in Tropical Oceanography* 5: 665-847.

Randall, J.E., K. Aida, T. Hibiya, N. Matsuura, H. Kamiya and Y. Hashimoto. 1971. Grammistin, the skin toxin of soapfishes, and its significance in the classification of the Grammistidae. *Publication of the Seto Marine Biological Laboratories* 10(2/3): 157-190.

Reed, L.Y. and H. Muench. 1938. A simple method for estimating fifty percent end points. *American Journal of Hygiene* 27: 493-497.

Rosen, M.W. and N.E. Cornford. 1971. Fluid friction of fish slimes. *Nature (London)* 234: 49.

Shai, Y., Y.R. Halari and A. Finkels. 1991. pH dependent pore formation properties of Pardaxin analogs. *Journal of Biological Chemistry* 266: 22346-23353.

Skeie, E.1962. The venom organs of the weeverfish (*Trachious draco* L.). Experimental studies on animals. *Meddelelser fra Danmarks Fiskeri-Og Havundersogelser* 3: 227-338.

Soong, T.W. and B. Venkatesh. 2006. Adaptive evolution of tetrodotoxin resistance in animals. *Trends in Genetics* 22: 621-626.

Spira, M., E. Klein, B. Hochner, Y. Yarom and M. Castel. 1976. Ultrastructural changes accompanying the disturbances of neuromuscular transmission caused by *Pardachirus* toxin. *Neuroscience* 1: 117-124.

Tachibana, K. and S.H. Gruber. 1988. Shark repellent lipophilic constituents in the defense secretion of the Moses Sole (*Pardachirus marmoratus*). *Toxicon* 26: 839-853.

Tachibana, K., M. Sakaitani and K. Nakanishi. 1984. Pavoninins: Shark repelling ichthyotoxins from the defense secretion of the pacific sole. *Science* 226: 703-705.

Tanu, M.B., Y. Mahmud, T. Takatani, K. Kawatsu, Y. Hamano, O. Arakawa and T. Noguchi. 2002. Localization of tetrodotoxin in the skin of a brackishwater puffer *Tetraodon steindachneri* on the basis of immunohistological study. *Toxicon* 40: 103-106.

Thompson, S.A., K. Tachibana, K. Nakanishi and I. Kubota. 1986. Mellitin-like peptides from the shark-repelling defense secretion of the sole *Pardachirus pavoninus*. *Science* 233: 341-343.

Thompson, S.A., K. Tachibana, I. Kubota and E. Zlotkin. 1988. Amino acid sequence of Pardaxin in the defense secretion of the Red Sea Moses sole. In: *Peptide Chemistry*. Protein Research Foundation, Osaka, T. Shiba and S. Sakakibura (eds.). pp. 127-132.

Venkatesh, B., S.Q. Lu, N. Dandona, S.L. See, S. Brenner and T.W. Soong. 2005. Genetic basis of tetrodotoxin resistance in pufferfishes. *Current Biology* 15: 2069-2072.

Wallett, T.S. 1972. *Shark Attack and Treatment of Victims in Southern African Waters*. Purnell and Sons, Cape Town.

Winn, H.E. and J.E. Bardach. 1959. Differential food selection by moray eels and a possible role of the mucus envelope of parrot fishes in reduction of predation. *Ecology* 40: 296-298.

Wolf, K. and M.C. Quimby. 1969. Fish cell and tissue culture. In: *Fish Physiology*, W.S. Hoar and D.C. Randall (eds.). Academic Press, London, pp. 253-301.

Yamamori, K., M. Nakamura and T.J. Hara. 1987. Gustatory responses to tetrodotoxin and saxitoxin in rainbow trout (*Salmo gairdneri*) and arctic char (*Salvelinus alpinus*): A possible biological defense mechanism. *Annals of New York Academy of Sciences* 510: 727-729.

Zaccone, G., B.G. Kapoor, S. Fasulo and L. Ainis. 2001. Structural, histochemical and functional aspects of the epidermis of fishes. In: *Advances in Marine Biology*, A.J. Southward, P.A.Tyler, C.M. Young and L.A. Fuiman (eds.). Academic Press, London, Vol. 40, pp. 253-348.

Zahuranec, B.J. 1975. *Shark Research: Present Status and Future Directions*. ONR report ACT-208, Arlington, VA.

Zlotkin, E. and Y. Barenholz. 1983. On the membranal action of Pardaxin. In: *Shark Repellents from the Sea*. AAAS Selected Symposium, T. Zahuranec (ed.). Westview Press, Boulder, 83, pp. 157-172.

Zlotkin, E. and S.H. Gruber. 1984. Synthetic surfactants: A new approach to the development of shark repellents. *Archives in Toxicology* 56: 55-58.

Defence Strategies of Opisthobranch Slugs against Predatory Fish

Arnaldo Marin

INTRODUCTION

The long-term evolution of species involved in a predator-prey interaction has been frequently regarded as an arms race (Janzen, 1980; Krebs and Davies, 1991). This has also been registered in the fossil record of molluscs (Vermeij, 1987). In the last 500 million years, the fossil record of shell-breaking families has increased, and so has the percentage of gastropods with narrow apertures and high-spired shells (Fig. 7.1). These shell characteristics are excellent defences against predators. At the same time as this was happening, weak shell designs gradually declined in number. At present, many fishes exist that eat hard-shelled prey by crushing them, but molluscs have probably evolved hard shells against all predators. There are many cases of specialized prey defences in molluscs, especially in

Author's address: Departamento de Ecología e Hidrología, Facultad de Biología, Universidad de Murcia, 30100-Murcia, Spain.

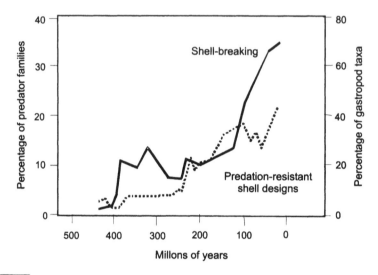

Fig. 7.1 Evolution of shell-breaking predators and gastropods with predation-resistant shell designs in the fossil record (data from Vermeij, 1987). Modified from Krebs and Davies (1991). Vertical axes are the number of shell-breaking families and the percentage of taxa with predation-resistant shell designs. Shells with narrow apertures, sculptured and high-spired gastropod shells are more difficult to break.

opisthobranchs, but there are hardly any fish species that specialize in shelled molluscs. One exception is the fish *Asemichthys taylori*, which possesses specialized teeth to puncture snails (Norton, 1988). In this chapter, we shall analyze the evolutionary defensive adjustments of shell-less gastropods, which have been necessary to maintain this arms race. Faulkner and Ghiselin (1983) proposed that the nudibranchs (opisthobranchs) evolved from shelled molluscs in a process that involved the gradual loss of the shell and detorsion of the visceral mass. Prior to replacing the shell as the basic defensive mechanism, the ancestral nudibranch had either developed a diet-derived chemical or a nematocyst-based defence. This hypothesis has been demonstrated in the Noah's ark (*Arca noae*), which shows a parallel evolution to some primitive shelled opisthobranchs that combine chemical and mechanical defence mechanisms (Marin and López Belluga, 2005). It is clear that such a parallel evolution is consistent with the hypothesis that predator escape and deterrence are primary factors in explaining the evolution of *Arca noae* and other chemically defended organisms, such as opisthobranchs. Mediterranean populations of the bivalve *Arca noae* are coated by the demo sponge *Crambre crambe*, and when fouled examples of Noah's ark were offered to starfish, snails and octopuses, they survived significantly more than aposymbiotic bivalves. Some of the observed benefits of this symbiosis are the enhanced survival of the bivalve, and the sponge being able to live in a site free of competitors. An aposymbiotic population of *Arca noae* lived in the Mar Menor, a hypersaline lagoon (SE Spain), until the opening of a channel connecting it with the Mediterranean Sea decreased the salinity from 50 to 42 psu, a decrease that allowed the biological invasion of the non-native snail *Hexaplex trunculus*. Noah's ark habits a broad range of salinity, while the symbiont sponge cannot grow in waters with high salinity. The extinction of this aposymbiotic population of *Arca noae* from the Mar Menor lagoon clearly illustrates the dramatic consequence of the loss of chemical defence. The probability of extinction of an unarmed species increases exponentially with the introduction of a new predator.

Molluscs constitute one of the largest phyla of animals, both in number of species and in number of individuals. They are characterized by soft bodies within a hard shell, although in some forms, the shell has been lost in the course of evolution (e.g., opisthobranchs, octopuses and squids). The gastropod subclass Opisthobranchia (from Latin opistho=behind and branch=gill) is divided into several orders: Cephalaspidea, Sacoglossa, Acochlidea, Anaspidea, Rhodopemorpha, Notaspidea, Thecosomata, Gymnosomata and Nudibranchia. In the Opisthobranchia, the general trend toward loss of the shell has been compensated for by various defensive adaptations involving chemical secretions or nematocyst-based defences, generally associated with conspicuous colouration (Edmunds, 1966; Ros, 1976, 1977; Fontana *et al.*, 1993; Gavagnin *et al.*, 1994a; Aguado and Marin, 2006). Some sea-slug nudibranchs are distinguished by a chromatic richness unusual in marine environments (Fig. 7.2). The coincidence between unpalatability and conspicuous colouration in nudibranchs is usually explained by the theory of warning coloration. One of the most surprising aspects of the biology of nudibranchs is the infrequency of their being preyed by other marine organisms (Thompson, 1960; Edmunds, 1991; Marin and Ros, 2004). Laboratory experiments offering a range of species to aquarium fish have largely provided negative results, although some species are finally consumed after repeated mouthing. Personal observations in the Mediterranean Sea showed the voracity of potential fish predators that repeatedly attacked to the umbrella of the dangerous jellyfish *Pelagia noctiluca* with broken tentacles.

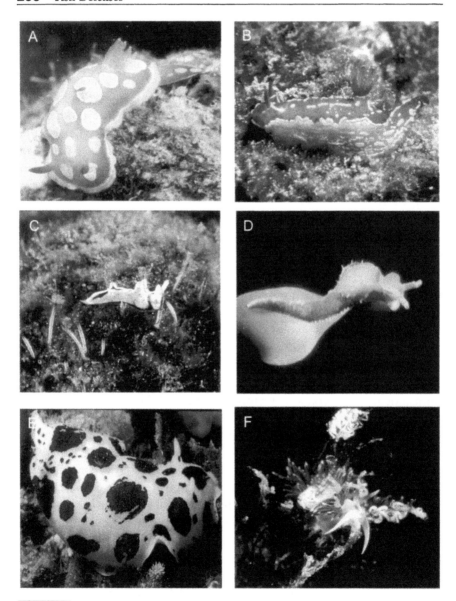

Fig. 7.2 In the Opisthobranchia, the general trend toward loss of the shell has been compensated for by various defensive adaptations involving chemical secretions or nematocyst-based defences, generally associated with conspicuous colouration. A, *Chromodoris luteorosea* (von Rapp, 1827); B, *Hypselodoris picta* (Schultz, 1836); C, *Elysia timida* (Risso, 1818); D, *Oxynoe olivacea* Rafinesque 1819; E, *Peltodoris atromaculata* Bergh, 1880; F, *Cratena peregrina* (Gmelin, 1791).

Offence and Defence

The opisthobranch defensive mechanisms have been classified into three basic categories: behavioural (autotomy, crypsis), morphological (spicules and kleptoplasty of cnidarian nematocysts), and chemical (direct use of toxins from prey and *in situ* synthesis of toxins as repugnatory fluids) (Todd, 1981), which may be employed in a hierarchical fashion in the different stages of fish predation (Fig. 7.3). Following the nomenclature of Endler (1986), the stages of fish predation and the corresponding anti-predator defences of opisthobranchs may be divided into six consecutive stages and the corresponding anti-predator defences:

(1) Encounter within a distance from which the predator can detect its prey. Reduces the random encounter rate between predator and prey (e.g., Rarity, apparent rarity, polymorphism).

(2) Detection of prey as objects that are distinct from the background. Reduces signal of prey in predator's sensory field (e.g., Crypsis, confusion).

(3) Identification as profitable or edible prey and decision to attack. Identification with inedible objects o distastefulness (e.g.,

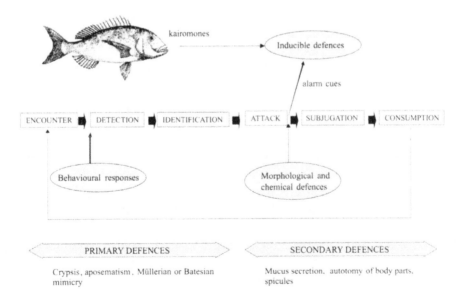

Fig. 7.3 Stages of predation and the corresponding anti-predator defences. A behavioural response interrupts the cycle before any encounters, whereas morphological changes interrupt between the attack and the capture. Modified from Brönmark and Hansson (2000).

Masquerade, Aposematism, Müllerian mimicry, Batesian mimicry).

(4) Attack. Prey reduces the approach of predator (e.g., Speed, predator saturation).

(5) Subjugation o prevent escape. Increase escapes possibility (e.g., Mucus secretion, autonomy of body parts, spicules).

(6) Consumption. Avoid digestion of prey through the gut (e.g., Safe passage through the gut, poisonous, lethal).

In visual predators such as fish, these anti-predator mechanisms may be employed in a hierarchical order and classified as 'primary' or 'secondary' defences. Thus, the primary line of defence to visual fish attack is either to evade detection, or having been detected to discourage the fish from initiating an attack. The most effective way to avoid predation is clearly to avoid encounters with predators in the first place. Many anti-predator adaptations involve crypsis (camouflage), polymorphism or mimicry of noxious species so that predators fail to distinguish potential prey from objects it does regard as food. Many opisthobranchs have shapes and pigmentation that closely match those of their host sponge (e.g., *Discodoris indecora* on the host prey, the sponge *Ircinia fasciculata*; Marin et al., 1997).

Some well-defended opisthobranchs are not just cryptic, but go so far as to advertise their unsuitability as food (e.g., species with bright or conspicuous warning colouration). Opisthobranchs display many different types of warning colouration, but the most common are bright colours such as orange, white and yellow, usually combined with dark blue, red brown or black to form a contrasting pattern. Prey use aposematic colouration to advertise noxious properties to potential predators. Guilford (1990a) provides four—but not necessarily mutually exclusive—hypotheses to explain the reason why prey use conspicuousness for warning. First, predators may learn to associate distastefulness with conspicuous colour pattern more rapidly than with a cryptic colour pattern. Second, certain specific colour patterns are easier to associate with distastefulness than others, and these are conspicuous. The third hypothesis is that new patterns are easier to learn. Fourth, conspicuous colouration will allow fewer recognition errors than crypsis. To the predator, crypsis is a 'non-signal', precisely the opposite of the warning signal of aposematism.

The widespread association between noxious qualities and warning colouration in animals appears to have arisen because predators more easily remember striking colours than cryptic ones (Gittelman et al., 1980). For example, conspicuous warning patterns such as the black-and-yellow banding common to many distasteful insects, can cause aversive or hesitant reaction in newly hatched chicks (Guilford, 1990b).

It is surely not a coincidence that many aposematic forms adopt similar patterns of colouration. Mimic species benefit from the presence of species with a similar colour-pattern, sharing the benefits of avoidance learning of common predators. Mimicry amongst species reduces the educational burden on predators, allowing them to learn to avoid unpalatable species with a sampling effort smaller than needed if the preys were not mimetic (Fisher, 1930). The two most common forms of mimicry are Batesian and Müllerian mimicry. In Müllerian mimicry, all species are noxious whereas in Batesian mimicry, a palatable species resembles an unpalatable species.

'Secondary' defence mechanisms are those mechanisms an animal brings into play when it encounters a predator. Some opisthobranch species rely entirely on avoidance after detection rather than investing in any 'primary' defence mechanisms. Numerous plants and animals exhibit plastic defences against herbivores and predators (Karban and Baldwin, 1997; Tollrian and Harvell, 1999) and, in many animals, the defensive traits are induced by chemical cues that are produced during predation events (Petranka et al., 1987; Chivers and Smith, 1998). These chemicals contain components from predators (termed kairomones) and components from injured prey (termed alarm pheromones). Molecules acting as alarm pheromones have been reported in the Pacific and Mediterranean Bullomorphs (e.g., Navanax inermis and Haminoea navicula) (Cimino and Sodano, 1989; Marin et al., 1999). In this chapter, neither kairomones nor alarm pheromones will be treated because of their limited contribution against fish predation.

Numerous marine invertebrates have developed defence mechanisms based on noxious or distasteful chemicals (Paul, 1992). Many chemicals are contained in the defensive exudates or in the external part of the body of several opisthobranchs. This fact represents circumstantial evidence that these molecules represent a deterrent for predators.

Although there have been several reviews of chemical and morphological defences in opisthobranchs and fish behaviour, there are no reviews of the ecological implications of the predator-prey relationship between opisthobranchs and fish. For heuristic reasons, we have subdivided the issue into five questions: (1) Can fish memorize the warning colouration of shell-less molluscs? (2) What is the perfect prey colour? (3) Do dorsal protuberances act as fish lure? (4) Why do some slug-like opisthobranchs miss their shell? (5) What are the ecological effects of fish predation?

CAN FISH MEMORIZE THE WARNING COLOURATION OF SHELL-LESS MOLLUSCS?

It is generally accepted that many marine species with warning colouration to deter predator fish use distasteful molecules. This is especially the case with opisthobranch gastropods, a class of mollusc with an evolutionary trend that involves reduction and loss of their shells and which display a broad variety of conspicuous patterns of colouration. Following the reduction and loss of their mechanical defence mechanisms (the shell), opisthobranchs have gradually increased their dependence on secondary defensive mechanisms (Faulkner and Ghiselin, 1983).

A series of organic molecules have been chemically characterized in these molluscs and a defensive role has been suggested for them based on the results of some bioassays not always performed following well-grounded ecological experiments. For this reason, laboratory and field experiments were conducted with marine fish, either in aquaria or in a marine habitat, provided with palatable food strips formed of carrageenan and pleasant food that sometimes contained noxious molecules. This protocol only served to show that opisthobranchs are unpalatable to marine fishes. However, the defensive ability of opisthobranchs against potential predators could be better investigated by reproducing behavioural factors that affect diet selection in marine fish. The conspicuous colouration of distasteful opisthobranchs suggests that some fish might learn to associate unpleasant taste with colour patterns. Decisions about whether or not to eat a particular food may be affected by previous experience with any poison or other harmful substance that may be present in such food (Broom, 1981). In order to demonstrate the ability of reef fishes to discriminate colours and to associate colours with noxious prey, Giménez-Casalduero et al. (1999) tested the acceptance of coloured (black, yellow and red) artificial nudibranch models by two different reef fish communities. They found differences in the initial colour preferences, one reef fish population showing a strong initial preference for red models, while the other population showed no preference.

Nudibranchs belonging to the family Hypselodoris are known to be specialized predators of sponges rich in secondary metabolites.

Hypselodoris picta (Fig. 7.2B) sequesters the furanosesquiterpenoids longifolin, nakafuran-9, and ent-furodysinin from its food sponges, *Dysidea fragilis, Microciona toxystila* and *Pleraplysilla spinifera* (Avila *et al.*, 1991). The ability of *Hypselodoris picta* to select sponges rich in furanosesquiterpenoids and to transfer the dietary metabolites to dorsal formations (mantle dermal formations) has been rigorously proved (García-Gómez *et al.*, 1990; Fontana *et al.*, 1993).

One way of examining the influence of the opisthobranch colour pattern and chemicals on fish predation and of determining the relative importance of each is by using models. In such an experiment, carrageenan and mussels were mixed and modelled according to the shape and colour of the natural molluscs (Fig. 7.4). The assays were conducted using 'aposematic' (colour-pattern of *Hypselodoris picta*) and 'control' (blue and without secondary metabolites) models, but with the same conditions of palatability, size, shape and texture. To determine whether fish could learn to avoid the colour pattern of *Hypselodoris picta*, we carried four field assays with artificial models. In a preliminary experiment, the 'control' and 'aposematic' models were offered to fish in the field, which showed no pre-existing aversion to any of their colour patterns. Thus, fish appeared initially to perceive both 'control' and 'aposematic' models as palatable and worthy of attack. However, when the 'aposematic' model was treated with furodysinin freshly extracted from the sponge just before the experiment, fish refused the food. During four consecutive sessions at half hourly intervals, both models were offered to fish, resulting in a loss of interest in the chemically protected models. The fish clearly improved their foraging efficiency with experience until 'aposematic' models were ignored (Fig. 7.4). Two hours after the last training session, the experiments were repeated, offering control and 'unarmed' aposematic models to fish. In the first case, all the models were eaten but, surprisingly, fish continued to refuse the aposematic models although they contained no toxin (Fig. 7.5). Learned avoidance of conspicuously coloured models also persisted even when the artificial models displayed a black-white pattern. It is obvious that the fish learnt through distasteful encounter and memorized warning colouration of the opisthobranch. Nevertheless, when the aposematic model was impregnated with toxin, the frequency of fish attack decreased. Thus, the defensive secretion ejected in alarm by *Hypselodoris picta* and many opisthobranchs could reduce the rates of fish predation and encourage a pre-existing aversion in the predator.

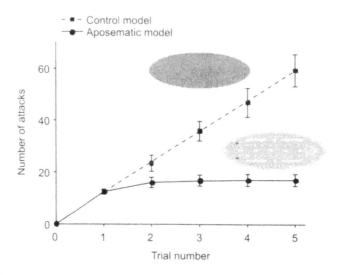

Fig. 7.4 Cumulative number of attacks by fish on distasteful or conspicuous and control models as a function of number of trials. Conspicuous models had the colour pattern of *Hypselodoris picta* and were impregnated with furodisynin.

BATESIAN AND MÜLLERIAN MIMICRY IN MARINE ECOSYSTEMS

The two most common phenomena of mimicry are Batesian and Müllerian mimicry. In Müllerian mimicry a less unpalatable species resembles a more unpalatable species, whereas in Batesian mimicry, a palatable species resembles an unpalatable species. Müllerian mimicry (Müller, 1879) is a classic example of an anti-predatory defence, where two or more aposematic species have sometimes strikingly similar warning patterns. The benefit of this resemblance is that if predators learn to avoid the warningly coloured prey from a fixed experience (Cott, 1940; Edmunds, 1974; Sherratt and Beatty, 2003), mimetic species (e.g., possessing adequate similarity) will have lower per capita mortality rates than dissimilar species for which predators have to learn each pattern separately (Müller, 1879; Rowe *et al.*, 2004). On the basis of this premise, Müllerian mimicry is considered an example of mutualism, where the cost of educating a predator is shared between similar prey types and the benefits of the resemblance can be calculated for each prey type by its frequency in a population (Müller, 1879; Joron and Mallet, 1998). Müllerian mimicry, like aposematism, is a strategy that is more beneficial the more common it is (Greenwood *et al.*, 1989; Kapan, 2001; Lindström *et al.*,

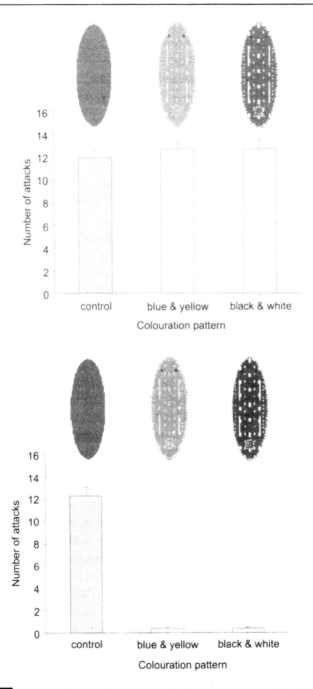

Fig. 7.5 Number of fish attacks before and after four sessions of training on distasteful models. Conspicuous models had the colour pattern of *Hypselodoris picta* but were not treated with furodisynin.

2001). This positive frequency dependence creates selection against dissimilar patterns, further promoting monomorphism in a warning pattern (Benson, 1972; Mallet and Barton, 1989; Kapan, 2001). Müllerian mimicry is an extension of the aposematic colouration strategy involving two or more species, where there is a strong stabilizing selection promoting monomorphism within species. Batesian mimicry is rather different in mechanism and evolutionary dynamics from Müllerian mimicry and aposematic colouration (Turner *et al.*, 1984). In Batesian mimicry, the innocuous mimic fools its predator by resembling a stinging or bad-tasting model that the predator has learned to avoid. Here, only the mimic benefits while both the unpalatable model and predator lose. Batesian mimicry can be unstable and potentially subject to rapid directional evolution. The occurrence of Batesian and Müllerian mimicry in terrestrial ecosystems is well established, but the possibility of its occurrence in marine species is more controversial (Edmunds, 1991; Ritland, 1991; Rudman, 1991).

Several authors have proposed that the colour-pattern convergence of chromodorids of the genera *Hypselodoris* and *Chromodorids* can follow a Müllerian or Batesian mimicry (Ros, 1976, 1977; Edmunds, 1987). Sympatric groups of similarly coloured chromodorid species occur in many parts of the world. Colour groups of chromodorids may be the result of a close genetic relationship, but frequently occur among clearly unrelated species (Rudman, 1991). There are two colour groups of chromodorids described in the Mediterranean, one belonging to the genus *Hypselodoris* (blue chromodorids) and the other to *Chromodoris* (gold chromodoris) (Ros, 1976, 1977). The blue forms are characterized by a general dark blue body colour with the notum edged with a white or yellow circle. Also, a white or yellow line, often broken, runs along the middle of the back. This basic colour-pattern changes slightly among species and individuals, so that it is sometimes difficult to distinguish between *Hypselodoris* species. Almost all *Hypselodoris* molluscs are brightly coloured, with a typical flat and oval shape. The *Hypselodoris* species sequester and store toxic, distasteful furanosesquiterpenoids from sponges (Fontana *et al.*, 1993). *Hypselodoris villafranca* contains the sponge-derived longiforin and nakafuran-9 in the mantle dermal formations, digestive gland and mucous secretion (Avila *et al.*, 1991). Some species such as *Hypselodoris orsini* transform the dietary metabolites of their sponge prey. *H. orsini* feeds on the sponge *Cacospongia mollior*, whose major metabolite is scalaradial.

H. orsini transforms scalaradial into deoxoscalarin and this latter into 6-ketodeoxoscalarin, which is stored in the mantle dermal formations (Cimino, *et al.*, 1993). *Hypselodoris tricolor* and *H. messinensis* feed on horny sponges (*Cacospongia* sp.) distributed along the western Atlantic coast and Mediterranean Sea (Cattaneo-Vietti *et al.*, 1990).

To explore the presence of Müllerian mimicry in *Hypselodoris* species models were prepared by mixing carrageenan and mussels modelled according to the colour pattern of *Hypselodoris tricolor*, *H. messinensis*, *H. villafranca* and *H. orsini* (Marin and Felipe, in preparation) (Fig. 7.6). To determine whether fish can learn to avoid the colour pattern of *Hypselodoris* species, we carried out four field assays with artificial models. Field assays were carried out as has been described previously with *H. picta*.

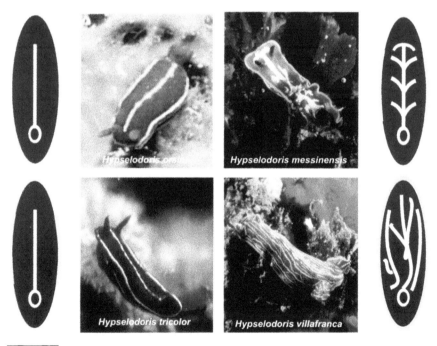

Fig. 7.6 Müllerian mimicry in the Mediterranean *Hypselodoris tricolor, H. messinensis, H. villafranca* and *H. orsini*. The blue forms are characterized by a general dark blue body colour with the notum edged with a white or yellow circle. Also, a white or yellow line, often broken, runs along the middle of the back. This basic colour-pattern changes slightly among species and individuals, so that it is sometimes difficult to distinguish between *Hypselodoris* species.

In a first experiment, field fish did not show differences in the number of attacks on *Hypselodoris tricolor, H. orsini, H. messinensis, H. villafranca, H. orsini* and control models when they were offered without toxin. However, when the *Hypselodoris tricolor* model was treated with furodysinin freshly extracted from the sponge during four consecutive training sessions, fish learned to avoid this model. In the next experiment, all models were offered without chemical defence. While fish profusely attacked control models, they completely ignored all *Hypselodoris* models. A second experiment was carried out with live *Hypselodoris* species of two different colour patterns, normal coloured and blue. The blue colour was obtained by dyeing *Hypselodoris picta* in Janus Green. The number of fish attacks on normal coloured patterns was lower than the blue-dyed opisthobranchs, suggesting a learned aversion toward its common colour pattern (Figs. 7.6 and 7.7). These last experiments demonstrate that fish cannot distinguish the colour pattern of *Hypselodoris* species which, therefore, may be termed Müllerian mimicry.

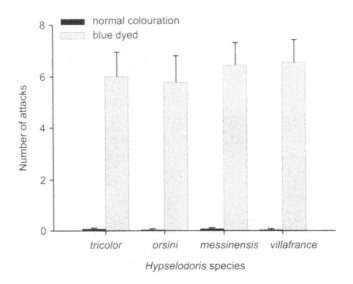

Fig. 7.7 Demonstration of Müllerian mimicry in *Hypselodoris tricolor, H. orsini, H. messinensis, H. villafranca* and *H. orsini*. Field assays with fish were carried out with models without chemical defence. Four training assays were carried out with *Hypselodoris tricolor* models impregnated with furodisynin.

EVOLUTION OF WARNING COLOURATION IN THE CONTEXT OF FISH DEFENCES

Marine ecologists have traditionally focused on competition and predation as determining factors in survival, but cooperation may play a greater role than previously believed (Michod, 1999). Shoaling by fish can lead to predator confusion and dilution of the attack and greater vigilance results in early predator detection (Csányi and Dóka, 1993). The herring *Clupea harengus* is adapted to a pelagic lifestyle and shows schooling behaviour (Axelsen *et al.*, 2000). During the spawning period, the school segregates vertically into a pelagic component that contracts into a tight ball and a demersal component that spreads out in a flat layer on the bottom (Axelsen *et al.*, 2000). Herrings stay in the pelagic ball in a suboptimal locality with respect to predation and food, waiting in the pelagic for the other fish in the school to spawn, supporting the suggestion that a school of herring generally makes collective decisions even if the optimal situation differs for individual fish (Fernö *et al.*, 1998). But, what is the ecological effect of fish schooling in the predator-prey relationship with aposematic opisthobranchs? The showing off warning colouration found in opisthobranchs is particularly effectiveness because potential fish predators frequently live in schools that may decrease the threat of other fish predators. When first encountering opisthobranchs or artificial models, individual fish leave the shoal and approach close to the potential prey. Field assays suggest that aposematic opisthobranchs can live in low densities because the negative signal transmission of these 'inspector' fish inhibits any attack by the shoal. Fish acquired a conditioned avoidance reaction against aposematic prey even without making direct contact with the prey, through observation of the consumption of other members of the shoal. The aposematic colouration of opisthobranchs may work as well for solitary species able to survive predator attacks as for gregarious species. The kin selection theory is an attractive way to explain how distasteful or poisonous species might have became conspicuously coloured, despite the fact that the most conspicuous animals are usually killed. Kin selection arguments have been applied to explain the evolution of warning colouration in terrestrial ecosystems but its application to explain aposematic colouration in marine species is controversial (Foukner and Ghiselin, 1983). However, many aposematic opisthobranchs do not live in family groups and, of course, they do not fulfil the necessary social conditions described in kin selection. Kin selection requires that members of a brood remain in close proximity so that the experience of a single predator is translated into the rejection of a sibling. Individual selection hypothesis indicates that if an aposematic animal survives the learning experience of predators, this individual gains a selective advantage. Edmunds (1991) suggested that in nudibranchs, kin selection should favour aposematic species with long life spans and non-planktonic development because this would enable the young to grow up close to the parent. Individual selection should be effective in nudibranchs with planktotrophic larvae. The planktotrophic development of many aposematic opisthobranchs forces one to think that individual selection, not kin selection, has been important in the evolution of the warning pattern of these opisthobranchs. Among butterflies and moths, warning colouration and aggregation of the broods of larvae are strongly associated (Sillen-Tullberg, 1988). This fact suggests that natural selection may promote many nudibranch species with the same pattern, as is often seen in Müllerian mimicry. Whatever the merits of such theoretical considerations, food dependency may have constrained the evolution of aposematic colouration in opisthobranchs. A substantial number of aposematic opisthobranchs, especially nudibranchs, obtain defensive metabolites (or nematocysts) from host food (sponges, algae, coelenterates, etc.) with a clamped dispersion, which, indirectly through their food, induces the free-swimming larvae to settle in the same area. A host food-associated kin selection rather that individual selection may explain the necessary conditions for the evolution of aposematic colouration in opisthobranchs.

WHAT IS THE PERFECT PREY COLOUR?

Since no defence is perfect and advertisement is potentially dangerous, natural selection on unpalatable prey should favour a combination of potentially aposematic and cryptic characters, which, when produced efficiently, should be alternative perceptions of the same structures or behaviours (Papageorgis, 1975). The marine ecosystems are a changing scenario. For example, the vertical light attenuation of marine ecosystems varies from hour to hour during the day, or the seaweed turfs grow seasonally in temperate seas. Apostatic selection suggests that predators prey upon the more common phenotypes or species (Holling, 1965). Gendron and Staddon (1983) showed that if we assume that the probability of detecting a prey is inversely related to search rate as well as to the degree of crypsis, then the optimum search rate represents a balance between encounter rate and degree of crypsis, which together determine the discovery rate. Specialist herbivores could be subjected to greater selective pressure by natural enemies of this sort than generalists living in the same environment. Thus, while specialization may perhaps offer refuge from some generalist predators, it will tend to increase danger from specialist natural enemies, or by generalists with learning ability (Rowell-Rahier and Pasteels, 1992).

MORPHOLOGICAL AND CHEMICAL CAMOUFLAGE

The transfer of fish deterrent metabolites from food to opisthobranch has been recognized in many species. Natural products originally obtained from algae, sponges, coelenterates, bryozoans and ascidians have been found in the defensive glands of many opisthobranchs. In all opisthobranchs, the digestive gland is the site of uptake of food into the rest of the body, and also of detoxification and other modifications of metabolites from the food. The metabolites that are found in the digestive gland are probably of dietary origin; those found in the skin but not the digestive gland are probably either secondarily modified or synthesized *de novo* (Cimino and Ghiselin, 1999).

The nudibranch *Discodoris indecora* shows perfect camouflage on the sponges *Ircinia variabilis* and *I. fasciculata*. The shape and colour of this nudibranch are remarkably similar to that of the sponge that is widespread in the shallow waters of the Mediterranean Sea (Marin *et al.*, 1997). The mantle is covered with large and rounded tubercles. In its sponge habitat, the foot and body are close to the sponge where the colour and shape of the mantle provide good camouflage. The nudibranch is able to transfer the sponge metabolites, sesterterpenoids palinurin and variabilin, from the digestive gland to the mantle glands located in the dorsal tubercles. When *Discodoris indecora* is handled roughly, the dorsal tubercles discharge a copious opalescent white secretion that contains the sponge metabolites. The sesterterpenoids palinurin and variabilin act as deterrent to the marine fish *Chromis chromis* and *Sparus auratus* (Marin *et al.*, 1997).

Nudibranchs may also obtain pigments from sponges in the diet. The retention of these pigments can be considered as a passive chemical defence mechanism. An example of such dietary camouflage, with pigments present in the sponge food, is found in the notaspidean *Tylodina perversa* and its prey the sponge *Aplysina aerophoba*. *Tylodina perversa* is a conspicuous yellow opisthobranch with a conical soft shell, the latter usually being colonised by seasonal seaweeds. *Tylodina perversa* is always found feeding on the yellow sponge *Aplysina aerophoba* from which they sequester brominated metabolites and the pigment uranidine (Ebel *et al.*, 1999). The yellow colour of *A. aerophoba* is due to the presence of the quinone pigment uranidine, which becomes black when exposed to air. Although *T. perversa* is partially covered by a soft shell, the brominated metabolites apparently provide protection from potential predators. When molested, the opisthobranch exudates a yellow mucus that is secreted from defensive glands in the mantle. This defensive mucus contains aerophobin-2, which was seen to act as a deterrent to the wrasse *Thalassoma pavo* (Ebel *et al.*, 1999).

An identical strategy has been described in *Oxynoe olivacea*, a green sacoglossan that lives camouflaged upon the alga *Caulerpa prolifera*, which transforms the major algal metabolite, caulerpenyne, to oxytoxin-1 and oxytoxin-2 by hydrolysis of the acetyl groups (Fig. 7.8). This process

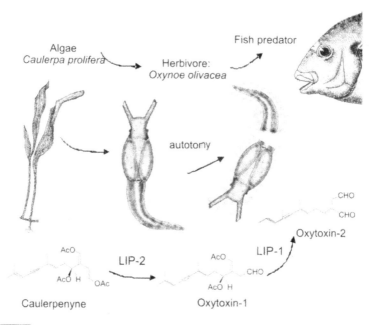

Fig. 7.8 Biotransformation of caulerpenyne in the sacoglossan *Oxynoe olivacea* from the algal food *Caulerpa prolifera*. Feeding assays with caulerpenyne demonstrated that this metabolite did not act as a deterrent towards marine fish, while the modified metabolites oxytoxin-1 and -2 did so. Modified from Marin and Ros (2004). The enzymatic transformation of caulerpenyne to oxytoxin-1 and oxytoxin-2 is due to two distinct hydrolytic enzymes, LIP-1 and LIP-2.

increases the toxicity of the algal metabolite (caulerpenyne) 100 times due to the activity of esterases Lip-1 and Lip-2 (Cutignano *et al.*, 2004).

In these three examples, we have seen that many opisthobranchs have a perfect camouflage on their host food, which also provide refuge and chemical defence against fish predators. However, crypsis may be costly because it prevents individuals from exploiting multiple habitats due to the need to match a particular background. This view changes the way that we think about adaptive colouration, so that we need to consider the costs and benefits of all types of colouration, not just warning signals.

HABITAT STABILITY

Hay *et al.* (1989) indicate that predation may be a major factor selected for feeding preferences and for the evolution of feeding specialization in small marine herbivores. Most studies of feeding preferences of fish herbivores have focused on the role of plant secondary metabolites and calcium carbonate. In some tropical seaweeds (e.g., *Halimeda goreauii*), the combination of calcium carbonate and secondary metabolites acts synergistically and deter feeding more than the sum of the effects of each component tested separately (Hay and Kappel, 1994). In many algae, calcification occurs as photosynthesis raises the pH of the water-filled spaces between algal cells and leads to the precipitation of calcium carbonate (Borowitzka, 1977). The nutritional quality and digestibility of plant foods is critical to herbivores. Herbivores usually select plant food according to its nutrient content (Gwynne and Bell, 1968).

Many sea slugs show a degree of feeding specialization similar to that seen in terrestrial insects. This is the case of a group of herbivore opisthobranchs, the sacoglossans, which retain functional chloroplasts from the algal diet (Marin and Ros, 1989). Most sacoglossans appear to feed on a very restricted number of closely related algal species (Jensen, 1994). In these specialized herbivores, seaweeds provide habitat as well as food. The herbivore populations depend on seasonal changes in food value, habitat stability, structure of living space and background colour. In this study we have chosen a monophagous opisthobranch, *Elysia timida*, that lives closely associated with the annual seaweed *Acetabularia acetabulum* (Marin and Ros, 1992, 1993). The green alga *Acetabularia acetabulum* produces calcium carbonate particles that discourage fish, sea urchins and others from eating the thallus in Mediterranean waters. The sacoglossan *E. timida* can feed suctorially on *Acetabularia acetabulum* every time that a portion of the stalk remains uncalcified. The mollusc retains functional chloroplasts from the food that are energy sources for the host, providing it with photosynthetic products (Marin and Ros, 1989). The host plant gradually changes in colour and in habitat complexity during its life cycle. Thus, the conspicuous white colour of *Elysia timida* in autumn provides a perfect camouflage in summer when the host algal turf is completely developed. *Elysia timida* contains a mixture of bioactive polypropionates also present in defensive secretion (Gavagnin *et al.*, 1994b). To evaluate whether the behaviour of fish changes with nasty encounters with the sacoglossan *Elysia timida*, two populations of wrasses *Thalassoma pavo* were collected from two different places, Palos Cape and Mazarrón Bay. In Palos Cape, *Elysia timida* is relatively abundant, whereas it is absent from Mazarrón Bay. We then offered white models reproducing the shape and colour of the mollusc and green models to the two fish populations. The inexperienced fish population immediately ate all models, whereas the experienced population refused the white model and showed a preference for the red and green models (Fig. 7.9).

But, what is the influence of habitat in all this? Let us look at two field experiments. In a first assay, rocky bottoms were selected which reproduced the background of the two extreme habitats of the food plant life. The white background and the more complex structure of calcified algal stalks in spring provide white models with a better protection from fish predation. In the second assay, the white models were treated with the mixture of polypropionates produced by *Elysia timida* and placed in the two algal habitats. In this case, the response from the inexperienced fishes was comparable with that of the fish population that cohabited with the opisthobranchs (Fig. 7.10).

These experiments suggest that aposematic colouration is the best defensive mechanism when meeting with predators and is inevitable in spite of individual loss is suffered during predator learning. On the contrary, when the seaweeds provide protection from predators, a cryptic colouration is the most economical method, while in seasonal and heterogeneous environments, both strategies may be complementary. The co-evolution of *Elysia timida* and *Acetabularia* has apparently favoured a white pattern that can be either aposematic or cryptic. This dual signal associated with the sacoglossan's colour-pattern could be reflective exploitation of a seasonal background.

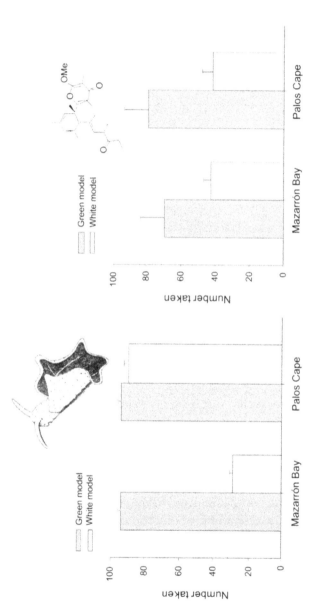

Fig. 7.9 The sacoglossan *Elysia timida* is able to biosynthesize *de novo* noxious polypropionates. The figure shows the frequency of green and white models eaten by two fish populations from two places. The fish population captured in a locality where no *Elysia timida* populations exist ate in both colour patterns, whereas the fish population that coexisted with the herbivore showed aversion to the white model. The assay suggests that herbivore colour pattern increased avoidance of all preys of the same species even in artificial coloured models. This is an important observation because it demonstrates that learned aversion exists in marine predators to some colour patterns present in opisthobranchs. When the experiment was carried out with the white model coated with polypropionates, the chemical defences reduced prey mortality.

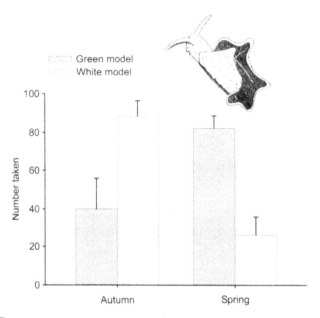

Fig. 7.10 *Elysia timida* is a specialized herbivore that feeds on the seaweed *Acetabularia acetabulum*. The population density of this herbivore is highly dependent on its algal food supply. This is controlled by the seasonal life cycle of the host plant, which grows from September to August, gradually calcifying the cellular wall and making it non-edible by the end of spring. The effects of algal architecture and unpalatatibility of *Elysia timida* are shown in this figure. The high habitat complexity reduces predation on this herbivore in the artificial models without polypropionates. The complex interactions among colour pattern, habitat and predator characters could give advantage to the presence of apparently contradictory signals (advertisement and concealment) in opisthobranchs.

Ecosystem Instability

Humans modify both the identities and numbers of species in ecosystems through disturbance, and the potential consequences of this modification for ecosystem functioning and services have received considerable attention during the last decade. Marine pollution induces changes in the composition of the benthic communities and favours the introduction of alien species, which is slowly altering predator-prey interactions and diversity. Shave *et al.* (1994) tested the anti-predator behaviour of a New Zealand freshwater crayfish (*Paranephrops zealandicus*) to the native long-finned eel (*Anguilla dieffenbachii*) and the introduced brown trout (*Salmo trutta*). Crayfish modified their behaviour in the presence of both trout and eels. However, a significantly greater number of defensive displays and swimming responses were made to eels than to trout. Crayfish were able to use chemical cues from skin mucus to detect eels but not trout. This may be a reflection of the different co-evolutionary histories crayfish have had with trout and eels so that crayfish may be at greater risk from the introduced predator because of their apparent inability to detect trout.

The intensive and unrelenting exploitation of the oceans has led to a progressive increase in the proportion of over fished stocks over the last 25 years. In the 1950s, most fisheries were undeveloped and very few were overexploited. Technological developments in gear and fish detection systems as well as huge investments in growing markets have changed that the scenario. Over 50% of all fish resources are currently either fished to full capacity or overexploited (GLOBEC, 2003). This means that the trophic structure of marine ecosystems has changed, and because the top control has been largely eliminated, the diversity of marine ecosystem is threatened.

Following severe stock collapses in many natural populations, efforts are now being invested in farming marine species, and in rehabilitating populations through controlled releases. Hatchery reared fishes show remarkable deficits in many aspects of their behavioural performance, resulting in high levels of mortality in the post-release phase (Suboski and Templeton, 1989; Brown and Smith, 1998; Brown and Day, 2002). It is commonly stated that anti-predator responses are weaker in hatchery reared than in wild fishes (Howell, 1994). Hatchery and wild fish grow up in very different environments; differential experience is likely to generate behavioural differences. Behaviour achieved early in life is likely to influence behaviour during later stages. Hence, deficiencies generated in early life are likely to affect later success. Offspring of wild trout *Salmo trutta* reared under hatchery conditions did not react to the presence of predatory brown trout (Avarez and Nicieza, 2003). Experience with predators during the first weeks of life probably explains the changes in the propensity to stay in refuges under predation threat. Several studies have indicated a negative effect of domestication on the predator avoidance behaviour of salmonids (Johnsson and Abrahams, 1991; Johnsson et al., 1996, 2001; Olla et al., 1998). Lack of experience with predators significantly changed the pattern and frequency of predator inspections in guppy *Poecilia reticulate* populations (Magurran, 1986; Magurran and Seghers, 1990). Malavasi et al. (2004) also found differences in anti-predator behaviour between hatchery reared and wild sea bass juveniles

Dicentrarchus labrax. An additional environmental problem is that farmed fish generally have a different genotype from that of local wild populations, and these artificially selected fish are typically bigger and more aggressive than wild fish. These differences may be problematic if farmed fish escape and begin breeding with local wild populations (Gro *et al.*, 2006).

DO DORSAL PROTUBERANCES ACT AS FISH LURE?

The localization of chemical defences in specific regions of an organism is a frequent characteristic in terrestrial plants and marine algae (Zangerl and Rutledge, 1996), and has been observed in some invertebrates, such as sponges (Schupp *et al.*, 1999), gorgonians (Harvell and Fenical, 1989), and molluscs (Cimino *et al.*, 1993). The allocation of defensive compounds to areas that are most vital for survival and fitness is a component of the optimal defence theory (Rhoades, 1979). In nudibranchs and sacoglossans where complete detorsion has occurred, the mantle cavity and gill have disappeared altogether. Respiration takes place through the general body surface. To help increase the body surface for this absorption, some have developed numerous projections called cerata, which are also utilised as a defensive structure. There is an evolutionary trend within opisthobranchs to develop dorsal papillae or cerata called 'aeolidisation' (Wägele and Willan, 2000). This trend for 'aeolidization' is apparent in many families of Nudibranchia, Sacoglossa and Acochlidiacea. A fascinating defensive mechanism of aeolid nudibranchs against fish is the storage and use of nematocysts from cnidarian prey (Edmunds, 1966). Nematocysts are ingested with other prey tissues and pass through the nudibranch's digestive system to the tips of the cerata, the dorsal extensions of the digestive gland. The nematocysts are stored in specialized structures, known as cnidosacs, at the tops of the cerata (Greenwood and Mariscal, 1984). The nematocysts are squeezed out from the cnidosacs when the aeolids are disturbed. The cerata are often brightly coloured in a manner that contrasts with the background colour of the mantle. Aeolids use dorsal protuberances as defensive lures. Predators may be attracted to the most unpleasantly flavoured part of the body, which they can bite and be repelled, without serious damage to the aeolid's body (Thompson, 1960; Edmunds, 1966). When attacked, an aeolid usually holds the cerata erect and may wave them towards the enemy. Predators are intercepted by the cerata, which avoid direct attack to the vital head and visceral mass of the aeolid.

Some Mediterranean aeolidacean nudibranchs such as *Cratena peregrina* exhibit warning colours. The cerata of *Cratena peregrina* has an iridescent blue and orange colouration with white areas on the upper parts that contrast with the white body. In addition, two conspicuous orange spots can easily be seen between the bases of the oral tentacles. Aguado and Marin (2006) analyzed the interaction between *Cratena peregrina* and predatory fish in laboratory and field assays with both live aeolids and artificial models. The first experiment was carried out with live *Cratena peregrina* of two different colour patterns, normal coloured and blue. The blue colour was obtained by dyeing *Cratena peregrina* in Janus Green. The number of attacks by fish was independent of the density of the prey, but the normal aeolids were attacked less than the blue ones. The fact that all normal *Cratena peregrina* survived while 12 to 20% of the blue aeolids died after fish attacks suggests that aposematic colouration provides a selective advantage against fish predators. Field and laboratory assays with artificial aeolids demonstrated that fish learned to avoid unpalatable models with the colour pattern of *Cratena peregrina*. The aeolid-like models were made with a cuttlefish-carrageenan mixture simulating the dorsal protuberances and the colour pattern of *Cratena peregrina*. After 3 to 4 training sessions with unpleasant models (impregnated of nematocysts), fish avoided palatable models with the same colour pattern. These results suggest that the colour pattern of *Cratena peregrina* combined with the presence of dorsal appendages and nematocysts encourage fish to avoid aeolids.

Defensive autotomy is a defence strategy widely used by opisthobranch molluscs as the last resource against fish predation (Stasek, 1967; Di Marzo *et al.*, 1993). This behaviour has adaptive significance in the opisthobranch when the autotomized structure: (1) is not essential for the continued existence of the prey itself (2) is the most frequently attacked part of the prey and (3) preferably contains the most potent deterrent substance (Todd, 1981). Regeneration of the autotomized parts occurs over different periods of time, varying from a few days to many weeks, depending on the species. Miller and Byrnea (2000) documented the regeneration of autotomized cerata in the aeolid nudibranch *Phidinna crassicornis*. Autotomized cerata exhibit a prolonged writhing response that may serve as a diversion to distract visual predators such as crabs and fishes. Four days after autotomy, regenerating cerata appeared as small protuberances. By day 24, the regenerates acquired their mature structural organisation and vivid colour. The cerata subsequently increased in length and diameter and were indistinguishable from surrounding cerata by 41 to 43 days after autotomy.

Dorsal protuberance or cerata are also formed in opisthobranchs capable of *de novo* biosynthesis of chemical defences. The biosynthesis of defensive metabolites is more costly than chemical constituents obtain from dietary sources. It is hardly surprising that species showing *de novo* biosynthesis of defensive metabolites use these metabolites for other purposes. The nudibranch *Tethys fimbria* is unique among living organisms in possessing great amounts of prostaglandin derivates (Marin *et al.*, 1991). Each side of the dorsum bears a series of cerata, which are readily autotomized when attacked by fish. These cerata may be found in different stages of regeneration. Autotomized cerata exhibited prolonged (up to 8 h) and irregular spontaneous contractions accompanied by extrusion of large quantities of slime. It is interesting that in the past authors described the body of the autotomized *Tethys fimbria* and its cerata (named *Phoenicurus*) as if they were different species (De Lacaze-Duthiers, 1885). Chemical analysis of the cerata led to the isolation of high levels of prostaglandin (PG) free acids and of PG-1,15 lactones of the E, A and F series, whereas the defensive secretion contained only PG-1,15 lactones of the E and A series (Fig. 7.11). The structural variety of the lactones and the data on their distribution in the body of *T. fimbria* suggest a range of different biological functions. PG free acids, derived from the opening of PG-1,15 lactones of the E series following the detachment of the cerata, are used *in vivo* to contract smooth muscle fibres. PG-1,15 lactones of the E series would participate in the chemical defensive mechanisms at two different levels: (1) directly, as the defence allomones of the ceratal secretion and (2) indirectly, as the precursors of bioactive PGs which, in turn, contract ceratal tissue, facilitating the secretion of the lactones themselves (Fig. 7.10). PGs also appear to be important in basic physiological functions of molluscs, including ion regulation, possible renal functions and reproductive biology (Cimino and Sodano, 1993).

The herbivorous sacoglossans *Cyerce cristallina* and *Ercolania funerea* secrete large amounts of slime, whose extracts displayed ichthyotoxic activity. *Cyerce cristallina* and *Ercolania funerea* undergo autotomy of the cerata followed by rapid (8 to 10 days) regeneration of dorsal appendages. Chemical analysis of the slime, mantle and cerata led to the isolation of polypropionate α- and γ-pyrones (Fig. 7.12). These secondary metabolites are synthesized *de novo* and possess structures that differ only in the degree

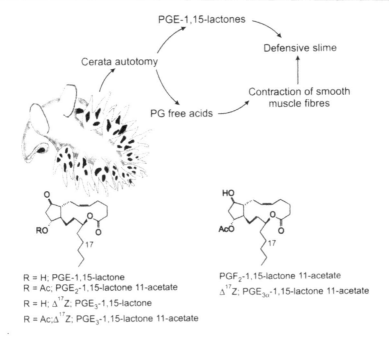

Fig. 7.11 Biological roles of prostaglandin derivates in the nudibranch *Tethys fimbria*. Prostagladin derivates are involved in the chemical defence and the spontaneous contractions of autotomized cerata.

of methylation and the geometry of double bonds of the side chain. The pyrones play a role either as defence allomones or as supportive inducers of cerata regeneration (Di Marzo *et al.*, 1993).

WHY DO SOME SLUG-LIKE OPISTHOBRANCHS MISS THEIR SHELL?

The lack of predation on sessile invertebrates such as sponges and coelenterates, including molluscs, by generalist predators has generally been attributed to their use of spicules and/or secondary metabolites as chemical defences (Thompson, 1960; Ros, 1976, 1977; Todd, 1981; McClintock, 1987). Many algae and invertebrates display redundant defence mechanisms against predation (Hay *et al.*, 1994; Schupp and Paul, 1994). Ascidians utilise both physical (spicules, tunic toughness) and chemical (secondary metabolites, acidity) defences and suffer relatively little predation by generalist predators. The ascidian *Cystodytes* (Polycitoridae) is widely distributed in both tropical and temperate waters.

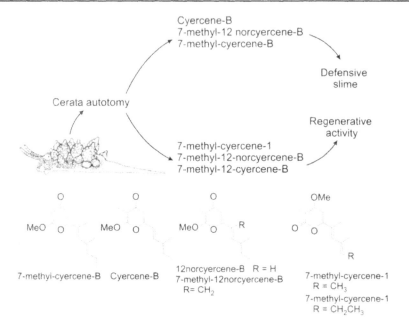

Fig. 7.12 Presence of polypropionate α- and γ-pyrones in the herbivorous sacoglossans *Ercolania funerea*. These secondary metabolites are synthesized *de novo* and play a role either as defence allomones as supportive inducers of cerata regeneration.

Cystodytes benefits from secondary metabolites (ascididemin), calcareous spicules and tunic acidity (pH<1). In field and laboratory assays conducted using artificial food, López-Legentil *et al.* (2006) found that all crude extracts and ascididemin significantly deterred fish predation, but not sea urchin predation. Calcareous spicules did not act as a deterrent in artificial food, and may only help protect the zooid in living colonies. Assays conducted using artificial food with sclerites incorporated showed that sclerites reduced feeding in fishes by 95% (Van Alstyne and Paul, 1992). This indicates that structural defences may play a role in determining an organism's ability to deter predators. Some sclerites are made up of calcium carbonate, which may serve as a structural as well as a chemical deterrence (Hay *et al.*, 1994; Schupp and Paul, 1994). Calcium carbonate will neutralise the low pH of some fish guts and the large amount of carbon dioxide released in the process can function as a chemical feeding deterrent (Hay *et al.*, 1994). However, the efficacy of the defensive role of structural defences has been doubted in some species. Chanas and Pawlik (1995) found no significant deterrence caused by siliceous spicules in sponges and suggested a deterrent role of sclerites of calcium carbonate related to an alteration of pH in the acidic gut of putative predators. Burns and Ilan (2003) found that deterrence in sponges was linked to spicule size, and only those spicules larger than 250 μm deterred predation. Several studies of sponges, gorgonians and ascidians (Pawlik et al., 1995; Koh et al., 2000; O'Neal and Pawlik, 2002; Puglisi et al., 2002) indicate that secondary metabolites are the primary means of defence against fish predators. In the sea fan Gorgonia ventalina, the sclerites incorporated into a carrageenan-based artificial diet reduced feeding in fishes by 95% (Van Alstayne and Paul, 1992). Sclerite concentration on this gorgonian ranged from 48.2% to 68.6% of the animal dry weight. Laboratory bioassays using greyhead wrasses, Halichoeres purpurescens, as well as field bioassays showed five gorgonian species from the family Ellisellidae and three from the family Plexauridae collected from Singapore reefs to be deterrent for fishes (Koh et al., 2000). Bioassays of fractions obtained from subsequent fractionation suggested synergistic or additive effects between the compounds present in gorgonians. Sclerites incorporated into fish feed in their natural concentrations were also tested for fish deterrence and were positive for only two gorgonian species from the family Ellisellidae. The results of this study suggest that inter-species variation in the shape and concentration of sclerites do not affect fish feeding. However, observed reductions in feeding might not be just due to concentration and morphology of sclerites alone, but to a reduction in food quality, since treatment pellets containing sclerites were quite probably of a lower nutritional quality than control pellets (Duffy and Paul, 1992; Koh et al., 2000).

Many nudibranchs have large quantities of endoskeletal calcareous spicules, which work as a flexible shell. The presence of skin spicules inside the mantle is generally considered as a defensive mechanism to make the mollusc unpalatable towards marine predators. Nudibranch defences against generalist predators include the inherently low nutritional value due to the high calcium carbonate content of the external tissues. The presence of large quantities of spicules suggests that nudibranchs are of poor nutritional values to fish predators. Several studies have indicated that chemical and physical defences commonly co-occur in opisthobranchs and can function either additively or synergistically to reduce susceptibility to consumers. In the nudibranch Doris verrucosa several defensive strategies occur at the same time. The nudibranch is perfectly mimetized in its habitat and its mantle is protected by spicules and by the presence of two toxic molecules, verrucosin-A and -B (Cimino and Sodano, 1989). The presence of large quantities of spicules, and their concomitant stiffening of an otherwise very soft body, may be of protective value to a nudibranch, which is subjected to repeated mouthing and rejection by a fish until all chemicals are exhausted. Increased tissue toughness due to spicules and the associated reduced damage from exploratory attacks by generalist predators may have been an important prerequisite for individual selection to favour chemical defence in nudibranchs (Penney, 2004). Interestingly, the phylogenetic pattern of spicules in dorid nudibranchs suggests that it is a primitive character for the group that is subsequently lost in taxa with the most effective chemical defences (Faulkner and Ghiselin, 1983; Cimino and Ghiselin, 1999). In fact, bioassays with artificial food indicated that spicules from the dorid *Cadlina luteomarginata* alone did not deter generalist crabs and anemones (Penney, 2006).

Peltodoris (=*Discodoris*) *atromaculata* bears, on the dorsal surface of their mantle, minute tubercles that are supported internally by calcareous spicules. The upper surface of the mantle is densely covered by spiculose tubercles of uniform size. The dorsal tubercles contain densely packed calcareous spicules of approximately 330 μm (Cattaneo-Vietti, 1993). The percentage of calcium carbonate in the mantle of the mollusc ranges from 58 to 68% of dry notum. The proportion of spicules oscillates between 3 and 4% of the mantle volume. To test whether mantle spicules of *Peltodoris atromaculata* deterred feeding by fishes, the isolated spicules were put into the carrageenan-based food and tested at 2% and 4% of the total volume. The diet was cut into small pieces of 0.5 ml^3 and offered to the gilthead bream *Sparus auratus*. The notal spicules of *Peltodoris atromaculata* seen to be significantly deterrent at the percentage volumes assayed. That is, notal spicules were strongly deterrent at the same concentration as found in living nudibranchs (Fig. 7.13). The predation of *Peltodoris atromaculata* on the sponge *Petrosia ficiformis* has been demonstrated by direct observations and gut analysis (Schmekel and Portman, 1982; Cattaneo-Vietti *et al.*, 1993), behavioural experiments and chemical analysis (Castiello *et al.*, 1980). The digestive gland of *Peltodoris atromaculata* contains chemical constituents of the sponge *Petrosia ficiformis*, mainly polyacetilenes (Castiello *et al.*, 1980) but they are not transported to the skin. The role of the digestive gland in protecting *Peltodoris atromaculata* is not clear, but

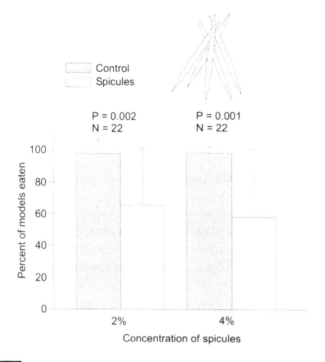

Fig. 7.13 Mean feeding rates of the fish *Sparus auratus* on artificial diet containing sclerites at 2% and 4% concentrations in paired feeding experiments. Vertical lines represent ± 1 SE.

its location within the body of the nudibranch is not consistent with an optimum defence location. There is some evidence that the sea hares store selected algal metabolites in the digestive gland and that these metabolites are more effective chemical repellents (Faulkner, 1992). The gut secretion of *Peltodoris atromaculata*, containing sponge polyacetilenes, may reinforce avoidance learning in fish predators. At worst, the defensive role of spicules is sufficient to explain avoidance learning in marine fishes. In addition, some large dorid species, notably *Peltodoris atromaculata*, present autotomy of the mantle margin. In fact, when *Peltodoris atromaculata* is rough handled, the mantle margin, containing densely packed calcareous spicules, is readily autotomized.

WHAT ARE THE ECOLOGICAL EFFECTS OF FISH PREDATION?

The roles of chemical or physical defences have not been widely considered in discussions of marine ecosystem structure or biogeochemical cycling. Feeding-deterrent properties exist across widely diverse plankton and benthos species (Ianora *et al.*, 2006). Defence mechanisms often act at other levels, such as fouling avoidance or space competition, and they may act in different life history stages (Stoecker, 1980; Becerro *et al.*, 1997; Pisut and Pawlik, 2002). At the same time, their multiple functions may constrain their evolution (Kubanek *et al.*, 2002), meaning that there is always scope for the evolution of specialist predators able to circumvent the defence mechanisms of a given species.

Hutchinson (1996) described the paradox of plankton and focused on the processes that support a high diversity of planktonic species in seemingly homogenous environments. In oceanic food webs, defences would have important implications for the regulation of material and energy, and such defences may help explain the abundance of co-existing species, all competing for similar resources, in a seemingly homogeneous habitat (McClintock *et al.*, 1996). The nature of grazing relationships may be able to modify the structure of prey populations, the creation of sinking particles, and the cycling of biogenic elements. The species diversity of planktonic ecosystems could likewise be increased by the diversity of unique defences and co-evolved specialized behaviours (McClintock *et al.*, 1996). The resource-partitioning model assumes that maximum species diversity will be obtained when full competitive resource partitioning has occurred. An alternative hypothesis suggests that species will be most diverse when competition is prevented. In this context, one of the causes of the high diversity of tropical and temperate seas is the diversity of predatory fish (Fig. 7.13). Defence studies have focused on single trophic levels, and there has been little attention paid as to how changing diversity at higher trophic levels may influence ecosystem processes. Further, all natural ecosystems contain communities with multiple trophic levels, and interactions between trophic levels are integral to the dynamics of most natural ecosystems, especially aquatic ones. Further studies should incorporate both bottom-up and top-down processes over several trophic levels to understand the mechanisms behind biodiversity effects on ecosystem functioning.

Predation is one of the major factors influencing community structure and ecosystem processes in aquatic systems through direct and indirect effects of prey density and prey behaviour. Many pelagic and benthic species are able to detect the presence of their fish predator by chemical cues alone. Pelagic planktivores, performing diel vertical migration, spend most of their time in near darkness, and hence the ability to detect predators by chemical cues alone is essential for their survival. A common anti-predator behaviour in many species is to decrease ingestion and their activity rate in response to predator cues, which reduces encounters with fish predators and the probability of being detected. Behaviour needs to be taken into account in food web studies, since much anti-predator behaviours, such as reduced feeding, results in effects on the lower trophic levels similar to those caused by direct fish predation (ghost predation). Ghost predation could explain the phenotypic plasticity of some species and, indirectly, species diversification through evolution. Conspecific populations living in habitats with different risks of predation often show phenotypic variation in defensive traits. Peckarsky et al. (2005) demonstrated that populations of mayflies living in fish and fishless streams are not genetically distinct, and are consistent with the hypothesis that traits associated with environments of different risk are phenotypically plastic. Traits of two species of mayflies (Baetidae: Baetis bicaudatus and Baetis sp. nov.) differ between populations living in fish and fishless streams in a high altitude drainage basin in western Colorado, USA. Mayflies, associated with the presence or absence of fish, are capable of responding to chemical cues associated with actively feeding trout, which trigger nocturnal feeding behaviour and accelerated development. This phenotypic plasticity enables larvae to adjust their behaviour and life history, thus completing development quickly in risky environments and maximizing survival, and extending the period of growth and thereby increasing fecundity in safer environments.

The complex predation (or grazing) interactions caused by the evolution of defensive properties relax predation pressure on chemically and/or physically defended species (Fig. 7.14). The co-evolution of specialized predators (host predation) that can feed despite the defences of their prey allows a range of niche diversification. If these specialized predators themselves receive defensive characteristics through the depredator-prey relationship, this further increases the complexity of the ecosystem. The opisthobranchs provide numerous examples of the acquisition of chemical defences derived from food.

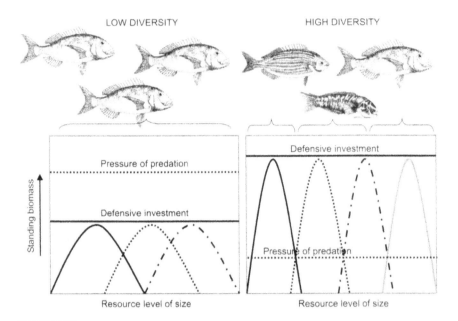

Fig. 7.14 Fish predation is one of the major factors influencing community structure and ecosystem processes in aquatic systems through the direct and indirect effects of prey density and prey behaviour. The complex predation (or grazing) interactions caused by the evolution of defensive properties relax the fish predation pressure on chemically and/or physically defended species. The coevolution of specialized predators that can feed despite the defences of their preys allows a range of niche diversification. If these specialized predators themselves receive defensive characteristics through the depredator-prey relationship, this further increases the complexity of the ecosystem.

References

Aguado, F. and A. Marín. 2006. Warning colouration associated with nematocyst-based defenses in aeolidacean nudibranchs. *Journal of Molecular Studies* 73: 23-28.

Avarez, D. and A.G. Nicieza. 2003. Predator avoidance behaviour in wild and hatchery-reared brown trout: the role of experience and domestication. *Journal of Fish Biology* 63: 1565-1577.

Avila, C., G. Cimino, A. Fontana, M. Gavagnin, J. Ortea and E. Trivellone. 1991. Defensive strategy of two Hypselodoris nudibranchs from Italian and Spanish coasts. *Journal of Chemical Ecology* 17: 625-636.

Axelsen, B.E., L. Nøttestad, A. Fernö, A. Johannessen and O.A. Misund. 2000. 'Await' in the pelagic: dynamic trade-off between reproduction and survival within a herring schoolsplitting vertically during spawning. *Marine Ecology Progress Series* 205: 259-269.

Becerro, M.A., X. Turon and M.J. Uriz. 1997. Multiple functions for secondary metabolites in encrusting marine invertebrates. *Journal of Chemical Ecology* 23: 1527-1547.

Benson, W.W. 1972. Natural selection for Müllerian mimicry in *Heliconius erato* in Costa Rica. *Science* 179: 936-939.

Borowitzka, M.A. 1977. Algal calcification. *Oceanography and Marine Biology Annual Review* 15: 189-223.

Brönmark, C. and L.A. Hansson. 2000. Chemical communication in aquatic systems: an introduction. *Oikos* 88: 103-109.

Broom, D.M. 1981. *Biology of Behaviour*. Cambridge University Press, Combridge.

Burns, E. and M. Ilan. 2003. Comparison of anti-predatory defenses of Red Sea and Caribbean sponges: II. Physical defense. *Marine Ecology Progress Series* 252: 115-123.

Cattaneo-Vietti, R., S. Angelini and G. Bavestrello. 1993. Skin and gut spicules in *Discodoris atromaculata* (Bergh, 1880) (Mollusca: Nudibranchia). *Bolletin Malacologico* 29: 173-180.

Castiello, D., G. Cimino, S. De Rosa, G. De Stefano and G. Sodano. 1980. High molecular weight polyacetylenes from nudibranch *Peltodoris atromaculata* and the sponge *Petrosia ficiformis*. *Tetrahedron Letters* 21: 5047.

Cattaneo-Vieti, R., R. Chemello and R. Giannuzzi-Savelli. 1990. *Atlas of Mediterranean Nudibranchs*. La Conchiglia, pp. 264.

Cimino, G., A. Fontana, F. Gimenez, A. Marin, E. Mollo, E. Trivellone and E. Zubia. 1993. Biotransformation of a dietary sesterterpenoid in the Mediterranean Nudibranch *Hypselodoris orsini*. *Experientia* 49: 582-586.

Cimino, G. and G. Sodano. 1989. The chemical ecology of Mediterranean opisthobranchs. *Chemica Scripta* 29: 389-394.

Chivers, D.P. and R.S. Mirza. 2001. Predator diet cues and the assessment of predation risk by aquatic vertebrates: A review and prospectus. In: *Chemical Signals in Vertebrates*, A. Marchlewska-Koj, J.J. Lepri and D. Müller-Schwarze (eds.). Kluwer Academic/Plenum Publishers, New York, Vol. 9, pp. 277-284.

Cott, H.B. 1940. *Adaptive Coloration in Animals*. Methuen, London.

Chanas, B. and J.R. Pawlik. 1995. Defenses of Caribbean sponges against predatory reef fish: II. Spicules, tissue toughness, and nutritional quality. *Marine Ecology Progress Series* 127: 95-211.

Csányi, V. and A. Dóka. 1993. Learning interactions between prey and predator fish. *Marine Behaviour and Physiology* 23: 63-78.

Cutignano, A., V. Notti, G. D'Ippolito, G. Cimino and A. Fontana. 2004. Lipase-mediated production of defensive toxins in the marine mollusc *Oxynoe olivacea*. *Organic and Biomolecular Chemistry* 2: 3167-3171.

De Lacaze-Duthiers, H. 1885. Sur le Phoenicure. C.r. hebd. Séanc. Acad. Sci., Paris CI: 30-35.

Di Marzo, V., A. Marin, R.R. Vardaro, L. De Petrocellis, G. Villani and G. Cimino. 1993. Histological and biochemical bases of defense mechanims in four species of Polybranchioidea ascoglossan molluscs. *Marine Biology* 117: 367-380.

Edmunds, M. 1966. Protective mechanisms in the Eolidacea (Mollusca: Nudibranchia). *Journal of Linean Society (Zoology)* 46: 27-71.

Edmunds, M. 1974. *Defence in Animals. A Survey of Anti-predator Defences*. Longman, New York.

Edmunds, M. 1987. Color in Opisthobranchs. *American Malacological Bulletin* 5: 185-196.

Edmunds, M. 1991. Does warning coloration occur in Nudibranchs? *Malacologia* 32: 241-255.

Ebel, R., A. Marin and P. Proksch. 1999. Organ-specific distribution of dietary alkaloids in the marine opisthobranch *Tylodina perversa*. *Biochem. Syt. Ecol.* 27: 769-777.

Endler, J.A. 1978. A predator's view of animal color pattern. In: *Evolutionary Biology*, M.K. Hecht, W.C. Steere and B. Wallace (eds.). Plenum Press, New York, Vol. 11, pp. 319-364.

Faulkner, D.J. 1992. Chemical defenses of marine molluscs. In: *Ecological Roles of Marine Natural Products*, V.J. Paul (ed.). Constock Publishing Associates, p. 119.

Faulkner, D.J. and M.T. Ghiselin. 1983. Chemical defence and evolutionary ecology of dorid nudibranchs and some other opisthobranch gastropods. *Marine Ecology Progress Series* 13: 295-301.

Fisher, R.A. 1930. *The Genetical Theory of Natural Selection*. Oxford University Press, Clarendon, Oxford.

Fontana, A., F. Gimenez, A. Marin, E. Mollo and G. Cimino. 1993. Transfer of secondary metabolites from the sponges *Dysidea fragilis* and *Pleraplysilla spinifera* to the mantle dermal formations (MDFs) of the nudibranch *Hypselodoris webbi*. *Experientia* 50: 510-516.

García-Gómez, J.C., A. Medina and R. Coveñas. 1991. Study on the anatomy and histology of the mantle dermal formations (MDFs) of *Chromodoris* and *Hypselodoris*. *Malacología* 32: 233-240.

Gavagnin, M., A. Marin, F. Castellucio, G. Villani and G. Cimino. 1994a. Defensive relationships between *Caulerpa prolifera* and its shelled sacoglossan predators. *Journal of Experimental Marine Biology and Ecology* 175: 197-210.

Gavagnin, M., A. Spinella, F. Castelluccio, G. Cimino and A. Marin. 1994b. Polypropionates from the Mediterranean mollusc *Elysia timida*. *Journal of Natural Products* 57: 298-304.

Gittleman, J.L., P.H. Harvey and P.J. Greenwood. 1980. The evolution of conspicuous coloration: some experiments in bad taste. *Animal Behaviour* 28: 897-899.

Guilford, T. 1990a. The secrets of aposematism: Unlearned responses to specific colours and patterns. *Tree* 5: 323.

Guilford, T. 1990b. The evolution of aposematism, In: *Insect Defenses: Adaptative Mechanisms and Strategies of Preys and Predators*, D.L. Evans and J.O. Schmidt (eds.). University of New York Press, New York.

Giménez-Casalduero, F., R.W. Thacker and V.J. Paul. 1999. Association of color and feeding deterrence by tropical reef fishes. *Chemoecology* 9: 33-39.

Greenwood, J.J.D., P.A. Cotton and D.A. Wilson. 1989. Frequency-dependent selection on aposematic prey: Some experiments. *Biological Journal of the Linnean Society* 36: 213-226.

Greenwood, P.G. and R.N. Mariscal. 1984. the utilization of cnidarian nematocysts by aeolid nudibranchs: Nematocyst maintenance and release in *Spurilla*. *Tissue and Cell* 15: 719-730

Gro, A., V. Salvanes and V. Braithwaite. 2006. The need to understand the behaviour of fish reared for mariculture or restocking. *ICES Journal of Marine Science* 63: 346-354.

Gwynne, M.O. and R.H.V. Bell. 1968. Selection of vegetation components by grazing ungulates in the Serengeti National Park. *Nature (London)* 220: 390-393.

Hard, I.H., W. Fenical, G. Cronin and M.E. Hay. 1996. Acutilols, potent herbivore feeding deterrents from the tropical brown alga, *Dictyota acutiloba*. *Phytochemistry* 43: 71-73.

Harvell, C.D. and W. Fenical. 1989. Chemical and structural defenses of Caribbean gorgonians (*Pseudopterogorgia* spp.): Intracolony localization of defense. *Limnology and Oceanography* 34: 382-389.

Hay, M.E., Q.E. Kappel and W. Fenical. 1994. Synergisms in plant defenses against herbivores: Interactions of chemistry, calcification, and plant quality. *Ecology* 75: 1714-1726.

Hutchinson, G.E. 1996. The paradox of the plankton. *American Naturalist* 95: 137-145.

Ianora1, A., M. Boersma, R. Casotti1, A. Fontana, J. Harder, F. Hoffmann, H. Pavia, P. Potin, S.A. Poulet and G. Toth. 2006. New trends in marine chemical ecology. *Estuaries and Coasts* 29: 531-551.

Janzen, D.H. 1980. When is it coevolution? *Evolution* 34: 611-612.

Jensen, K.R. 1994. Behavioural adaptations and diet specificity of sacoglossan opisthobranchs. *Ethology, Ecology and Evolution* 6: 87-101.

Joron, M. and J. Mallet. 1998. Diversity in mimicry: Paradox or paradigm? *Trends in Ecology and Evolution* 13: 461-466.

Johnsson, J.I. and M.V. Abrahams. 1991. Interbreeding with domestic strain increases foraging under threat of predation in juvenile steelhead trout (*Oncorhynchus mykiss*): An experimental study. *Canadian Journal of Fisheries and Aquatic Sciences* 48: 243-247.

Johnsson, J.I., J. Höjesjö and I.A. Fleming. 2001. Behavioural and heart rate responses to predation risk in wild and domesticated Atlantic salmon. *Canadian Journal of Fisheries and Aquatic Sciences* 58: 788-794.

Johnsson, J.I., E. Petersson, E. Jönsson, B. Björnsson and T. Järvi. 1996. Domestication and growth hormone alter antipredator behaviour and growth patterns in juvenile brown trout, *Salmo trutta*. *Canadian Journal of Fisheries and Aquatic Sciences* 53: 1546-1554.

Kapan, D.D. 2001. Three-butterfly system provides a field test of Müllerian mimicry. *Nature (London)* 409: 338-340.

Karban, R. and I.T. Baldwin. 1997. *Induced Responses to Herbivory*. University of Chicago Press, Chicago, IL.

Koh, L.L., N.K.C. Goh, L.M. Chou and Y.W. Tan. 2000. Chemical and physical defenses of Singapore gorgonians (Octocorallia: Gorgonacea). *Journal of Experimental Biology and Ecology* 251: 103-115.

Krebs, J.R. and N.B. Davies. 1991. *Behavioural Ecology. An Evolutionary Approach*. Blackwell Scientific Publications, Oxford.

Kubanek, J., K.E. Whalen, S. Engel, S.R. Kelly, T.P. Henkel, W. Fenical and J.R. Pawlik. 2002. Multiple defensive roles for triterpene glycosides from two Caribbean sponges. *Oecologia* 131: 125-136.

Lindström, L., R.V. Alatalo and J. Mappes. 1997. Imperfect Batesian mimicry: The effects of the frequency and the distastefulness of the model. *Proceedings of the Royal Society of London* B264: 149-153.

López-Legentil, S., X. Turon and P. Peter Schupp. 2006. Chemical and physical defenses against predators in Cystodytes (Ascidiacea). Journal of Experimental Marine Biology and Ecology 332: 27-36.

Norton, S. 1988. Role of gastropod shell and operculum in inhibiting predation by fish. Science 241: 92-94.

Magurran, A.E. 1986. Predator inspection behaviour in minnow shoals: Differences between populations and individuals. Behavioural Ecology and Sociobiology 19: 267-273.

Magurran, A.E. and B.H. Seghers. 1990. Population differences in predator recognition and attack cone avoidance in the guppy Poecilia reticulata. Animal Behaviour 40: 443-452.

Malavasi, S., V. Georgalas, M. Lugli, P. Torricelli and D. Mainardi. 2004. Differences in the pattern of antipredator behaviour between hatchery-reared and wild European sea bass juveniles. Journal of Fish Biology A65: 143-155.

Mallet, J. and N.H. Barton. 1989. Strong natural selection in a warning-colour hybrid zone. Evolution 43: 421-431.

Marin, A. and J.D. Ros. 1989. The chloroplast-animal association in four Iberian Sacoglossan Opisthobranchs: Elysia timida, Elysia translucens, Thuridilla hopei and Bosellia mimetica. Scientia Marina 53: 429-440.

Marin, A. and J.D. Ros. 1992. Dynamics of a peculiar plant-herbivore relationship: The photosynthetic sacoglossan Elysia timida and the chlorophycean Acetabularia acetabulum. Marine Biology 112: 677-682.

Marin, A. and J.D. Ros. 1993. Ultrastructural and ecological aspects of the development of chloroplast retention in the sacoglossan gastropod Elysia timida. Journal of Molecular Studies 59: 95-104.

Marín, A., L.A. Alvarez, G. Cimino and A. Spinella. 1999. Chemical defence in cephalaspidean gastropods: Origin, anatomical location and ecological roles. Journal of Molecular Studies 65: 121-131.

Marín, A. and J.D. Ros. 2004. Chemical defenses in sacoglossan opisthobranchs: taxonomic trends and evolutive implications. Scientia Marina 68: 227-241.

Marín, A. and M.D. López Belluga. 2005. Sponge coating decreases predation on the bivalve Arca noae. Journal of Molecular Studies 71: 1-6.

Marin, A., V. Di Marzo and G. Cimino. 1991. A histological and chemical study of the cerata of the opisthobranch mollusc Tethys fimbria. Marine Biology 111: 353-358.

Marin, A., M.D. López Belluga, G. Scognamiglio and G. Cimino. 1997. Morphological and chemical camouflage of the Mediterranean nudibranch Discodoris indecora on the sponges Ircinia variabilis and Ircinia fasciculata. Journal of Molecular Studies 63: 431-439.

McClintock, J.B. 1987. Investigation of the relationship between invertebrates predation and biochemical composition, energy content, spicule armament and toxicity of benthic sponges at McMurdo Sound, Antarctica. Marine Biology 94: 479-487.

McClintock, J.B., D.P. Swenson, D.K. Steinberg and A.A. Michaels. 1996. Feeding-deterrent properties of common oceanic holoplankton from Bermudian waters. Limnology and Oceanography 41: 798-801.

Michod, R.E. 1999. *Darwinian Dynamics. Evolutionary Transitions in Fitness and Individuality*, Princeton University Press, Princeton.

Jennifer, A., J.A. Miller and M. Byrnea. 2000. Ceratal autotomy and regeneration in the aeolid nudibranch *Phidiana crassicornis* and the role of predators. *Invertebrate Biology* 119: 167-176.

Müller, F. 1879. *Ituna* and *Thyridia*: A remarkable case in butterflies. *Proceedings of the Entomological Society of London* 1879: 20-29.

Olla, B.L., M.W. Davis and C.H. Ryer. 1998. Understanding how the hatchery environment represses or promotes the development of behavioural survival skills. *Bulletin of Marine Sciences* 62: 531-550.

O'Neal, W. and J.R. Pawlik. 2002. A reappraisal of the chemical and physical defenses of Caribbean gorgonian corals against predatory fishes. *Marine Ecology Progress Series* 240: 117-126.

Papageorgis, C. 1975. Mimicry in Neotropical butterflies. *American Scientist* 63: 522-532.

Paul, V.J. 1992. *Ecological Roles of Marine Natural Products*. Comstock Publishing Associates, Ithaca, New York.

Pawlik, J.R., B. Chanas, R.J. Toonen and W. Fenical. 1995. Defense of Caribbean sponges against predatory reef fish: I. Chemical deterrency. *Marine Ecology Progress Series* 17: 183-194.

Peckarsky, B.L., J.M. Hughes, P.B. Mather, M. Hillyer and A.C. Encalada. 2005. Are populations of mayflies living in adjacent fish and fishless streams genetically differentiated? *Freshwater Biology* 50: 42-51.

Penney, B.K. 2002. Lowered nutritional quality supplements nudibranch chemical defense. *Oecologia* 132: 411-418.

Penney, B.K. 2006. Morphology and biological roles of spicule networks in *Cadlina luteomarginata* (Nudibranchia, Doridina). *Invertebrate Biology* 125: 222-232.

Petranka, J.W., L.B. Kats and A. Sih. 1987. Predator-prey interactions among fish and larval amphibians: use of chemical cues to detect predatory fish. *Animal Behaviour* 35: 420-425.

Pisut, D.P. and J.R. Pawlik. 2002. Anti-predatory chemical defenses of ascidians: secondary metabolites or inorganic acids? *Journal of Experimental Marine Biology and Ecology* 270: 203-214.

Puglisi, M.P., V.J. Paul, J. Biggs and M. Slattery. 2002. Co-occurrence of chemical and structural defenses in gorgonian corals from Guam. *Marine Ecology Progress Series* 239: 105-114.

Ritland, D.R. and L.P. Brower. 1991. The viceroy is not a Batesian mimic. *Nature* (*London*) 350: 497-498.

Ros, J.C. 1976. Sistemas de defensa en los opistobranquios. *Oecol. Aquat.* 2: 41-77.

Ros, J.C. 1977. La defensa en los opistobranquios. *Investigación y Ciencia* 12: 48-60.

Rowe, C., L. Lindström and A. Lyytinen. 2004. The importance of pattern similarity between Müllerian mimics on predator avoidance learning. *Proceedings of the Royal Society of London* B271: 407-413.

Rhoades, D.F. 1979. Evolution of plant chemical defenses against herbivores. In: *Herbivores: Their Interaction with Secondary Plant Metabolites*, G.A. Rosenthal (ed.). Academic Press, Orlando, pp. 4-55.

Rudman, W.B. 1991. Purpose in pattern: The evolution of colour in chromodorid nudibranchs. *Journal of Molecular Studies* 57: 5-21.

Shave, C.R., C.R. Townsend and T.A. Crowl. 1994. Anti-predator behaviours of a freshwater crayfish (*Paranephrops zealandicus*) to a native and an introduced predator. *New Zealand Journal of Ecology* 18: 1-10.

Sherratt, T.N. and C.D. Beatty. 2003. The evolution of warning signals as reliable indicators of prey defense. *American Naturalist* 162: 377-389.

Schupp, P.J. and V.J. Paul. 1994. Calcification and secondary metabolites in tropical seaweeds: variable effects on herbivorous fishes. *Ecology* 75: 1172-1185.

Schupp, P., C. Eder, V. Paul and P. Proksch. 1999. Distribution of secondary metabolites in the sponge *Oceanapia* sp. and its ecological implications. *Marine Biology* 135: 573-580.

Schmekel, L. and A. Portmann. 1982. *Opisthobranchia des Mittelmeeres*. Springer-Verlag, Berlin.

Sillen-Tullberg, B. 1985. The significance of coloration per se, independent of background, for predator avoidance of aposematic prey. *Animal Behaviour* 33: 1382-1384.

Stasek, C.R. 1967. Autotomy in the Mollusca. *California Academy of Sciences, Occasional Papers* 61: 1-44.

Stoecker, D. 1980. Chemical defenses of ascidians against predators. *Ecology* 61: 1327-1334.

Rudman, W.B. 1991. Purpose in patterns: The evolution of colours in Chromodorids Nudibranchs. *Journal of Molecular Studies* 57: 5-21.

Todd, C.D. 1981. Ecology of nudibranch molluscs. *Marine Biology Annual Review* 19: 141-234.

Tollrian, R. and C.D. Harvell. 1999. *The Ecology and Evolution of Inducible Defenses*. Princeton University Press, Princeton.

Thompson, T.E. 1960. Defensive adaptations in opisthobranchs. *Journal of Marine Biological Association of the United Kingdom* 39: 123-134.

Turner, J.R., G.E.P. Kearney and L.S. Exton. 1984. Mimicry and the Monte Carlo predator: the palatability spectrum and the origins of mimicry. *Biological Journal of the Linnean Society* 23: 247-268.

Van Alstyne, K.L. and V.J. Paul. 1992. Chemical and structural defenses in the sea fan *Gorgonia ventalina*: Effects against generalist and specialist predators. *Coral Reefs* 11: 155-159.

Wägele, H. and R.C. Willan. 2000. Phylogeny of the Nudibranchia. *Zoological Journal of the Linnean Society*: 130: 83-181.

Zangerl, A.R. and C.E. Rutledge. 1996. The probability of attack and patterns of constitutive and induced defense: A test of optimal defense theory. *American Naturalist* 147: 599-607.

Behavioural Defenses in Fish

Jörgen I. Johnsson

INTRODUCTION

In the best of all possible worlds, fish would not need any behavioural defenses (Voltaire, 1759). However, fish do not live in the best of all possible worlds. Instead, many species live under harsh environmental conditions, sharing habitats with predators, parasites and competitors and only the individuals that deal most efficiently with these problems will pass on their genes to future generations (Dawkins, 1976). Such efficiency requires appropriate behavioural responses. It is not sufficient just to avoid all threats, because fish need not only to stay alive, but must also generally grow to be able to reproduce. Therefore, many behavioural actions reflect trade-offs between growth, survival and reproduction. In addition, the nature of these trade-offs often changes during the lifespan of an individual, for example, as sexual maturity is approaching (Ludwig and Rowe, 1990).

The aim of this chapter is to provide an insight into the great diversity of behavioural defenses in fish and the mechanisms shaping this diversity.

Author's address: Department of Zoology, Göteborg University, SE-405 30 Göteborg, Sweden.
E-mail: jorgen.johnsson@zool.gu.se

In a broader context, behavioural defenses could be defined as any action defending an individual or its resources. However, I have restricted the scope of this chapter to defenses against predators and parasites. While it was still not possible to cover all aspects in depth, I have suggested other relevant literature for further reading. Although the focus is on behaviour, some morphological defenses are also discussed, as these are often co-adapted with behaviour (Marshall, 2000). The main part of the chapter focuses on the diverse anti-predator tactics exhibited by fishes. Less is known about behaviour serving to mitigate parasites and other pathogens, but recent progress in this research field is also discussed, followed by some suggestions for further research to complete the chapter. Hopefully, it will become clear to the reader that fish very rarely live in the best of all possible worlds but, nevertheless, have come up with an intriguing diversity of solutions to cope with the challenges of life.

DEFENSE AGAINST PREDATORS

All behavioural actions are the end results of complex interactions between genetic and environmental factors (Pigliucci, 2001). Moreover, constraints imposed by phylogeny, physiology and morphology will set limits to the behavioural options available to a specific species or individual. For example, a poor swimmer is not likely to outswim a predator and may, therefore, have to rely on hiding, or on morphological defense. Due to morphological constraints, crucian carp (*Carassius carassius*) are poor swimmers, and thus have limited ability to outswim attacking pike (*Esox lucius*). This condition has likely contributed to the evolution of an induced morphological defense in this species. In the presence of pike crucian carp grow a 'hunched back' which makes them more difficult to handle and swallow by gape limited predators (Brönmark and Miner, 1992). Conversely, morphological defense can influence behavioural responses to predation threat. For example, fish species with more body armor are less inclined to flee from approaching predators compared to less armored species (McLean and Godin, 1989).

Nevertheless, fish are often remarkably plastic in their behaviour, altering strategies in response to a changing environment where learning ability often plays a crucial role in the development of anti-predator behaviour (Smith, 1997; Shettleworth, 2001). On a longer time scale, variation in predation pressure can result in the evolution of interpopulation differentiation in behavioural defense tactics (Giles and

Huntingford, 1984). For example, inexperienced European minnows (*Phoxinus phoxinus*) from high-predation sites show more predator inspection behaviour and form larger shoals in the presence of predatory pike than do conspecifics from low-predation sites (Magurran and Pitcher, 1987). In addition, minnows from high-predation sites are more able to alter anti-predator behaviour in response to experience (Magurran, 1990). Thus, variable predation pressure seems to have differentiated both the basic innate anti-predator response in inexperienced fish and the (innate) capacity for learning and, thereby, the plasticity in anti-predator behaviour in minnows. Variation in predation pressure may not only select for interpopulation differences in anti-predator behaviour, but can also influence other behaviours, like spatial orientation, foraging behaviour and aggression (Endler, 1995; Brown and Braithwaite, 2005).

Figure 8.1 shows possible outcomes of interactions between predators and prey at different steps in the predator-prey encounter sequence (Lima and Dill, 1990). This sequence is used as a framework and the reader can go back to it from the text to identify the position of a certain behavioural defense in the predator-prey encounter sequence. The text is organized after the encounter sequence, starting with strategies adopted to prevent an encounter taking place at all, continuing with strategies intended to detect and recognize predators, and tricks employed to deter them. If this is not possible, fish have to flee, but even if they are captured there may still be last-resort tactics available to increase the chance of survival.

Avoiding Encounter

Habitat choice

The spatial and temporal distribution of prey and predators will determine the probability that a predator-prey encounter occurs (Fig. 8.1). For example, fish may avoid habitats with increased predation risk even if these have higher prey densities than safer habitats. In a classic experiment, Gilliam and Fraser (1987) tested the hypothesis that animals chose foraging habitats in order to minimize the ratio of mortality rate to energy intake rate (Fig. 8.2). Juvenile creek chubs (*Semotilus atromaculatus*) were allowed to choose between foraging arenas that differed in resource densities and mortality hazards and their preferences agreed well with theoretical predictions. Two years later Abrahams and Dill (1989) used a titration technique to calculate the amount of extra food needed to balance an increased risk of predation in guppy (*Poecilia*

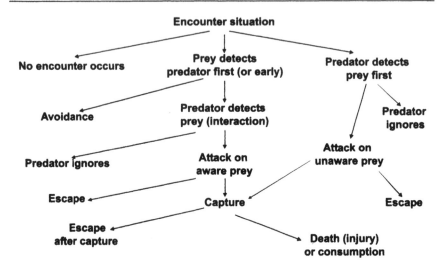

Fig. 8.1 The possible outcomes of an encounter between prey and predator, modified after Lima and Dill (1990). Prey behaviours at the different steps of the encounter sequence are discussed in the text.

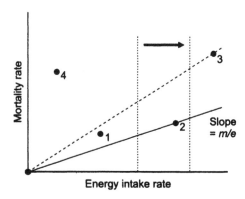

Fig. 8.2 Habitat selection under predation risk. Points 1-4 indicate habitats with different prey density and predation risk. The dotted vertical lines indicate minimum energetic requirements for survival. The solid line indicates the optimal habitat choice according to the Gilliam and Fraser (1987) rule: minimize the ratio of mortality rate to energy intake rate. The dashed line indicates an alternative choice of the habitat with highest prey density and predation risk. Selection of foraging habitat can be influenced by energetic demands (as illustrated by the arrow) which may increase due to starvation, parasitic infection (Barber *et al.*, 2000), domestication selection (Johnsson, 1993) or genetic modification for increased growth potential (Abrahams and Sutterlin, 1999) (Figure adapted from Gilliam and Fraser, 1987).

reticulata). Many studies since then have confirmed that fish are able to adjust their habitat choice trading off foraging benefits against the risk of predation. For example, juvenile coho salmon (*Oncorhynchus kisutch*) select habitats by trading off access to protective cover against food availability (Grand and Dill, 1997). Habitat choice in fish is often size-dependent as larger individuals generally are less vulnerable to gape-limited predators (Godin, 1997), and therèfore can switch to more food-rich habitats. Such size-dependent ontogenetic habitat shifts have been demonstrated in a number of species, for example the bluegill sunfish, *Lepomis macrochirus* (Werner *et al.*, 1983; Werner and Hall, 1988).

Predation risk may also induce shifts in vertical position within the habitat. For example, fish may avoid feeding in the surface, lowering their vertical position to reduce the risk of aerial predation. Consequently, simulated attacks from aerial predators (e.g., herons) result in lowered vertical position in sticklebacks (Giles and Huntingford, 1984) and salmonids (Jönsson *et al.*, 1996). The relation between predation risk and vertical position raises the possibility that aerial predation selects against the evolution of air (surface) breathing in fishes. This hypothesis was tested in a predation experiment involving green heron (*Butorides striatus*), preying on six bimodal and six water-breathing species of fish. Consistent with the hypothesis, the bimodal species suffered higher predation mortality than the water-breathing species (Kramer *et al.*, 1983).

From the previous examples it should be clear that predation is an important selective factor influencing habitat choice and other anti-predator adaptations. The most extensively studied species in this respect is probably the guppy where predation intensity is known to affect a number of anti-predator traits, which in turn affects guppy ecology, population biology and differentiation (Endler, 1995). Artificial selection can also alter anti-predator behaviour as shown in experiments comparing wild-type and domesticated salmonids (Weber and Fausch, 2003). For example, domesticated brown trout (*Salmo trutta*) are more willing to risk exposure to predators during foraging compared to wild counterparts (Johnsson *et al.*, 1996). A possible explanation may be that selection for fast growth in a predator-free (hatchery) environment selects for a high-gain/high-risk behavioural phenotype (Johnsson, 1993; see Fig. 8.2). However, it is still unclear to what extent behavioural effects of domestication are due to relaxed predation selection or directional selection for fast growth.

Activity patterns

In cases where risky habitats cannot be avoided, the probability of an encounter can still be reduced by ceasing or adjusting activity. Fish commonly avoid predators by hiding in refuges in the habitat. For example, many species of flatfish burrow into the substrate (Ryer et al., 2004). However, as safety from predators has to be balanced against food intake and/or reproductive opportunities there is strong selection on the ability to assess predation risk accurately (Burrows, 1994).

Predator regimes can influence activity patterns of the prey both on a daily and seasonal basis. Trinidian guppies (*Poecilia reticulata*) are nocturnal feeders in the absence of predation risk but cease nocturnal feeding when the predator (*Hoplias malabaricus*) is present. This results in higher growth rate and increased courtship activity in the absence of the nocturnal predator, highlighting the link between predation risk, behaviour, growth and reproductive success (Fraser et al., 2004). Juvenile Atlantic salmon (*Salmo salar*) are mainly nocturnal during winter, but switch to 24-h activity in the summer. As ectotherms they have low metabolic rate in cold water during winter, which reduces swimming speed and their ability to escape from endothermic predators (Webb, 1978). It has therefore been suggested that fish should reduce their risk of predation in winter by hiding during the day, when the risk from endothermic predators is highest (Fraser et al., 1993).

Due to foraging demands it may not be optimal to cease activity completely, but fish may adjust their prey choice to reduce activity and, thereby, the probability of being detected and preyed upon. Consistently, attack distances and thereby the consumption of larger flies were reduced in juvenile coho salmon (*Oncorhynchus kisutch*) when foraging in the presence of a rainbow trout (*Oncorhynchus mykiss*) predator model (Dill and Fraser, 1984). Prey choice can also be influenced by risks posed by conspecifics. Northern pike frequently choose prey that are smaller than the size that would maximise net energy intake. Could this reflect a response to threats from other pike that are frequently cannibalistic and, in addition, often steal prey from each other? Researchers from Lund University addressed this question and experimentally showed that attacks on larger crucian carp increased handling time, and thereby the risk of cannibalistic attacks and stealing attempts from other pike (Nilsson and Brönmark, 1999). Thus, prey choice in pike appears to balance energy gain against the risk of mortality or loss of captured prey.

Crypsis

In this section adaptations to reduce detection by predators are briefly summarized. For a more thorough discussion on the theoretical basis and empirical evidence for crypis, see Ruxton et al. (2004). Many fishes, with flatfishes as striking examples, avoid predators by matching the background and can change colour rapidly. Immobility is often, but not always, an important component of crypsis. Although movement will break crypsis against a static background, the leaf-fish (Monocirrhus polyacanthus) is masquerading against moving backgrounds by exhibiting swimming movements that resemble a drifting leaf (Randall and Randall, 1960). This type of masquerading is not limited to leaf-fish. At least seventeen genera of fishes have been reported to resemble plants (Breder, 1946).

Another way of reducing the probability of detection by predators is to conceal the body outline. The idea of disruptive colouration was first formalised by Cott (1940). Although the idea is widely accepted, empirical evidence is rare and evolutionary patterns are unclear since markings that function disruptively may also function as background matching. For example, the saddle pattern of alternative light and dark stripes of several benthic freshwater fishes may help fish match the background of large dark stones on light substrate, as well as breaking up the fish body outline into smaller units (Armbruster and Page, 1996). This type of crypsis may provide acceptable protection against a range of backgrounds, reducing the risk of detection when foraging in benthic habitats with variable structure.

Many mid-water fish (e.g., herring and salmon) are predominantly laterally silvered and ventrally silvered or whitish, but darkened dorsally, which is likely an adaptation to provide background matching both from above and from below. Indeed, some catfishes (family Mochokidae), which swim upside-down when feeding and breathing at the surface at night, show reverse countershading with a light dorsa and dark ventra (Chapman et al., 1994). One of these species (Synodontis nigriventris) is uniformly coloured during the day, when it avoids the surface, but shifts to reverse countershading as it rises to the surface at night (Nagaishi et al., 1989). It is unclear whether countershading colouration results from selection for background matching or self-shadow concealment (Ruxton et al., 2004).

Crypsis can also be achieved by transparency and silvering, which is a common feature in pelagic fish species. Transparency is more common in juvenile or planktonic stages, whereas silvering seems largely confined to larger and/or adult stages of fish. Transparency probably becomes impossible for larger fish that need to move quickly and must have large amounts of opaque respiratory and connective tissues associated with their muscle mass (Ruxton et al., 2004). The different depth distributions of larvae and adults may also influence the relative efficiency of transparency and silvering, since the latter is more effective at greater depths where the light field is more homogenous (Denton et al., 1972).

Detecting Predators

Vigilance

When predator encounters cannot be avoided it is important to detect them as soon as possible to increase the likelihood of escape. One way of achieving this is to adjust foraging technique. Head-down foraging postures have been found to reduce the probability of escape from attacking predators (Krause and Godin, 1996). Consistently, convict cichlids (Archocentrus nigrofasciatus) subjected to conspecific chemical alarm cues switch from substrate foraging to head-up foraging postures in the water column (Foam et al., 2005; Fig. 8.3). Prey density may also affect vigilance if fish foraging in a dense swarm of prey cannot pay sufficient attention to detect approaching predators. If given a choice fish should choose lower density prey swarms when their expected predation risk is increased. This hypothesis was addressed in an experiment where sticklebacks (Gasterosteus aculeatus) were allowed a foraging choice among water flea swarms (Daphnia magna) of different densities. Half of the fish were subjected to a simulated predator attack from a model kingfisher (Alcedo atthis) before the foraging trial. As predicted from the hypothesis, sticklebacks subjected to elevated predation risk attacked lower prey densities than unfrightened conspecifics (Milinski and Heller, 1978).

Predation risk can also be reduced by avoiding involvement in escalated aggressive interaction, since this may reduce vigilance. For example, cichlids (Nannacara anomala) detect approaching predators later when involved in escalated mouth-fighting, whereas the predator is detected at larger distance if it approaches during low-level interactions like broadside display (Jakobsson et al., 1995). However, fish may face a dilemma if they need to fight to secure access to protective cover in habitats with predators. Indeed, brown trout parr previously subjected to (simulated) predator attacks defended territories with protective cover

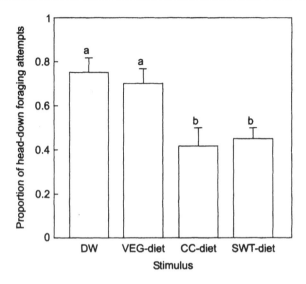

Fig. 8.3 Mean±SE foraging attempts directed towards the horizontal food patch (head-down posture) for convict cichlids exposed to: distilled water (DW), predators fed vegetable diet (VEG-diet), predators fed cichlids (CC-diet) or predators fed swordtails (SWT-diet). Different letters denote significant differences (P<0.05) (Adapted from Foam *et al.*, 2005).

more aggressively than trout not previously exposed to predation risk (Johnsson *et al.*, 2004).

Shoaling

Another strategy to ensure early predator detection is to form shoals, larger shoals being more efficient in predator detection (the 'many eyes' hypothesis). This hypothesis was confirmed in an experiment on European minnows (*Phoxinus phoxinus*) showing that minnows in larger shoals reduced their foraging sooner, but remained feeding on the patch for longer when approached by a model pike, *Esox lucius* (Magurran *et al.*, 1985). As groups persist over time individuals become familiar and their social hierarchy stabilizes which may allow attention to be switched from costly aggressive interactions to feeding and vigilance against predators. Indeed, in familiar groups of brown trout parr aggression was lower, food intake higher and escape responses to predator attacks faster compared with unfamiliar groups (Griffiths *et al.*, 2004). For a more thorough discussion of the anti-predator benefits of schooling, see Krause and Ruxton (2002).

Predator selection has contributed to the evolution of a rich variety of sensory adaptations in fish (Pitcher, 1993). Sensory detection of predators can be facilitated by appropriate behavioural responses. For example, chemical detection of approaching predators in streams may be enhanced by choosing foraging sites downstream of the location where the predator was last encountered. Furthermore, if coexisting fish species differ in their sensory abilities, the less equipped species may exploit the abilities of the other. This seems to be the case in rays where cowtail stingrays (*Pastinachus stephen*) prefer to rest with whiprays (*Himantura uarnak*) rather than joining conspecific groups (Semeniuk and Dill, 2006). Ray tails are equipped with mechanoreceptors capable of detecting predators, and the authors suggest that cowtails benefit from associating with the longer-tailed whiprays by exploiting their superior predator detecting abilities (Fig. 8.4).

Alarm signaling

Alarm substance (AS, sometimes called Schreckstoff) is released when the epidermis is injured in several species belonging to the superorder Ostariophysi (minnows, catfish, etc.). Fish that can detect AS react with a number of specific antipredator behaviours, and in the field an area in which AS was released can be avoided for up to 12 hours (Mathis and Smith, 1992; Chivers and Smith, 1994). Experiments on fathead minnows, *Pimephales promelas*, suggest that they can use AS as a chemical cue to label the dangerousness of predatory pike (Mathis and Smith, 1993a,b,c), and that this AS warning reduces the probability of predation for minnows receiving the signal (Mathis and Smith, 1993a, b, c). Recent studies show that AS can be detected also by species that do not produce the chemical themselves, including some salmonids (Brown and Smith, 1998) and sticklebacks (Brown and Godin, 1997). It is debated whether AS has evolved primarily for signal function (as a pheromone), or for some other function. Thus, the question is whether fish have evolved the ability to produce the chemical, or to detect it (Abrahams, 2006). It is challenging to explain how the chemical could evolve as a pheromone, since a predatory-injured sender is unlikely to benefit from the signal. A commonly proposed mechanism is based on kin selection, that the signal would increase survival of kin and thereby the inclusive fitness of the sender (Hamilton, 1964). An alternative explanation is that AS attracts additional predators to the place of capture, so that the resulting fight over

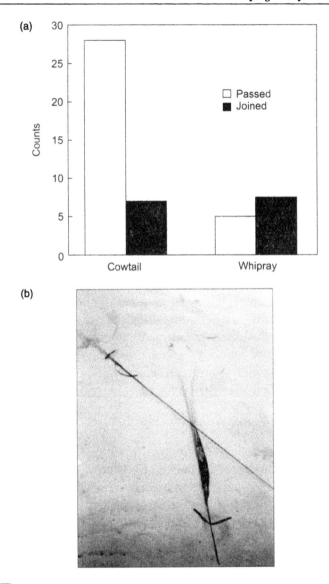

Fig. 8.4 a. The number of resting cowtail stingrays and whiprays passed and joined by cowtails searching for resting opportunities (n=12). Adapted from Semeniuk and Dill (2006). b. A mixed-species group of three rays showing the difference in relative tail length: cowtails centre and top right, whiptail top left. Photograph by Christina Semeniuk.

the prey increases its chance of escape (Chivers *et al.*, 1996). However, the pheromone function of AS has recently been questioned as fish seem to become less responsive to the chemical as their environment approaches natural conditions (Magurran et al., 1996; Irwing and Magurran, 1997). In addition, recent experiments suggest that the response to AS is dependent on the level of predation risk as well as the reliability of visual information (Hartman and Abrahams, 2000; Smith and Belk, 2001), suggesting that AS is a cue, not a pheromone. For a more comprehensive review of the debated role of AS in predator defense, see Abrahams (2006).

Avoiding Attack after Detection

In this section I shall discuss secondary defenses, which are tactics employed to avoid attack after detection, usually by making the attack seem unprofitable to the predator (Edmunds, 1974). First, however, fish may need to find out whether the approaching animal really is a predator. Failing to recognize a predator may result in death, whereas identifying a harmless animal as a predator may only result in lost time for foraging (the death versus dinner dilemma). Fish should therefore initially fear all large unfamiliar animals, learn which species are dangerous and should be avoided, and habituate to species that do not attack (Smith, 1997). Such learning responses have been confirmed experimentally in several fish species, including the humbug damsel fish *Dascyllus aruanus* (Coates, 1980), *Chromis caerulus* (Karplus et al., 1982) and the paradise fish *Macropodus opercularis* (Csányi, 1985).

Predator inspection

Predator inspection, swimming towards a potential predator, has been observed in several fish species, but the function of the behaviour is still debated. Approaching predators may serve several functions including information gathering (inspection), alarm signalling and pursuit deterrence (Smith, 1997). The paradise fish, *Macropodus opercularis*, only needs to approach a non-predatory fish once to learn and remember that it is harmless, whereas habituation to real predators or realistic models take longer time (Csányi, 1985). By approaching a predator fish may also obtain information about its motivational state. For example, by visual inspection only, guppies can distinguish between hungry and satiated predators (Licht, 1989). Some species can also identify predators by smell, and sniffing the predator for prey pheromones during inspection may aid in the assessment of its dangerousness (Smith, 1997). Inspection may also act as an alarm signal by showing the predator's location to other prey.

The detection signalling hypothesis

Perhaps the inspection simply signals to the predator that it has been detected? Could the prey really benefit by providing this information? Yes, if the chance of capturing an alert prey is reduced, the predator may benefit by aborting an energetically and/or time consuming attack with little chance of success. If the signal causes the predator to cancel its attack the signaller also benefits by saving the cost of fleeing from the predator and avoiding risk of capture. Since both predator and prey benefit, a detection signal has potential to evolve (Ruxton et al., 2004). However, this scenario may provide opportunities for prey to cheat, signalling that they have detected the predator when they have not. If cheating prey increase in frequency, predators should be selected to ignore the signal and, hence, the signal could break down. This scenario was modeled by Bergstrom and Lachmann (2001), suggesting that there is a set of conditions necessary to produce a stable signaling system where prey should only signal when they are reasonably sure that they have detected the predator signal, and that the predator response by canceling its attack. The conditions are that: (1) there is a cost to signalling; but (2) the cost is not so high that signalling is never profitable; (3) prey can assess the likelihood of predator presence with some accuracy, if not perfect; (4) Prey that suspects predator presence is harder to capture than less cautious individuals; and (5) the cost of the predator attacking is not too high.

The quality signalling hypothesis

An alternative (non-mutually exclusive) benefit from predator inspection may be to signal to the potential predator that the signaller is difficult to catch. Again, theoretical work suggests that honest signalling of intrinsic 'quality' and thus, catchability, may be evolutionarily stable under certain conditions (Vega-Redondo and Hansson, 1993). Predatory cichlids (*Aequidens pulcher*) are less likely to attack and kill inspecting guppies than fish that do not inspect (Godin and Davis, 1995), which could be explained either by the 'detection signalling' hypothesis, or by the 'quality hypothesis'. Godin and Davis suggested that quality signalling may be a plausible explanation since previous findings suggest that larger, better armoured and high condition prey are more likely to inspect (Godin and Davis, 1995). However, the interpretation of Godin and Davis' findings have been subject to further discussion (Godin and Davis, 1995; Milinski and Boltshauser, 1995), one problem being that inspection behaviour may be correlated with some other general measure of 'quality', for example, colouration, utilized by the predator. If so, an inspecting individual would not receive fewer predator attacks than a non-inspecting individual of similar quality. This situation was experimentally simulated by Milinski and colleagues by mimicking inspection behaviour using dead sticklebacks (*Gasterosteus aculeatus*) and live predatory pike. 'Inspecting' dead sticklebacks did not receive fewer attacks than non-inspecting ones, suggesting that predators may use other cues than inspection behaviour itself when considering an attack (Milinski *et al.*, 1997). Characin fish (*Hemigrammus erythrozonus*) start to flick their fins in response to alarm pheromones emitted when conspecific skin is damaged (Brown *et al.*, 1999). The authors suggested that this behaviour functions as a predator deterrent, but again alternative explanations, like startling effects cannot be ruled out (Ruxton *et al.*, 2004).

Deterrence, deceit and deflection

Deterrence: Honest warning signals

Aposematic warning displays that inform the predator about the unpalatability of its potential prey are common in many animal taxa. Warning displays have interested biologists since the time of Wallace (1867), Darwin (1887) and a thorough review of the theories about their function and evolution is provided by Ruxton *et al.* (2004). Whereas warning displays are apparently widespread in fish, experimental evidence for aposematic function is rare. However, interesting work has been conducted on some species from the families Trachinidae and Uroscopidae (weeverfishes and stargazers). They all have a poisonous first dorsal fin and hide by burying themselves in the sand. If experiencing predation threat, however, they rapidly raise their first dorsal fin exposing a black spot. The aposematic function of the suddenly appearing spot is supported by the observation that the dorsal fin is not raised in response to non-predator stimuli, for example, when a prey species is approaching (Bedini, *et al.*, 2003).

Deceit: Batesian mimicry

Interestingly, some sole species (Soleidae) that live sympatrically with these weeverfish and stargazers also possess a dark spot, but on their nontoxic right pectoral fin, which is raised only when a predator is threatening (Bedini *et al.*, 2003). The authors suggest that the black spot display in these soles functions as a Batesian mimic, allowing the non-poisonous mimics to benefit by resembling their poisonous models (Bates, 1862). While this is a plausible interpretation, a more convincing example of Batesian mimicry is perhaps provided by the plesiopid fish *Calloplesiops altivelis* with its caudal colouring patterns resembling the head of a toxic moray eel (*Gymnothorax meleagris*). When attacked by the predator the fish flies to a crevice, but leaves its moray eel-like tail exposed (McCosker, 1977; Fig. 8.5). Similarly, resemblances between non-venomous and venomous blennies likely have evolved through Batesian mimicry (Losey, 1972).

Fig. 8.5 Behaviour-facilitated Batesian mimicry in a fish. The caudal region of the plesiopid fish *Calloplesiops altivelis* resembles the head of a toxic moray eel (*Gymnothorax meleagris*). When attacked by a predator the fish flies to a crevice, but leaves its moray eel-like tail exposed. Photograph by J Randall.

Deflection: False eyespots

Many tropical fish species carry dark spots at the posterior end of their bodies, which has been suggested to deflect the predator attack away from the head, thereby increasing the survival chances of the prey. A couple of studies have addressed this hypothesis by adding dark posterior spots to species not featuring them. Overall, the results support the deflection hypothesis although the experimental details and results are insufficient to draw firm conclusions (McPhail, 1977; Dale and Pappantoniu, 1986). In butterflyfishes (*Chaetodon* spp.) false eyespots occurs concomitantly with eye camouflage in 41 out of 90 species, which is consistent with a deflective function (Neudecker, 1989). Furthermore, observations in the wild of butterflyfishes recovering fully despite missing as much as 10% of the posterial body region (likely due to predator attacks) suggest that predator deflection may allow escape and survival even when the predator attack is 'partially' successful (Neudecker, 1989).

Fleeing from Predators

Should I stay or should I go?

This section provides a brief summary of the present understanding of flight behaviour in fish. The subject has been treated in detail by Godin (1997) and his review is recommended for further reading. Once a prey becomes aware of an approaching predator, several alternative responses may occur. The choice of action depends on a number of factors, including the perceived risk of predation, as well as the energetic and reproductive state of the prey (Milinski, 1993; Rodewald and Foster, 1998). The threatened prey may freeze, perform inspection, or use other tactics to discourage an attack, as discussed previously, but in many cases flight may be the only alternative available to avoid being killed. For example, in red drum larvae (*Sciaenops ocellatus*) flight responses to attacks from longnose killifish (*Fundulus similis*) are critical for successful escape (Fuiman *et al.*, 2006). At the same time flight means lost energy and foraging opportunities to the prey. So, when should the prey flee? This adaptive dilemma, the economics of fleeing, has been addressed in depth by Ydenberg and Dill (1986) who discussed various factors influencing the initiation of prey flight.

Flight initiation distance

The flight initiation distance (FID) of a fish theoretically depends on the relative costs of fleeing and remaining. Costs of fleeing include energetic losses and lost foraging or mating opportunities, whereas remaining increases the probability of death or injury. From this economical theory a number of predictions of FID can be generated and empirically tested. For example, FID should increase with increasing predator size and approach velocity, predictions that has been confirmed in studies on several animal taxa (Ydenberg and Dill, 1986), including fish (Dill, 1974; Helfman, 1989). Also, more armoured phenotypes should be less likely to flee from predators, as has been shown for brook stickleback, *Culaea inconstans* (Reist, 1983). It should be pointed out, however, that several studies have failed to show within-species correlations between body armour and flight behaviour (e.g., McLean and Godin, 1989).

As the individual risk of predation diminishes with group size (the dilution effect (Turner and Pitcher, 1986)), FID should decrease with increasing group size. Consistently, when approached by a model predatory pike, European minnows in larger groups remain longer on the feeding patch, whereas minnows in smaller groups leave at larger distance from the predator model. The story is a bit more complex, however, as minnows in larger shoals reduced foraging sooner although remaining on the patch (Magurran *et al.*, 1985). This suggests that minnows in larger groups detect the predator sooner (the 'many eyes' hypothesis), but do not 'calculate' the risk as sufficiently high to induce flight immediately. In addition to the dilution hypothesis, which we started out with, there are at least two other possible explanations for the behavioural differences between minnows in small and large shoals. (1) Late detection may reduce opportunities for calculating risk and the best strategy may therefore be to suppose the highest risk and fly immediately (Milinski, 1993). (2) Smaller shoals cannot rely on the confusion effect of larger shoals making it difficult for a predator to single out a prey for pursuit (Neill and Cullen, 1974). This example illustrates both the strength and difficulties of using the cost-benefit approach to predict anti-predator responses: the final behavioural action is often determined by the combined effect of several interacting mechanisms.

Flight tactics

Fish do not only need to decide when to flee. In addition, the trajectory, speed and duration of the flight will affect the probability of escape and, thus, fitness. The kinematic optimality model developed by Weihs and Webb (1984) predicts the optimal turning angles that maximise the instantaneous distance separating prey and predator, depending on their relative speed. According to their model, a prey which is faster than, or as fast as the predator should flee in direct line with the predator's attack path. When the predator is faster than the prey, however, the prey should deviate from the predator attack path, with an optimal turning angle of less than 21°. Laboratory experiments suggest that escape responses commonly are directed away from the predator stimulus, but the responses are highly variable (Godin, 1997). For example, the escape angles of angelfish, *Pterophyllum eimekei*, were found to be highly variable among individuals, but the majority of fish escaped at about 130-180° away from the direction of the stimulus (Domenici and Blake, 1993). The escape trajectory is also influenced by the initial body form at the start of the acceleration phase, which is commonly C-shaped, but in some species the body is bent into an S-shape (Webb, 1976).

In Atlantic herring larvae (*Clupea harengus*) the probability of escaping predator attacks from juvenile herring increases with body size as facilitated by morphological development and increased swimming speed (Fuiman, 1993). Furthermore, Fuiman's results suggest that escape angle and flight duration is of limited importance to predict escape success of larvae whereas a late fright response followed by rapid acceleration is more critical. Thus, the successful herring flight tactic is similar to that of a matador dodging a bull (Blaxter and Fuiman, 1990). Fuiman (1993) emphasizes that herring larvae encounter a diversity of predators with different hunting tactics, which may necessitate predator-specific escape tactics. Thus, researchers studying different fish species and ecosystems should expect to find a variety of optimal flight tactics.

As pointed out earlier, morphological constraints can influence the behavioural options available to a species. Juvenile rock sole (*Lepidopsetta polyxystra*) show typical flatfish defense mechanisms relaying on immobility, burial and crypsis. Pacific halibut (*Hippoglossus stenolepsis*), on the other hand, has a less developed ability to mimic sediments, but a deeper/narrower body allowing higher swimming speed than by rock sole. In a predation experiment using larger halibut as predators, small halibut were more likely than rock sole to flush and flee on the approach of a predator. Also, fleeing halibut were more likely to escape than fleeing rock sole. Overall, however, predation rates on halibut were higher than on rock sole, likely due to the superior camouflage of the latter species (Ryer *et al.*, 2004).

As touched on before, variation in predation pressure can result in the evolution of interpopulation differences in antipredator tactics, including flight behaviour. This is exemplified by variation in escape behaviour among trinidian guppy populations subjected to different predator regimes

(Seghers, 1974). Similarly, artificial selection may alter innate flight behaviour, likely as a response to reduced predation pressure in the protected hatchery environment. For example, juvenile Atlantic salmon from a domestic strain react to predation threat with flights of shorter duration than their wild progenitors (Johnsson et al., 2001).

Although the knowledge about flight behaviour in fish is increasing, information is still scarce about the fitness consequences of different flight tactics. Future research should pursue this question using realistic scenarios and a holistic approach, integrating other co-variables of the flight response, for example timing and speed (Godin, 1997).

Escape after Capture

'It is not over until the fat lady sings'. Although this saying is unlikely to be appreciated by a perch stuck between the jaws of a pike, fish do sometimes escape predation after capture (Smith and Lemly, 1986; Reimchen, 1988). A striking example is the aforementioned observation of butterflyfishes (e.g., *Chaetodon citrinellus* and *Chaetodon atrimaculatus*) recovering after predation events despite missing substantial parts of their body (Neudecker, 1989). Similarly, stream-living salmonid parr frequently escape attacks from larger fish, birds and mammalian predators, and as much as 10% of the individuals can have bite marks from mink, *Mustela vison* (Heggenes and Borgström, 1988). Marine fish larvae also sometimes escape after being stung by the medusa *Aurelia arita* (Bailey and Batty, 1984). This section provides a brief discussion on such last-resort escape tactics, a research field that unfortunately has received limited attention from researchers up to date.

The relative size of the predator and prey is not only critical for the likelihood of the prey being caught (Godin, 1997), but may also affect the probability of escaping after capture. The fact that larger prey generally requires longer handling time facilitates escape simply by prolonging the time window during which flight is possible. In addition, larger prey are stronger and may therefore be more able to break loose during manipulation. For example, the ability of juvenile plaice (*Pleuronectes platessa*) to escape after being captured by a predatory shrimp (*Crangon crangon*) increases with size (Gibson et al., 1995). Many predatory fish predominantly attack and grasp the fish from the side. They then need to turn larger prey to a headfirst position to allow swallowing (Reimchen, 1991), which may provide opportunities for escape. As mentioned previously, increased handling time of larger prey increases the risk of attacks from kleptoparasitic or cannibalistic conspecifics in pike. These interactions also resulted in higher escape probability of captured crucian carp (Nilsson and Brönmark, 1999).

The fact that prey handling can attract additional predators opens up the scene for prey adaptations to signal the predation event in order to escape in the resulting tumult, a function that also has been proposed to explain the distress calls given by some birds following predator capture (e.g., Högstedt, 1983). Could alarm substance (AS) in ostariophysian fishes have evolved for this function (Smith, 1992)? This idea is consistent with the fact that fathead minnow alarm substances attracts predators like pike and diving beetles, *Colymbetes sculptilis* (Mathis *et al.*, 1995; see also Tester, 1963). In addition, a subsequent study showed that attraction of additional pike increases the probability of minnows escaping after capture (Chivers *et al.*, 1996). Thus, AS may confer a selective advantage to the involuntary sender. It is still unclear, however, if the substance originally evolved for this function (see Abrahams, 2006). More research is clearly needed to clarify the evolution and importance of AS and other adaptations enhancing escape after capture.

DEFENSE AGAINST PARASITES

A major difference between predator-prey relations and parasite-host relations, which we now turn to, is the relative effects on the prey and the host. Whereas failure to avoid predators generally results in death, parasite effects on hosts can vary from lethal to more or less insignificant. Parasites frequently utilise fish as hosts and it is rare to find totally non-infected individuals in natural environments. Parasites can have a wide range of effects on host behaviour including alterations of anti-predator behaviour, which may potentially affect the outcomes at all stages in the predator-prey encounter sequence (for a thorough discussion, see review by Barber *et al.*, 2000).

Whether parasites benefit from increased predation on their host or not depends on a number of factors, for example if the parasite has a direct life cycle transmitting from one definite host, or an indirect life cycle using at least one intermediate host harbouring sexually immature forms. In the definitive host, parasites need to reach sexual maturity to reproduce and are therefore generally not expected to benefit from increased predation, especially since this generally kills the parasite as well (Smith Trail, 1980).

However, in populations with unnaturally high levels of infection (which may occur due to pollution or artificial rearing) host behaviour may be affected resulting in increased susceptibility to predation (Herting and Witt, 1967).

In parasites with an indirect life cycle, the situation is different as it may be adaptive for the parasite to manipulate host behaviour in order to increase the transmission rate from the intermediate to the definitive host. However, in fish only a few studies have clearly demonstrated increased susceptibility of intermediate hosts to predation. The best evidence comes from a study by Lafferty and Morris (1996) showing increased predation by the definitive hosts, herons and egrets, on the intermediate host California killifish (*Fundulus parvipinnis*) infected by the brain-encysting trematode *Euhaplorchis californiensis*.

Considering the potentially negative effects of parasites on host fitness we would expect hosts to evolve adaptations to reduce the probability of parasite infection, as well as the severity of acquired infections. Behavioural avoidance is a first line of defense against parasites, which if successfully employed will reduce the demands on the immune system (Hart, 1994). Some examples of such behavioural adaptations are discussed below.

Avoiding Infection

Avoiding infected prey

If ingestion of infected prey carries costs that exceed benefits from energy intake, we would expect hosts to avoid infected prey. Two experiments on sticklebacks have been conducted to test this hypothesis, but both studies failed to show any avoidance of infected prey (Urdal *et al.*, 1995; Wedekind and Milinski, 1996). In the latter study sticklebacks actually preferred infected copepods, probably due to their impaired escape performance. More studies are definitely needed in this area, not least on the ability of fish to recognize infected prey.

Avoiding infected individuals

Infections from directly transmitted parasites may be prevented simply by avoiding infected individuals or groups. Avoidance of parasitized groups may not only be adaptive to avoid the infection *per se*, as groups with infected individuals may experience reduced anti-predator benefits due to reduced vigilance, increased risk taking and increased conspicuousness (Barber *et al.*, 2000; Krause and Ruxton, 2002). Three-spined sticklebacks avoid conspecific schools infected by *Argulus canadensis* (Dugatkin *et al.*, 1994), a parasite shown to reduce growth and increase mortality of infected individuals. Sticklebacks also avoid schools infected with the cestode *Schistocephalus solidus* (Barber *et al.*, 1998). Fish may recognize conspicuous parasites directly, as seems to be the case in the killifish (*Fundulus diaphanous*) where experimental simulation of the black spots caused by the trematode *Crassiphiala bulboglossa*, induces discrimination (Krause and Godin, 1996). In other cases, like in the previously mentioned study on sticklebacks and *Argulus*, discrimination appears to be based on the altered host behaviour induced by the parasite (Dugatkin *et al.*, 1994).

Parasite recognition also influences mate choice in several species, for example in the sex-role reversed pipefish *Sygnathus typhle* where males avoid mating with females that have black spots caused by the trematod *Cryptocotyle*. Males benefit from discrimination because parasitized females have reduced fecundity (Rosenqvist and Johansson, 1995). Similarly, female three-spined sticklebacks prefer brighter red males reducing their probability of mating with males infected by 'whitespot' *Ichtyopthirius multifilis* (Milinski and Bakker, 1990). In many cases, however, fish appear to have limited abilities to recognize parasitized individuals, as well as their own infection status (Krause and Godin, 1996), which in turn can limit the evolution of anti-parasite adaptations (Barber *et al.*, 2000).

Avoiding infectious habitats

An alternative strategy to stay uninfected is to avoid microhabitats with increased risk of infection. Support for this comes from a study on three-spined and black-spotted sticklebacks (*Gasterosteus wheatlandi*), which reduced their proximity to substrate and vegetation when the ectoparasite *Argulus canadensis* was added to an experimental tank. The behaviour appears adaptive since proximity to substrate increases infection risk (Poulin and Fitzgerald, 1989). Similarly, rainbow trout leave protective shelters invaded by the parasitic eye fluke *Diplostomum spathaceum*. A rapid reaction to parasite presence is critical since the likelihood of infection is higher for individuals with longer response time (Karvonen *et al.*, 2004; Fig. 8.6). Earlier hatching can also be considered as an anti-infection strategy, if it reduces time spent in an infectious habitat and/or

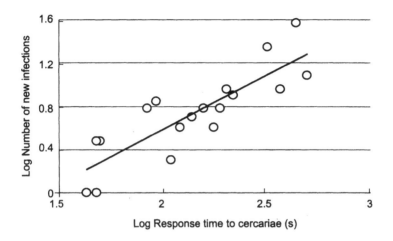

Fig. 8.6 Number of new *Diplostomum spathaceum* infections in rainbow trout as a function of response time to the parasite cercariae (log scale). Adapted from Karvonen *et al.*, 2004.

life stage. Whitefish eggs (*Coregonus* spp.) accelerate hatching in response to water-borne cues from other eggs infected by a virulent egg parasite (*Pseudomonas fluorescence*). This response increases survival as the earlier hatching allows larvae to swim away from the infected eggs (Wedekind, 2002). The exact nature of these waterborne cues and the evolution of the response mechanism are still unknown.

Reducing Infection

Parasite removal

Even after infection fish may be able to reduce negative impacts using appropriate behavioural actions. A common observation, both in wild populations and in aquaculture, is individuals attempting to remove ectoparasites by scraping their body against available substrate (Urawa, 1992). It is still unclear to what extent this behaviour reduces parasite load, and how any such benefits are balanced against costs such as increased conspicuousness to predators and physical damage causing secondary infections.

The best-studied infection-reducing mechanism is probably the visitation of cleaning stations where station-holding species pick ectoparasites and necrotic tissue from client fish (Gorlick *et al.*, 1978). This system has attracted interest in aquaculture where wrasses (Labridae) and other cleaner species have been successfully used as biological parasite control systems (Cowell et al., 1993; Treasurer, 2002). Recent research has, however, revealed that cleaner-host relations in nature are dynamic and not always perfectly symbiotic.

Although a number of studies have shown that cleaner fish indeed can reduce parasite load significantly (e.g., Grutter, 1999), they do also frequently cheat by tearing away scales and other healthy tissue from the host. In this system, ectoparasite abundance is one important factor influencing the frequency of cheating behaviour in cleaners (Cheney and Côté, 2005). Cleaner wrasse (*Labroides dimidiatus*) also adjust their behaviour depending on whether the host species is predatory, in which case cheating may increase mortality risk (Bshary and Wuerth, 2001). Interestingly, cleaner-client relations involving cleaner wrasse also possess features of a social control game where clients are more likely to team up with cleaners known to be cooperative (Bshary and Grutter, 2006). Another interesting twist is that the cleaner-client system itself can be 'parasitized'. The aggressive bluestriped fangblenny fish (*Plagiotremus rinorhynchus*) deceptively mimics the cleaner wrasse *Labroides dimidiatus*, and can thereby ambush and tear away tissue and scales from the deceived host fish. As if this was not enough, the fangblenny can also totally alter its external appearance to blend into schools of small reef fish, a fascinating example of opportunistic facultative mimicry (Côté and Cheney, 2005).

Prophylactic feeding

It has been suggested that animals may counteract parasitic infections by selectively foraging on prophylactic and/or anti-parasitic food items (Lozano, 1991). Such behaviour has indeed been demonstrated for birds where starlings (*Sturnus vulgaris*) line their nest with herbaceous plants that appear to increase nestling resistance against parasite infections (Gwinner et al., 2000). However, such preferences are yet to be described in fishes. Carotenoid pigments are interesting candidates for such selective feeding in fish. These compounds, which are precursors of vitamin A, possess both antioxidant and immunostimulant properties (Christiansen, 1995) and are also an important component of sexual colouration in many fish species. In arctic charr (*Salvelinus alpinus*), the colour intensity of adults have been found to reflect immune status (Skarstein and Folstad, 1996), and in guppies female preferences for carotenoid-pigmented males have been suggested to result from a sensory bias originally evolved to detect carotenoid-rich fruits (Grether *et al.*, 2004). The evolutionary mechanisms linking feeding preferences, infection resistance and mate choice are likely to be complex, but progress in our knowledge is expected due to the rapid development of this research field.

SUGGESTIONS FOR FURTHER RESEARCH

Hopefully this chapter, to some extent, has reflected our present understanding of behavioural defense strategies in fishes. Nevertheless, there are considerable knowledge gaps in many areas. For example, most studies to date have been conducted in restricted laboratory environments. Extrapolating these results to wild conditions should be made with caution, due to the physical and biological complexity of most natural environments. The environmental sensitivity of the effects of alarm substance, discussed in this chapter, provides an illustration of the problem (Abrahams, 2006).

More long-term studies are needed to reveal how behavioural actions affect individual fitness, and how individual variation in behavioural defense tactics influence life history and population growth. Such information is not acquired without considerable effort, combining controlled laboratory studies and selection experiments with long-term field surveys, like in the extensive research on Trinidadian guppies (e.g., Endler, 1995). Extensive behavioural observation in the wild can provide valuable and novel information, the recent studies of cleaner-client systems on Indonesian reefs providing an excellent example (Côté and Cheney, 2005). More comprehensive behavioural studies in the wild should be possible in the future as facilitated by rapid technical progress. For example, modern data storage tags now allow tracking of movements of many fish species in their natural environment (e.g., Hunter *et al.*, 2006). It is important that such techniques are not only used descriptively, but also utilised in manipulative experiments, for example to test predictions of predator impact on movement patterns in the wild. Another promising and rapidly developing field is behavioural genetics where researchers now are developing molecular techniques to reveal the genetic regulation of complex behavioural variation (Sneddon *et al.*, 2005).

No modern technique can, however, relieve our dependence on theoretical progress, new ways of thinking about problems. For example, in most studies to date predators have been treated as abstract or invariable sources of risk to which prey respond, rather than dynamic partners in a predator-prey interaction. Considering and focusing also on the strategic responses of predators to prey behaviour can yield novel predictions and new insight in future studies on predator-prey interactions, as pointed out by Lima (2002).

Finally, fish behaviour is receiving increasing attention in many applied contexts, including welfare and performance in aquaculture (Huntingford and Adams, 2005), hatchery-release programmes (Brown and Laland, 2001), effects of fishing (De Robertis and Wilson, 2006), habitat degradation (Polte and Asmus, 2006), pollution (Zhou and Weis, 1999) and biological risk assessment (Devlin et al., 2006). Nevertheless, critical behavioural aspects are still often ignored in fish management and conservation (Shumway, 1998). Ensuring that up-to-date scientific knowledge is appropriately used for managing and conserving fish populations should be a stimulating challenge for future fish behaviour researchers worldwide.

Acknowledgements

JIJ was financed by the Swedish Research Council for Environment, Agricultural sciences and Spatial Planning. Special thanks to Professor Ian Fleming for providing space and a stimulating working environment at the Ocean Science Centre, Memorial University of Newfoundland, for the completion of this chapter. I also thank Neil Metcalfe and Fredrik Sundström for valuable comments on the manuscript.

References

Abrahams, M.V. 2006. The physiology of antipredator behaviour: what you do with what you've got. In: *Behaviour and Physiology of Fish*, K.A. Sloman, R.W. Wilson and S. Balshine (eds.). Academic Press, San Diego, pp. 79-108.

Abrahams, M.V. and L.M. Dill. 1989. A determination of the energetic equivalence of the risk of predation. *Ecology* 70: 999-1007.

Abrahams, M.V. and A. Sutterlin. 1999. The foraging and antipredator behaviour of growth-enhanced transgenic Atlantic salmon. *Animal Behaviour* 58: 933-942.

Armbruster, J.W. and L.M. Page. 1996. Convergence of a cryptic saddle pattern in benthic freshwater fishes. *Environmental Biology of Fishes* 45: 249-257.

Bailey, K.M. and R.S. Batty. 1984. Laboratory study of predation by *Aurelia aurita* on larvae of cod, flounder, plaice and herring: Development and vulnerability to capture. *Marine Biology* 83: 287-291.

Barber, I., L.C. Downey and V.A. Braithwaite. 1998. Parasitism, oddity and the mechanism of shoal choice. *Journal of Fish Biology* 53:1365-1368.

Barber, I., D. Hoare and J. Krause. 2000. Effects of parasites on fish behaviour: a review and evolutionary perspective. *Reviews in Fish Biology and Fisheries* 10: 131-165.

Bates, H.W. 1862. Contributions to an insect fauna of the Amazon valley. Lepidoptera: Heliconidae. *Transactions of the Linnean Society of London* 23: 495-566.

Bedini, R., M.G. Canali and A. Bedini. 2003. True and false threatening visual cues in some Mediterranean fish. *Journal of the Marine Biological Association of the United Kingdom* 83: 265-270.

Bergstrom, C.T. and M. Lachmann. 2001. Alarm calls as costly signals of antipredator vigilance: the watchful babbler game. *Animal Behaviour* 61: 535-543.

Blaxter, J.H.S. and L.A. Fuiman. 1990. The role of the sensory systems of herring larvae in evading predatory fishes. *Journal of the Marine Biological Association of the United Kingdom* 70: 413-427.

Brönmark, C. and J.G. Miner. 1992. Predator-induced phenotypical change in body morphology in crucian carp. *Science* 258: 1348-1350.

Brown, C. and K. Laland. 2001. Social learning and life skills training for hatchery reared fish. *Journal of Fish Biology* 59: 471-493.

Brown, C. and V.A. Braithwaite. 2005. Effects of predation pressure on the cognitive ability of the poeciliid *Brachyraphis episcopi*. *Behavioral Ecology* 16: 482-487.

Brown, G.E. and J.-G. J. Godin. 1997. Anti-predator responses to conspecific and heterospecific skin extract by three-spine sticklebacks: Alarm pheromones revisited. *Behaviour* 134: 1123-1134.

Brown, G.E. and J.F. Smith. 1998. Acquired predator recognition in juvenile rainbow trout (*Oncorhynchus mykiss*): conditioning hatchery-reared fish to recognize chemical cues of a predator. *Canadian Journal of Fisheries and Aquatic Sciences* 55: 611-617.

Brown, G.E., J.-G. J. Godin and J. Pedersen. 1999. Fin-flicking behaviour: a visual antipredator alarm signal in a characin fish, *Hemigrammus erythrozonus*. *Animal Behaviour* 58: 469-475.

Bshary, R. and M. Wurth. 2001. Cleaner fish *Labroides dimidiatus* manipulate client reef fish by providing tactile stimulation. *Proceedings of the Royal Society of London* B268: 1495-1501.

Bshary, R. and A.S. Grutter. 2006. Image scoring and cooperation in a cleaner fish mutualism. *Nature (London)* 441: 975-978.

Burrows, M.T. 1994. An optimal foraging and migration model for juvenile plaice. *Evolutionary Ecology* 8: 125-149.

Chapman, L.J., L. Kaufman and C.A. Chapman. 1994. Why swim upsidedown? A comparative study of two mochokid catfishes. *Copeia* 1994: 130-135.

Cheney, K.L. and I.M. Côté. 2005. Mutualism or parasitism? The variable outcome of cleaning symbioses. *Biology Letters* 1: 162-165.

Chivers, D.P. and R.J.F. Smith. 1994. Intra- and inter-specific avoidance of areas marked with skin extract from brook sticklebacks (*Culaea inconstans*) in a natural habitat. *Environmental Biology of Fishes* 49: 89-96.

Chivers, D.P., G.E. Brown and J.F.S. Smith. 1996. The evolution of chemical alarm signals: attracting predators benefits alarm signal senders. *American Naturalist* 148: 649-659.

Christiansen, R., J. Glette, O. Lie, O.J. Torrissen and R. Waagbo. 1995. Antioxidant status and immunity in Atlantic salmon, *Salmo-salar* L., fed semi-purified diets with and without astaxanthin supplementation. *Journal of Fish Diseases* 18: 317-328.

Coates, D. 1980. The discrimination and reactions towards predatory and non-predatory species of fish by humbug damselfish, *Dascyllus auranus* (Pisces, Pomacentridae). *Zeitschrift für Tierpsychologie* 52: 347-354.

Côté, I.M. and K.L. Cheney. 2005. Animal mimicry: Choosing when to be a cleaner-fish mimic — A dangerous fish can discard a seemingly harmless disguise to suit its circumstances. *Nature (London)* 433: 211-212.

Cott, H.B. 1940. *Adaptive Coloration in Animals*. Methuen, London.

Cowell, L.E., W.O. Watanabe, W.D. Head, J.J. Grover and J.M. Sheaker. 1993. Use of tropical cleaner fish to control the ectoparasite *Neobenedenia-Melleni* (Monogenea, capsalidae) on Seawater-cultured Florida Red Tilapia. *Aquaculture* 113: 189-200.

Csányi, V. 1985. Ethological analysis of predator avoidance by the paradise fish (*Macropodus opercularis* L.). 1. Recognition and learning of predators. *Behaviour* 92: 227-240.

Dale, G. and A.P. Pappantoniu. 1986. Eye-picking behavior of the cutlips minnow, *Exoglossum maxillingua*: applications to studies of eyespot mimicrys. *Annals of the New York Academy of Sciences* 463: 177-178.

Darwin, C. 1887. *The Life and Letters of Charles Darwin: Including an Autobiographical Chapter,* edited by his son Francis Darwin, Murray, London.

Dawkins, R. 1976. *The Selfish Gene*. Oxford University Press, Oxford.

Denton, E.J., J.B. Gilpin-Brown and P.G. Wright. 1972. The angular distribution of light produced by some mesopelagic fish in relation to their camouflage. *Proceedings of the Royal Society of London* B182: 145-158.

De Robertis, A. and C.D. Wilson. 2006. Walleye pollock respond to trawling vessels. *ICES Journal of Marine Science* 63: 514-522.

Devlin, R.H., L.F. Sundström and W.M. Muir. 2006. Interface of biotechnology and ecology for environmental risk assessments of transgenic fish. *Trends in Biotechnology* 24: 89-97.

Dill, L.M. 1974. The escape response of the zebra danio (*Brachydanio rerio*). I. The stimulus for escape. *Animal Behaviour* 22: 711-722.

Dill, L.M. and A.H.G. Fraser. 1984. Risk of predation and the feeding behavior of juvenile coho salmon (*Oncorhynchus kisutch*). *Behavioural Ecology and Sociobiology* 16: 65-71.

Domenici, P. and R.W. Blake. 1993. Escape trajectories in angelfish (*Pterophyllum eimekei*). *Journal of Experimental Biology* 177: 253-272.

Dugatkin, L.A., G.J. Fitzgerald and J. Lavoie. 1994. Juvenile 3-spined sticklebacks avoid parasitized conspecifics. *Environmental Biology of Fishes* 39: 215-218.

Edmunds, M. 1974. *Defence in Animals: A Survey of Anti-predator Defenses*. Longman, Harlow, Essex.

Endler, J.A. 1995. Multiple-trait coevolution and environmental gradients in guppies. *Trends in Ecology and Evolution* 10: 22-29.

Foam, P.E., M.C. Harvey, R.S. Mirza and G.E. Brown. 2005. Heads up: juvenile convict cichlids switch to threat-sensitive foraging tactics based on chemosensory information. *Animal Behaviour* 70: 601-607.

Fraser, D.F., J.F. Gilliam, J.T. Akkara, B.W. Albanese and S.B. Snider. 2004. Night feeding by guppies under predator release: Effects on growth and daytime courtship. *Ecology* 85: 312-319.

Fraser, N.H.C., N.B. Metcalfe and J.E. Thorpe. 1993. Temperature-dependent switch between diurnal and nocturnal foraging in salmon. *Proceedings of the Royal Society of London* B252: 135-139.

Fuiman, L.A. 1993. Development of predator evasion in Atlantic herring, *Clupea harengus* L. *Animal Behaviour* 45: 1101-1116.

Fuiman, L.A., K.A. Rose, J.H. Cowan and E.P. Smith. 2006. Survival skills required for predator evasion by fish larvae and their relation to laboratory measures of performance. *Animal Behaviour* 71: 1389.

Gibson, R.N., M.C. Yin and L. Robb. 1995. The behavioral basis of predator-prey size relationships between shrimp (*Crangon crangon*) and juvenile plaice (*Pleuronectes platessa*). *Journal of the Marine Biological Association of the United Kingdom* 75: 337-349.

Giles, N. and F.A. Huntingford. 1984. Predation risk and inter-population variation in anti-predator behaviour in three-spined stickleback, *Gasterosteus aculeatus* L. *Animal Behaviour* 32: 264-275.

Gilliam, J.F. and D.F. Fraser. 1987. Habitat selection under predation hazard: test of a model with foraging minnows. *Ecology* 68: 1856-1862.

Godin, J.-G. J. 1997. Evading predators. In: *Behavioural Ecology of Teleost Fishes*, J.-G. J. Godin (ed.). Oxford University Press, Oxford, pp. 191-236.

Godin, J.-G. J. and S.A. Davis. 1995a. Who dares, benefits: predator approach behaviour in the guppy (*Poecilia reticulata*) deters predator pursuit. *Proceedings of the Royal Society of London* B259: 193-200.

Godin, J.-G. J. and S.A. Davis. 1995b. Boldness and predator deterrence: a reply to Milinski and Boltshauser. *Proceedings of the Royal Society of London* B262: 107-112.

Gorlick, D.L., P.D. Atkins and G.S. Losey, Jr. 1978. Cleaning stations as water holes, garbage dumps, and sites for the evolution of reciprocal altruism? *American Naturalist* 112: 341-353.

Grand, T.C. and L.M. Dill. 1997. The energetic equivalence of cover to juvenile coho salmon (*Oncorhynchus kisutch*): Ideal-free distribution theory applied. *Behavioural Ecology* 8: 437-447.

Grether, G.F., S. Kasahara, G.R. Kolluru and E.L. Cooper. 2004. Sex-specific effects of carotenoid intake on the immunological response to allografts in guppies (*Poecilia reticulata*). *Proceedings of the Royal Society of London* B271: 45-49.

Griffiths, S.W., S. Brockmark, J. Höjesjö and J.I. Johnsson. 2004. Coping with divided attention: the advantage of familiarity. *Proceedings of the Royal Society of London* B271: 695-699.

Grutter, A.S. 1999. Cleaner fish really do clean. *Nature (London)* 398: 672-673.

Gwinner, H., M. Oltrogge, L. Trost and U. Nienaber. 2000. Green plants in starling nests: effects on nestlings. *Animal Behaviour* 59: 301-309.

Hamilton, W.D. 1964. The genetical theory of social behaviour, I, II. *Journal of Theoretical Biology* 7: 1-52.

Hart, B.L. 1994. Behavioral defense against parasites — interaction with parasite invasiveness. *Parasitology* 109: S139-S151.

Hartman, E.J. and M.V. Abrahams. 2000. Sensory compensation and the detection of predators: the interaction between chemical and visual information. *Proceedings of the Royal Society of London* B267: 571-575.

Heggenes, J. and R. Borgstrom. 1988. Effect of mink, *Mustela vison* Schreber, predation on cohorts of juvenile Atlantic salmon, *Salmo salar* L., and brown trout, *S. trutta* L., in three small streams. *Journal of Fish Biology* 33: 885-894.

Helfman, G.S. 1989. Threat-sensitive predator-avoidance in damselfish-trumpetfish interactions. *Behavioural Ecology and Sociobiology* 24: 47-58.

Herting, G.E. and A. Witt. 1967. The role of physical fitness of forage fishes in relation to their vulnerability to predation by bowfin (*Amia calva*). *Transactions of the American Fisheries Society* 96: 427-430.

Hunter, E., F. Berry, A.A. Buckley, C. Stewart and J. D. Metcalfe. 2006. Seasonal migration of thornback rays and implications for closure management. *Journal of Applied Ecology* 43: 710-720.

Huntingford, F. and C. Adams. 2005. Behavioural syndromes in farmed fish: implications for production and welfare. *Behaviour* 142: 1207-1221.

Högstedt, G. 1983. Adaptation unto death: function of fear screams. *American Naturalist* 121: 562-570.

Irwing, P.W. and A.E. Magurran. 1997. Context-dependent fright reactions in captive European minnows: The importance of naturalness in laboratory experiments. *Animal Behaviour* 53: 1193-1201.

Jakobsson, S., O. Brick and C. Kullberg. 1995. Escalated fighting behaviour incurs increase predation risk. *Animal Behaviour* 49: 235-239.

Johnsson, J.I. 1993. Big and brave: selection affects foraging under risk of predation in juvenile rainbow trout, *Oncorhynchus mykiss*. *Animal Behaviour* 45: 1219-1225.

Johnsson, J.I., J. Höjesjö and I.A. Fleming. 2001. Behavioural and heart rate responses to predation risk in wild and domesticated Atlantic salmon. *Canadian Journal of Fisheries and Aquatic Sciences* 58: 788-794.

Johnsson, J.I., A. Rydeborg and L.F. Sundström. 2004. Predation risk and the territory value of cover: an experimental study. *Behavioural Ecology and Sociobiology* 56: 388-392.

Johnsson, J.I., E. Petersson, E. Jönsson, B.T. Björnsson and T. Järvi. 1996. Domestication and growth hormone alter antipredator behaviour and growth patterns in juvenile brown trout, *Salmo trutta*. *Canadian Journal of Fisheries and Aquatic Sciences* 53: 1546-1554.

Jönsson, E., J.I. Johnsson and B.T. Björnsson. 1996. Growth hormone increases predation exposure of rainbow trout. *Proceedings of the Royal Society of London* B263: 647-651.

Karplus, I., M. Goren and D. Algom. 1982. A preliminary experimental analysis of predator face recognition by *Chromis caerulus* (Pisces, Pomacentridae). *Zeitschrift für Tierpsychologie* 58: 53-65.

Karvonen, A., O. Seppala and E.T. Valtonen. 2004. Parasite resistance and avoidance behaviour in preventing eye fluke infections in fish. *Parasitology* 129: 159-164.

Kramer, D.L., D. Manley and R. Bourgeois. 1983. The effect of respiratory mode and oxygen concentration on the risk of aerial predation in fishes. *Canadian Journal of Zoology* 61: 653-665.

Krause, J. and J.-G. J. Godin. 1996a. Influence of prey foraging posture on flight behavior and predation risk: predators take advantage of unwary prey. *Behavioural Ecology* 7: 264-271.

Krause, J. and J.G. J. Godin. 1996b. Influence of parasitism on shoal choice in the banded killifish (*Fundulus diaphanus*, Teleostei, Cyprinodontidae). *Ethology* 102: 40-49.

Krause, J. and G.D. Ruxton. 2002. *Living in Groups*. Oxford University Press, Oxford.

Lafferty, K.D. and A.K. Morris. 1996. Altered behavior of parasitized killifish increases susceptibility to predation by bird final hosts. *Ecology* 77: 1390-1397.

Licht, T. 1989. Discriminating between hungry and satiated predators: the response of guppies (*Poecilia reticulata*) from high and low predation sites. *Ethology* 82: 238-243.

Lima, S.L. 2002. Putting predators back into behavioral predator-prey interactions. *Trends in Ecology and Evolution* 17: 70-75.

Lima, S.L. and L.M. Dill. 1990. Behavioral decisions under the risk of predation: a review and prospectus. *Canadian Journal of Zoology* 68: 619-640.

Losey, G.S. 1972. Predation protection in the poison-fangblenny, Meiacanthus atrodorsalis, and its mimics, *Escenius bicolor* and *Runula laudadus* (Blenniidae). *Pacific Science* 26: 129-139.

Lozano, G.A. 1991. Optimal foraging theory — a possible role for parasites. *Oikos* 60: 391-395.

Ludwig, D. and L. Rowe. 1990. Life-history strategies for energy gain and predator avoidance. *American Naturalist* 135: 686-707.

Magurran, A.E. 1990. The inheritance and development of minnow anti-predator behaviour. *Animal Behaviour* 39: 834-842.

Magurran, A.E. and T.J. Pitcher. 1987. Provenance, shoal size and the sociobiology of predator-evasion behavior in minnow shoals. *Proceedings of the Royal Society of London B* 229: 439-465.

Magurran, A.E., W. Oulton and T.J. Pitcher. 1985. Vigilant behaviour and shoal size in minnows. *Zietschrift für Tierpsychologie* 67: 167-178.

Magurran, A.E., P. W. Irwing and P. A. Henderson. 1996. Is there a fish alarm pheromone? A wild study and critique. *Proceedings of the Royal Society of London B* 263: 1551-1556.

Marshall, N.J. 2000. Communication and camouflage with the same 'bright' colours in reef fishes. *Philosophical Transactions of the Royal Society of London* B 355: 1243-1248.

Mathis, A. and R.J.F. Smith. 1992. Avoidance of areas marked with a chemical alarm substance by fathead minnows (*Pimephales promelas*) in a natural habitat. *Canadian Journal of Zoology* 70: 1473-1476.

Mathis, A. and R.J.F. Smith. 1993a. Chemical alarm signals increase the survival time of fathead minnows (*Pimephales prophales*) during encounters with northern pike (*Esox lucius*). *Behavioural Ecology* 4: 260-265.

Mathis, A. and R.J.F. Smith. 1993b. Chemical labelling of northern pike (*Esox lucius*) by the alarm pheromone of fathead minnows (*Pimephales promelas*). *Journal of Chemical Ecology* 19: 1967-1979.

Mathis, A. and R.J.F. Smith. 1993c. Fathead minnows (*Pimephales promelas*) learn to recognise pike (*Esox lucius*) as predators on the basis of chemical stimuli from minnows in the pike's diet. *Animal Behaviour* 46: 645-656.

Mathis, A., D.P. Chivers and R.J.F. Smith. 1995. Chemical alarm signals: predator deterrents or predator attractants? *American Naturalist* 145: 994-1005.

McCosker, J.E. 1977. Fright posture of a plesiopsid fish *Calloplesiops altivelis*: Example of Batesian mimicry. *Science* 197: 400-401.

McLean, E. and J.-G. Godin. 1989. Distance to cover and fleeing from predation in fish with different amounts of defensive armour. *Oikos* 55: 281-290.

McPhail, J.D. 1977. A possible function of the cadual spots in characid fishes. *Canadian Journal of Zoology* 55: 1063-1066.

Milinski, M. 1993. Predation risk and feeding behaviour. In: *Behaviour of Teleost fishes*, T.J. Pitcher (ed.). Chapman & Hall, London, pp. 285-305.

Milinski, M. and R. Heller. 1978. Influence of a predator on the optimal foraging behaviour of sticklebacks (*Gasterosteus aculeatus* L.). *Nature (London)* 275: 642-666.

Milinski, M. and T.C.M. Bakker. 1990. Female sticklebacks use male coloration in mate choice and hence avoid parasitized males. *Nature (London)* 344: 330-333.

Milinski, M. and P. Boltshauser. 1995. Boldness and predator deterrence: a critique of Godin and Davis. *Proceedings of the Royal Society of London* B262: 103-105.

Milinski, M., J.H. Lüth, R. Eggler and G.A. Parker. 1997. Cooperation under predation risk: experiments on costs and benefits. *Proceedings of the Royal Society of London* B264: 831-837.

Nagaishi, H., H. Nishi, R. Fujii and N. Oshima. 1989. Correlation between body colour and behaviour in the upside-down catfish, *Synodontis nigriventis*. *Comparative Biochemistry and Physiology* A 92: 323-326.

Neill, S.R.S.J. and J.M. Cullen. 1974. Experiments on whether schooling of prey affects hunting behaviour of cephalopods and fish predators. *Journal of Zoology* 172: 549-569.

Neudecker, S. 1989. Eye camouflage and false eyespots — Chaetodontid responses to predators. *Environmental Biology of Fishes* 25: 142-157.

Nilsson, P.A. and C. Bronmark. 1999. Foraging among cannibals and kleptoparasites: effects of prey size on pike behavior. *Behavioural Ecology* 10: 557-566.

Pigliucci, M. 2001. *Phenotypic Plasticity: Beyond Nature and Nurture*. John Hopkins Press, Baltimore.

Pitcher, T.J. 1993. *Behaviour of Teleost Fishes*. Chapman & Hall, London.

Polte, P. and H. Asmus. 2006. Intertidal seagrass beds (*Zostera noltii*) as spawning grounds for transient fishes in the Wadder. Sea. *Marine Ecology Progress Series* 312: 235-243.

Poulin, R. and G.J. Fitzgerald. 1989. Risk of parasitism and microhabitat selection in juvenile sticklebacks. *Canadian Journal of Zoology* 67: 14-18.

Reimchen, T.E. 1988. Inefficient predators and prey injuries in a population of giant stickleback. *Canadian Journal of Zoology* 66: 2036-2044.

Reimchen, T.E. 1991. Evolutionary attributes of headfirst prey manipulation and swallowing in piscivores. *Canadian Journal of Zoology* 69: 2912-2916.

Reist, J.D. 1983. Behavioral variation in pelvic phenotypes of brook stickleback, *Culaea inconstans*, in response to northern pike, *Esox lucius*. *Environmental Biology of Fishes* 8: 255-267.

Rodewald, A.D. and S.A. Foster. 1998. Effects of gravidity on habitat use and antipredator behaviour in three-spined sticklebacks. *Journal of Fish Biology* 52: 973-984.

Rosenqvist, G. and K. Johansson. 1995. Male avoidance of parasitized females explained by direct benefits in a pipefish. *Animal Behaviour* 49: 1039-1045.

Ruxton, G.D., T.N. Sherratt and M.P. Speed. 2004. *Avoiding Attack, the Evolutionary Ecology of Crypsis, Warning Signals, and Mimicry*. Oxford University Press, New York.

Ryer, C.H., A.W. Stoner and R.H. Titgen. 2004. Behavioral mechanisms underlying the refuge value of benthic habitat structure for two flatfishes with differing anti-predator strategies. *Marine Ecology Progress Series* 268: 231-243.

Seghers, B.H. 1974. Geographic variation in the responses of guppies (*Poecilia reticulata*) to aerial predators. *Oecologia* 14: 93-98.

Semeniuk, C.A.D. and L.M. Dill. 2006. Anti-predator benefits of mixed-species groups of cowtail stingrays (*Pastinachus stephen*) and whiprays (*Himantura uarnak*) at rest. *Ethology* 112: 33-43.

Shettleworth, S.J. 2001. Animal cognition and animal behaviour. *Animal Behaviour* 61: 277-286.

Shumway, C.A. 1999. A neglected science: applying behavior to aquatic conservation. *Environmental Biology of Fishes* 55: 183-201.

Skarstein, F. and I. Folstad. 1996. Sexual dichromatism and the immunocompetence handicap: An observational approach using Arctic charr. *Oikos* 76: 359-367.

Smith, M.E. and M.C. Belk. 2001. Risk assessment in western mosquitofish (*Gambusia affinis*): Do multiple cues have additive effects? *Behavioural Ecology and Sociobiology* 51: 101-107.

Smith, R.J.F. 1992. Alarm signals in fishes. *Reviews in Fish Biology and Fisheries* 2: 33-63.

Smith, R.J.F. 1997. Avoiding and deterring predators. In: *Behavioural Ecology of Teleost Fishes*, J.-G. J. Godin (ed.). Oxford University Press, Oxford, pp. 163-190.

Smith, R.J.F. and A.D. Lemly. 1986. Survival of fathead minnows after injury by predators and its possible role in the evolution of alarm signals. *Environmental Biology of Fishes* 15: 147-149.

Smith Trail, D.R. 1980. Behavioural interactions between parasites and hosts: host suicide and the evolution of complex life cycles. *American Naturalist* 116: 77-91.

Sneddon, L.U., J. Margareto and A.R. Cossins. 2005. The use of transcriptomics to address questions in behaviour: Production of a suppression subtractive hybridisation library from dominance hierarchies of rainbow trout. *Physiological and Biochemical Zoology* 78: 695-705.

Tester, A.L. 1963. The role of olfaction in shark predation. *Pacific Science* 17: 145-170.

Treasurer, J.W. 2002. A review of potential pathogens of sea lice and the application of cleaner fish in biological control. *Pest Management Science* 58: 546-558.

Turner, G.F. and T.J. Pitcher. 1986. Attack abatement: A model for group protection by combined avoidance and dilution. *American Naturalist* 128: 228-240.

Urawa, S. 1992. *Trichodina-Truttae* Mueller, 1937 (Ciliophora, Peritrichida) on juvenile chum salmon (*Oncorhynchus keta*) — Pathogenicity and host-parasite interactions. *Fish Pathology* 27: 29-37.

Urdal, K., J.F. Tierney and P.J. Jakobsen. 1995. The tapeworm *Schistocephalus solidus* alters the activity and response, but not the predation suceptibility of infected copepods. *Journal of Parasitology* 81: 330-333.

Wallace, A.R. 1867. *Proceedings of the Entomological Society of London*, March 4th, IXXX-IXXXi.

Wedekind, C. 2002. Induced hatching to avoid infectious egg disease in whitefish. *Current Biology* 12: 69-71.

Wedekind, C. and M. Milinski. 1996. Do three-spined sticklebacks avoid consuming copepods, the first intermediate host of *Schistocephalus solidus?* An experimental analysis of behavioural resistance. *Parasitology* 112: 371-383.

Vega-Redondo, F. and O. Hansson. 1993. A game-theoretic model of predator-prey signalling. *Journal of Theoretical Biology* 162: 309-319.

Webb, P.W. 1976. The effect of size on the fast-start performance of rainbow trout *Salmo gairdneri* and consideration of piscivorous predator-prey interactions. *Journal of Experimental Biology* 65: 157-177.

Webb, P.W. 1978. Temperature effects on acceleration in rainbow trout (*Salmo gairdneri*). *Journal of the Fisheries Research Board of Canada* 35: 1417-1422.

Weber, E.D. and K.D. Fausch. 2003. Interactions between hatchery and wild salmonids in streams: differences in biology and evidence for competition. *Canadian Journal of Fish Aquatic Sciences* 60: 1018-1036.

Weihs, D. and P.W. Webb. 1984. Optimal avoidance and evasion tactics in predator-prey interactions. *Journal of Theoretical Biology* 106: 189-206.

Werner, E.E. and D.J. Hall. 1988. Ontogenetic habitat shifts in bluegill: the foraging rate-predation risk trade-of. *Ecology* 659: 1352-1366.

Werner, E.E., J.F. Gilliam, D.J. Hall and G.G. Mittelbach. 1983. An experimental test of the effects of predation risk on habitat use in fish. *Ecology* 64: 1540-1548.

Voltaire, F.-M.A. 1759. *Candide*, Various Publishers.

Ydenberg, R.C. and L.M. Dill. 1986. The economics of fleeing from predators. *Advances in Study of Behaviour* 16: 229-249.

Zhou, T. and J.S. Weis. 1999. Predator avoidance in mummichog larvae from a polluted habitat. *Journal of Fish Biology* 54: 44-57.

Defense against Pathogens and Predators during the Evolution of Parental Care in Fishes

Jason H. Knouft

INTRODUCTION

Parental care is a fundamental aspect of the life histories of numerous vertebrate taxa, including fishes, amphibians, reptiles, birds, and mammals (Clutton-Brock, 1991; Reynolds *et al.*, 2002). In the broadest sense, parental care includes any pre-zygotic or post-zygotic behavior by a parent that can potentially increase the survivorship of offspring (Trivers, 1972; Clutton-Brock, 1991). While the ultimate objective of parental investment in offspring care is to maximize survival of progeny, differences in modes of parental care among species are numerous and presumed to arise from tradeoffs in the benefits for offspring survival versus the costs to the parent (Clutton-Brock, 1991; Sargent, 1997). Although the costs and benefits of parental care are often determined by contemporary selective

Author's address: Department of Biology, Saint Louis University, 3507 Laclede Avenue, St. Louis, Missouri, 63103-2010, USA.
E-mail: jknouft@slu.edu

factors, forms of parental care are ultimately constrained by evolutionary history. Considering this aspect, the evolution of parental care can be assumed to be driven by the relative costs and benefits of providing care to offspring within a set of constraints dictated by lineage specific traits (Maynard Smith, 1977; Clutton-Brock, 1991).

As noted by Dobzhansky (1973), 'Nothing in biology makes sense except in the light of evolution', and the study of parental care is no exception. Fishes are the most diverse group of vertebrates and display a wide variety of parental care behaviors (Breder and Rosen, 1966; Perrone and Zaret, 1979; Gross and Sargent, 1985). In some species, parental care is provided solely by the male, and in others, solely by the female. Additionally, there is a biparental version of parental care exhibited by species where both the male and female contribute towards caring for the offspring. These different strategies are, in some cases, found in related groups allowing for the investigation of factors driving evolutionary transitions in parental care. In fact, the study of parental care in fishes in an evolutionary context has provided the most resolved understanding of the costs and benefits associated with transitions between care types in any vertebrate group (Gross and Sargent, 1985; Balshine-Earn and Earn, 1998; Goodwin et al., 1998; Lindstrom, 2000; St Mary et al., 2001).

The singular benefit of parental care is increased offspring survival. This enhanced survival improves the fitness of parents by increasing the likelihood that their genes will be passed on to future generations. However, the costs to the parents can be numerous and include decreased future mating, decreased future survivorship and decreased future fertility (Clutton-Brock, 1991; Sargent 1997; Reynolds et al., 2002). Because the primary benefit of parental care is relatively consistent and the costs can vary, the evolution of forms of parental care is often driven by optimisation of offspring survival versus the varying costs associated with male, female, or biparental offspring care.

While significant pre-zygotic contributions are made by fishes to ensure offspring survival, most often by females in the form of egg provisioning, parental care against pathogens and predators is generally a post-zygotic parental responsibility. The evolutionary transitions in the parental responsibilities after fertilization are well documented for fishes. A lack of post-zygotic parental care is the most common characteristic among species of fishes and is presumed to be the ancestral state (Gross and Sargent, 1985; Reynolds et al., 2002). Based on analyses of the phylogenetic relationships among fishes, male care, derived from territoriality behaviors, is likely to be the next most common evolutionary step. In cases where longer care time is required, often related to increases in egg size or environmental changes, females will join the male at the nest, thus creating a biparental care situation (Gross and Sargent, 1985). In cases of biparental care, males have been shown to desert the female when other mating prospects are available (Keenleyside, 1983; Balshine-Earn and Earn, 1998). This scenario has been proposed to facilitate an evolutionary transition to female only care. Female care can then lead to the reemergence of no care. A general review of fish life histories indicates that uniparental male care is the most common behavior among families of fishes exhibiting care (49%), followed by biparental care (13%), and uniparental female care (7%) (Gross and Sargent, 1985). The consistencies of these patterns will be revealed with the application of well-resolved phylogenetic information to species life history data; however, these evolutionary transitions appear to be relatively consistent in externally fertilising species (Gross and Sargent, 1985; Reynolds et al., 2002).

NEST GUARDING IN FISHES

Although some fishes exhibit highly derived forms of parental care in which offspring are protected inside the body of the parent, for example mouthbrooding (Koblmuller et al., 2004) and placental viviparity (Reznick et al., 2002), the majority of care among species occurs while eggs are developing in the aquatic environment. In these cases, nest guarding is the most common parental behavior exhibited among families of fishes, with males more likely than females to be the guarding sex (Blumer, 1979). This bias may be due to the greater net fitness advantage to males resulting from guarding (Gross and Sargent, 1985). This parental presence at the nest has been hypothesized to inhibit predation, decrease the amount of debris on the eggs, and provide oxygen and assist in the removal of waste (Moyle and Cech, 1996; Knouft and Page, 2004; Green and McCormick, 2005).

Considering the wide variety of potential nest predators in aquatic systems, including other fishes, amphibians, and macro-invertebrates (Knouft and Page, 2004), the benefit of a parental presence at the nest seems obvious. However, a detailed investigation of the evolution of parental care in fishes has resulted in the suggestion that the presence of a parent at a nest does not automatically ensure that parental care will be provided to the offspring (Gross and Sargent, 1985). In the case of male nest guarding, the evolutionary transition preceding this behavior may include an intermediate stage in which the male gains some benefit from remaining at the nest, such as attracting additional females, while the offspring apparently realise no benefit from the male's presence (Gross and Sargent, 1985; Clutton-Brock, 1991). Considering the prevalence of nest guarding among fishes, the ubiquity of predators in the aquatic environment, and the possibility that nest guarding does not necessarily indicate the existence of parental care, it is interesting that relatively little research has been directed at the relationship between nest guarding and nest predation in freshwater fishes. Nevertheless, studies have shown that the presence of parents at the nest inhibits predation on eggs by other fishes and macro-invertebrates (McKaye, 1984; Rahel, 1989; Mol, 1996; Knouft and Page, 2004).

MICROBIAL DEFENSES IN FISHES

While the predatory potential of vertebrates and macro-invertebrates has been documented, the predatory potential of microscopic organisms remains relatively unknown (Paxton and Willoughby, 2000; Knouft et al., 2003). Predation is defined by the acquisition and consumption of one individual by another. From this perspective, fertilized eggs deposited into the aquatic environment are subject to predation by heterotrophic microbes. Considering the fact that nearly all substrate surfaces in aquatic systems are covered by biofilms composed of heterotrophic microbes, the potential for egg predation by these microbes is potentially as high, if not higher than the threat posed by macro-predators (Knouft et al., 2003).

Recent studies have identified the presence of antimicrobial compounds in the epidermal mucus of a variety of fishes (e.g., *Oncorhynchus mykiss* (Austin and McIntosh, 1988); *Cyprinus carpio* (Cole et al., 1997); *Pleuronectes americanus* (LeMaitre et al., 1997); *Morone saxatilis* × *M. chrysops* hybrid (Silphaduang and Noga, 2001)). These species are members of different orders (Salmoniformes, Cypriniformes, Pleuronectiformes and Perciformes, respectively), suggesting that the presence of antimicrobial compounds in fish mucus has evolved multiple times in distinct lineages. It has been suggested that these compounds, believed to be ubiquitous in fishes, may serve as a first line of defense against microbial pathogens (Boman, 1995).

Although a considerable amount of effort has been directed towards understanding reproductive behavior in fishes (Breder and Rosen, 1966;

Perrone and Zaret, 1979; Baylis, 1981; Gross and Sargent, 1985), limited experimental work has been directed at understanding the relationship between parental care, microbial infection and development of embryos. This is noteworthy considering that most surfaces in aquatic systems are covered by biofilms that are composed primarily of heterotrophic bacteria and fungi (Lock *et al.*, 1984). Localized growth inhibition of these microbes should be essential to survival during all life stages of aquatic organisms, particularly developing eggs.

The threat posed to fish eggs by aquatic microbes, particularly fungi and water molds, has long been recognized as a concern in aquaculture (Kitancharoen *et al.*, 1997). Unfertilized eggs appear to be the most susceptible to fungal colonization (Paxton and Willoughby, 2000), although the reason for this is not clear. Once present, the fungal colonizers can then impact developing eggs and cause death by direct infection or by suffocation of adjacent eggs (Pottinger and Day, 1999). Because of the density dependent aspect of this mode of colonization, the threat posed by microbial pathogens should be greatest in species that cluster eggs in continuous masses (as opposed to species that lay discrete unattached eggs).

PARENTAL CARE AND MICROBIAL INFESTATION

Empirical studies have shown that eggs covered in debris and/or microbes do not develop normally, and consequently do not result in viable offspring (e.g., Knouft *et al.*, 2003; Green and McCormick, 2005). Guarding parents may act to clean debris from eggs and increase oxygen levels by fanning the eggs with fins. During this behavior, the parent will fan or brush the eggs with a fin and generate a mild flow of water over the eggs. Green and McCormick (2005) demonstrated that this behavior increases the level of dissolved oxygen around the eggs, and the parent will modify fanning rates in response to varying oxygen levels. While this behavior can increase offspring survivorship by maintaining sufficient oxygen levels, there is no indication that this behavior decreases or removes attached microbes on the surface of eggs, as has been suggested (Bart and Page, 1991). Indeed, it is difficult to imagine that a mild flow of water could be responsible for dislodging microbes from the surface of an egg.

Filial cannibalism, the consumption of eggs by parents, has been documented in some species (Lindstrom, 1997; Klug and St Mary, 2005). This behavior has been hypothesized as a method for removing infected eggs from the nest; further, there may be fitness benefits due to increases survival of remaining eggs, or energy gains by parents that can be reinvested into future reproduction (Klug *et al.*, 2006). While egg consumption has been inferred as a potential mechanism to reduce microbial infection in nests, there is no experimental evidence that fishes preferentially consume infected eggs or that guarding parents can even identify infected eggs in the nest.

Recent work has indicated that parents may contribute antimicrobial compounds inhibiting microbial growth on fish eggs. *Perca fluviatilis*, a species exhibiting no post-zygotic parental care, reproduces by dispersing a contiguous, folded egg mass surrounded in a gelatinous matrix. Paxton and Willoughby (2000) noted that microbial growth by *Aphanomyces* spp. and *Saprolegnia* spp. occurred within dead *P. fluviatilis* eggs, but did not spread to adjacent live eggs, suggesting an antimicrobial component in the egg mass. Experimental manipulations indicated that developing eggs that were exposed to multiple species of *Saprolegnia* did not experience a greater level of infestation and mortality than unexposed eggs (Paxton and Willoughby, 2000). This result was suggested to be, in part, due to antifungal properties of the *P. fluviatilis* egg mass. The decrease in colonization rates on developing eggs could also suggest an innate immune response by developing embryos that inhibits microbial pathogen growth on eggs.

Do parents guarding the nest provide antimicrobial care to developing eggs? In species that provide parental care in the form of nest guarding, eggs are often deposited directly onto the substrate that generally supports heterotrophic biofilms. Knouft *et al.* (2003) described experiments on *Etheostoma crossopterum* (Percidae), the Fringed Darter, examining whether the presence of a guarding male inhibits microbial colonization of eggs. *Etheostoma crossopterum* is a small, benthic stream fish that is native to North America. Adults are sexually dimorphic during the breeding season, with males exhibiting larger body size and bolder patterning on the body. During the reproductive period from March through May, a male establishes a territory in the cavity under a stone in the stream (Page, 1983). Multiple females will sequentially attach eggs to the underside of the stone that are simultaneously fertilised by the male. This process results in a single layer of up to 1500 eggs deposited on the nest stone directly above the male. The male remains at the nest until the eggs hatch, which generally occurs between five and ten days depending on water temperature (Page, 1983).

In some cases males will abandon nests for unknown reasons. When this occurs, eggs quickly become covered in fungus, bacteria, and water molds (*Saprolegnia* spp.), and the entire nest can become non-viable within as little as a day. Knouft *et al.* (2003) experimentally removed males from the nest and demonstrated that microbial infection is not likely to be the cause of male abandonment, but infection appears to occur after the male has left the nest. Thus, male presence at the nest appears to inhibit microbial colonization.

Breeding male *Etheostoma crossopterum* are active under the nest and frequently rub against the eggs with their nape. The nape in breeding males is noticeably swollen and contains an increased concentration of mucus secreting cells relative to non-breeding males and females (Bart and Page, 1991) (Fig. 9.1). Consequently, the male is frequently applying mucus to the eggs. Knouft *et al.* (2003) demonstrated cytotoxic activity in the mucus to bacteria (*Salmonella typhimurium*) and a complete inhibitory effect against *Saprolegnia* spp. growth. Although the compound was not identified, this was the first documentation of an antimicrobial form of parental care in vertebrates.

A similar form of antimicrobial care was identified in two species of marine blennies (*Ophioblennius atlanticus* and *Salaria pavo*) (Giacomello *et al.*, 2006). Males in these species, as well as other species of Blenniidae, guard nests of eggs deposited on the substrate. Males in these species also possess anal glands on the anterior anal fin rays that are well developed in nest-guarding males relative to males engaged in opportunistic breeding

Fig. 9.1   Breeding male *Etheostoma crossopterum*. Arrow indicates nape area apparently used to apply antimicrobial compounds to eggs. (Photo provided by L.M. Page.)

behaviors (Gonçalves *et al.*, 1996; Neat *et al.*, 2003; Giacomello and Rasotto, 2005). During spawning, males rub these glands over the nest area and continue this process while tending the developing eggs (Giacomello *et al.*, 2006). The glands contain high concentrations of cells that produce mucus secretions that have been demonstrated to exhibit bacteriolytic activity (Giacomello *et al.*, 2006). While the evolutionary origin of the glands is unclear, these structures may have become adapted to provide antimicrobial protection of the developing eggs.

Evidence for antimicrobial parental care in fishes is not restricted to species that guard eggs deposited on the substrate. Members of the Seahorse genus *Hippocampus* exhibit a paternal-based form of care in which a female deposits her eggs into a male's brood pouch. The eggs are then fertilized by the male and the offspring develop for several weeks until they are released from the brood pouch. Melamed *et al.* (2005) constructed a cDNA library from the tissue lining the male's pouch to enhance the understanding of the functioning of the male brood pouch. Among several genes encoding a variety of proteins, Melamed *et al.* (2005) identified genes encoding C-type lectins. Initial results indicate that these compounds, which are secreted in significant amounts into the male pouch, inhibit bacterial growth. Consequently, the male not only appears to provide physical concealment of developing larvae, but also an environment that minimizes the chances of microbial infection.

In the freshwater Discus (*Symphysodon aequifasciata*), both male and female parents feed developing larvae with epidermal mucus secretions (Kishida and Specker, 1994). This resource is apparently crucial, as larval mortality increases in offspring separated from parents (Hildemann, 1959; Schutz and Barlow, 1997). Chong *et al.* (2005) applied a proteomics approach to investigate the potential contribution of parental mucus to developing offspring in this species. Several proteins were found to be uniquely expressed in parental mucus; including compounds facilitating cell repair and stress mediation in the adults. Additionally, Chong *et al.* (2005) identified a C-type lectin expressed in parental mucus that is hypothesised to enhance antimicrobial resistance for both parents and developing fry.

The identification of antimicrobial parental care in fishes is a recent discovery (Knouft *et al.*, 2003). Accordingly, very little is known about the extent of this general tactic among different taxonomic groups of fishes as well as the variety of mechanisms used by different species to protect developing offspring from microbial pathogens. However, even with the limited number of studies on antimicrobial parental care in fishes, a relatively diverse array of tactics has already been revealed (Knouft et al., 2003; Chong et al., 2005; Melamed et al., 2005; Giacomello et al., 2006). Moreover, the species in these studies all represent different Families (Percidae, Blenniidae, Cichlidae and Sygnathidae), suggesting that these types of behaviors may be widely distributed among taxonomic groups of fishes and have independent evolutionary origins.

ASSESSING THE EVOLUTION OF ANTIMICROBIAL PARENTAL CARE

The study of parental care in fishes in an evolutionary context has yielded a clearer understanding of the potential factors responsible for transitions to male, female, and biparental care from ancestors exhibiting no parental care (Perrone and Zaret, 1979; Gross and Sargent, 1985). Examination of the evolution of antimicrobial components of parental care potentially offers the opportunity for a more detailed understanding into all stages of the evolution of parental care in fishes while gaining insights into the evolution of innate immunity. At this time, the small number of documented cases of antimicrobial parental care may seem to limit the possibilities for the study of the evolution of this reproductive tactic. However, numerous distantly related fish species produce antimicrobial compounds, suggesting that this innate immunity is not phylogenetically constrained to a particular group, and potentially widespread among all fishes. With nest guarding being the most common form of parental care in fishes, the possibility also exists that species have integrated this innate immunity into parental care tactics. From an evolutionary perspective, the wealth of phylogenetic studies on fishes provides the historical framework to examine the relationship between parental care and innate immunity. Finally, advances in proteomics allow the characterization of the protein content of fish mucus, thus providing the data for comparative evolutionary analyses.

Previous work has demonstrated that male parental care usually arises from lineages exhibiting no care. This pattern provides a potentially useful framework for understanding the evolutionary origins of antimicrobial parent care. As discussed previously, *Perca fluviatilis* (Percidae), a species which provides no active parental care and occupies a relatively basal evolutionary position in Percidae (Sloss et al., 2004), appears to protect developing eggs by producing a gelatinous matrix around a dispersed egg mass (Paxton and Willoughby, 2000). *Etheostoma crossopterum* (Percidae), a species which provides male parental care and occupies a relatively derived evolutionary position in Percidae (Sloss *et al.*, 2004), applies antimicrobial compounds in the male's mucus to developing eggs (Knouft *et al.*, 2003). This system provides a potentially appropriate, yet unexplored, opportunity to understand the origins of antimicrobial parental care. Two scenarios can be proposed to explain the origin of antimicrobial parental care. The antimicrobial compounds employed by *E. crossopterum* for parental care may be similar to those in *P. fluviatilis*. This result would indicate that these compounds have been conserved and can be traced to a single evolutionary origin, or that multiple independent origins of antimicrobial compounds have led to convergent evolution. Alternatively, *E. crossopterum* and *P. fluviatilis* may not produce similar antimicrobial compounds, suggesting that antimicrobial parental care has arisen as a novel response to microbial conditions associated with nests constructed on a biofilm covered substrate.

The protein dispensability hypothesis provides a framework to assess the evolution of individual compounds and suites of compounds important to parental care among numerous closely related species (Wilson *et al.*, 1977). This hypothesis predicts that proteins that contribute positively to individual fitness should be conserved over evolutionary time (Wilson *et al.*, 1977; Hirsh and Fraser, 2001). For example, *E. crossopterum* is a member of the subgenus *Catonotus* which contains 18 species that all exhibit the same form of male nest guarding behavior (Page, 1983, 1985; Porterfield *et al.*, 1999). When viewed in an evolutionary context, proteins that are potentially important to offspring survival should be shared by breeding males among species. Moreover, more closely related species should have more similar proteomic profiles than distantly related species. While this hypothesis has not been tested, it provides a potentially useful framework for the understanding of the evolution of antimicrobial parental care and local adaptation to the microbial environment by species exhibiting parental care.

CONCLUSIONS

Parental care is a fundamental component of the life histories of many species. The evolutionary processes that are responsible for different forms of parental care exhibited among species of fishes are likely influenced by the potential effect of predators on offspring survival. While these include the obvious macro-predators, such as other fishes and invertebrates, microbial pathogens have recently been identified as significant potential sources of mortality for developing eggs. Multiple species have evolved mechanisms to inhibit microbial growth on developing eggs and embryos and incorporated these mechanisms into parental care behaviors (Knouft et al., 2003; Chong et al., 2005; Melamed et al., 2005; Giacomello et al., 2006). While the general evolution of parental care in fishes has received much attention, little is known about the evolution of factors associated with antimicrobial parental care. Nevertheless, recent work has indicated that microbes are a significant concern for developing eggs and embryos and likely influence modes of parental care in fishes. Fortunately, the opportunity to integrate the vast amounts of available information on the evolutionary relationships of fishes with proteomic techniques offers a novel approach to the understanding of this important component of parental care, and more generally, the evolution of innate immunity in fishes.

Acknowledgements

I thank Allison Miller for extremely helpful comments on a draft of this manuscript and Larry Page for providing the photograph used in Figure 9.1.

References

Austin, B. and D. McIntosh. 1988. Natural antibacterial compounds on the surface of rainbow trout. *Journal of Fish Diseases* 11: 275-277.

Balshine-Earn, S. and D.J.D. Earn. 1998. On the evolutionary pathway of parental care in mouth-brooding cichlid fish. *Proceedings of the Royal Society of London* B265: 2217-2222.

Bart, H.L., Jr. and L.M. Page. 1991. Morphology and adaptive significance of fin knobs in egg-clustering darters. *Copeia* 1991: 80-86.

Baylis, J.R. 1981. The evolution of parental care in fishes, with reference to Darwin's rule of male sexual selection. *Environmental Biology of Fishes* 6: 223-251.

Blumer, L.S. 1979. Male parental care in the bony fishes. *Quarterly Review of Biology* 54: 149-161.

Boman, H.G. 1995. Peptide antibiotics and their role in innate immunity. *Annual Review of Immunology* 13: 61-92.

Breder, C. and D. Rosen. 1966. *Modes of Reproduction in Fishes*, T.F.H. Publications, Inc., Neptune City, N.J.

Chong, K., S. Joshi, L.T. Jin and C.S. Shu-Chien. 2005. Proteomics profiling of epidermal mucus secretion of a cichlid (*Symphysodon aequifasciata*) demonstrating parental care behavior. *Proteomics* 5: 2251-2258.

Clutton-Brock, T.H. 1991. *The Evolution of Parental Care*. Princeton University Press, Princeton.

Cole, A.M., P. Weis and G. Diamond. 1997. Isolation and characterization of pleurocidin, an antimicrobial peptide in the skin secretions of winter flounder. *Journal of Biological Chemistry* 272: 12008-12013.

Dobzhansky, T. 1973. Nothing in biology makes sense except in the light of evolution. *American Biology Teacher* 35: 125-129.

Giacomello, E., D. Marchini and M.B. Rasotto. 2006. A male sexually dimorphic trait provides antimicrobials to eggs in blenny fish. *Biology Letters* 2: 330-333.

Gonçalves, E.J., V.C. Almada, R.F. Oliveira and A.J. Santos. 1996. Female mimicry as a mating tactic in males of the blenniid fish *Salaria parvo*. *Journal of the Marine Biological Association of the United Kingdom* 76: 529-538.

Goodwin, N.B., S. Balshine-Earn and J.D. Reynolds. 1998. Evolutionary transitions in parental care in cichlid fish. *Proceedings of the Royal Society of London* B265: 2265-2272.

Green, B.S. and M.I. McCormick. 2005. O-2 replenishment to fish nests: males adjust brood care to ambient conditions and brood development. *Behavioral Ecology* 16: 389-397.

Gross, M.R. and R.C. Sargent. 1985. The evolution of male and female parental care in fishes. *American Zoologist* 25: 807-822.

Hildemann, W.H. 1959. A cichlid fish (*Symphysodon discus*) with unique nurture habits. *American Naturalist* 93: 27-34.

Hirsh, A.E. and H.B. Fraser. 2001. Protein dispensability and rate of evolution. *Nature (London)* 411: 1046-1049.

Keenleyside, M.H.A. 1983. Mate desertion in relation to adult sex-ratio in the biparental cichlid fish *Herotilapia multispinosa*. *Animal Behavior* 31: 683-688.

Kishida, M. and J.L. Specker. 1994. Vitellogenin in the surface mucus of tilapia (*Oreochromis mossambicus*): possibility for uptake by the free-swimming embryos. *Journal of Experimental Zoology* 268: 259-268.

Kitancharoen, N., K. Hatai and A. Yamamoto. 1997. Aquatic fungi developing on eggs of salmonids. *Journal of Aquatic Animal Health* 9: 314-316.

Klug, H., K. Lindstrom and C.M. St Mary. 2006. Parents benefit from eating offspring: density-dependent egg survivorship compensates for filial cannibalism. *Evolution* 60: 2087-2095.

Klug, H. and C.M. St Mary. 2005. Reproductive fitness consequences of filial cannibalism in the flagfish, *Jordanella floridae*. *Animal Behaviour* 70: 685-691.

Knouft, J.H. and L.M. Page. 2004. Nest defense against predators by the male Fringed Darter (*Etheostoma crossopertum*). *Copeia* 2004: 915-918.

Knouft, J.H., L.M. Page and M.J. Plewa. 2003. Antimicrobial egg cleaning by the fringed darter (Perciformes: Percidae: *Etheostoma crossopterum*): implications of a novel component of parental care in fishes. *Proceedings of the Royal Society of London* B270: 2405-2411.

Koblmuller, S., W. Salzburger and C. Sturmbauer. 2004. Evolutionary relationships in the sand-dwelling cichlid lineage of Lake Tanganyika suggest multiple colonization of rocky habitats and convergent origin of biparental mouthbrooding. *Journal of Molecular Evolution* 58: 79-96.

LeMaitre, C., N. Orange, P. Saglio, N. Saint, J. Gagnon and G. Molle. 1997. Characterization and ion channel activities of novel antibacterial proteins from the skin mucosa of carp (*Cyprinus carpio*). *European Journal of Biochemistry* 240: 143-149.

Lindstrom, K.B. 1997. Food access, brood size and filial cannibalism in the fantail darter, *Etheostoma flabellare*. *Behavioral Ecology and Sociobiology* 40: 107-110.

Lindstrom, K.B. 2000. The evolution of filial cannibalism and female mate choice strategies as resolutions to sexual conflict in fishes. *Evolution* 54: 617-627.

Lock, M.A., R.R. Wallace, J.W. Costerton, R.M. Ventullo and S.E. Charleton. 1984. River epilithon: toward a structural-functional model. *Oikos* 42: 10-22.

Maynard Smith, J. 1977. Parental investment: a prospective analysis. *Animal Behavior* 25: 1-9.

McKaye, K.R. 1984. Behavioral aspects of cichlid reproductive strategies: patterns of territoriality and brood defense in Central American substratum spawners and African mouth brooders. In: *Fish Reproduction: Strategies and Tactics*, G.W. Potts and R.J. Wootton (eds.). Academic Press, London, pp. 245-273.

Melamed, P., Y. Xue, J.F.D. Poon, Q. Wu, H. Xie, J. Yeo, T.W.J. Foo and H. K. Chua. 2005. The male seahorse synthesizes and secretes a novel C-type lectin into the brood pouch during early pregnancy. *FEBS Journal* 272: 1221-1235.

Mol, J.H. 1996. Impact of predation on early stages of the armoured catfish *Hoplosternum thoractum* (Siluriformes-Callychthyidae) and implications for the syntopic occurrence with other related catfishes in a neotropical multi-predator swamp. *Oecologia* 107: 395-410.

Moyle, P.B. and J.J. Cech. 1996. *Fishes: An Introduction to Ichthyology*. Prentice-Hall, Englewood Cliffs, New Jersey.

Neat, F.C., L. Locatello and M.B. Rasotto. 2003. Reproductive morphology in relation to alternative male reproductive tactics in *Scartella cristata*. *Journal of Fish Biology* 62: 1381-1391.

Page, L.M. 1983. *Handbook of Darters*. T.F.H. Publications, Inc., Neptune City, N.J.

Page, L.M. 1985. Evolution of reproductive behaviors in percid fishes. *Bulletin of the Illinois Natural History Survey* 33: 275-295.

Paxton, C.G.M. and L.G. Willoughby. 2000. Resistance of perch eggs to attack by aquatic fungi. *Journal of Fish Biology* 57: 562-570.

Perrone, M. and T.M. Zaret. 1979. Parental care patterns of fishes. *American Naturalist* 113: 351-361.

Porterfield, J.C., L.M. Page and T.J. Near. 1999. Phylogenetic relationships among fantail darters (Percidae: *Etheostoma: Catonotus*): Total evidence analysis of morphological and molecular data. *Copeia* 1999: 551-564.

Pottinger, T.G. and J.G. Day. 1999. A *Saprolegnia parasitica* challenge system for rainbow trout: assessment of Pyceze as an anti-fungal agent for both fish and ova. *Diseases of Aquatic Organisms* 36: 129-141.

Rahel, F.J. 1989. Nest defense and aggressive interactions between a small benthic fish (*Etheostoma nigrum*) and crayfish. *Environmental Biology of Fishes* 24: 301-306.

Reynolds, J.D., N.B. Goodwin and R.P. Freckleton. 2002. Evolutionary transitions and live bearing in vertebrates. *Philosophical Transactions of the Royal Society of London* B2002: 269-281.

Reznick, D.N., M. Mateos and M.S. Springer. 2002. Independent origins and rapid evolution of the placenta in the fish genus *Poeciliopsis*. *Science* 298: 1018-1020.

Sargent, R.C. 1997. Parental care. In: *Behavioural Ecology of Teleost Fishes*, J.J. Godin (ed.). Oxford University Press, Oxford, pp. 290-313.

Schutz, M. and G.W. Barlow. 1997. Young of the Midas Cichlid get biologically active nonnutrients by eating mucus from the surface of their parents. *Fish Physiology and Biochemistry* 16: 11-18.

Silphaduang, U. and E.J. Noga. 2001. Peptide antibiotics in mast cells of fish. *Nature* (*London*) 414: 268-269.

Sloss, B.L., N. Billington and B.M. Burr. 2004. A molecular phylogeny of the Percidae (Teleostei, Perciformes) based on mitochondrial DNA sequence. *Molecular Phylogenetics and Evolution* 32: 545-562.

St Mary, C.M., C.G. Noureddine and K. Lindstrom. 2001. Environmental effects on male reproductive success and parental care in the Florida Flagfish (*Jordanella floridae*). *Ethology* 107: 1035-1052.

Trivers, R.L. 1972. Parental investment and sexual selection. In: *Sexual Selection and the Descent of Man*, B. Campbell (ed.). Aldine, Chicago, pp. 136-179.

Wilson, A.C., S.S. Carlson and T.J. White. 1977. Biochemical evolution. *Annual Review of Biochemistry* 46: 573-639.

The Nose Knows: Chemically Mediated Antipredator Defences in Ostariophysans

Reehan S. Mirza

INTRODUCTION

Fishes have a variety of ways of defending themselves against environmental stressors. These stressors can be biotic (predators, pathogens, etc.) or abiotic (contaminants, temperature, oxygen levels, etc.) Exposure to any environmental stressor can elicit an inducible defence. Inducible defences are temporary and are evoked when a preset threshold of stress is exceeded and subside once the stressor falls below threshold response levels (Havel, 1987; Harvell, 1990; Tollrian and Harvell, 1999). The exposures to stressors are intermittent, and thus permanent defences are not favoured due to the biological cost involved. Inducible defences involve a wide variety of responses, including changes in chemical/biochemical, physiological, morphological, life-historical and behavioural processes. Each one of these responses acts on a different

Author's address: Department of Biology, Nipissing University, North Bay, ON, Canada P1B 8L7.

E-mail: reehanm@nipissingu.ca

timescale from seconds to minutes (physiological, chemical/biochemical, behavioural) to days and weeks or longer (morphological, shifts in life-history traits). The particular induced response elicited is most likely correlated with the length of exposure to the environmental stressor.

This chapter focuses on the inducible defences evoked by chemical cues associated with predation. Predation is a strong selective force that shapes many behavioural, morphological and life-historical responses. Animals must be able to assess local predation risk accurately; otherwise the probability of surviving an encounter with a predator is low. To ensure accurate assessment of predation risk, animals are constantly sampling their environments because predation pressure is not constant and fluctuates on a daily basis (Lima and Bednekoff, 1999; Sih, 2000). In aquatic systems, we can group sensory information into visual, chemical, mechanical and electrical stimuli (Smith, 1992). Consequently, fishes exhibit a wide array of sensory structures to detect and process this information.

Transfer of information from one individual to another is the basis of communication. However, communication tends generally is defined as having three components: a sender, a signal and a receiver. The interaction between the signaller and receiver results in an exchange of information that leads to fitness benefits both for the senders that are manipulating receivers, and for the receivers through their responses to the information provided (Johnstone, 1997). Under this definition, there is intent from the sender to signal the receiver. Selection should favour the shaping of specific signals sent by the signaller, but at the same time, they shape the reception capabilities of the receiver to maximize reception (Bradbury and Vehrencamp, 1998; Wisenden and Chivers, 2006). However, this definition does not account for all sensory information exchanged between individuals. Receivers may benefit from detecting information, but with no benefit to the sender. This form of information transfer has been termed public information (Danchin et al., 2004; Wisenden and Chivers, 2006). The concept of public information fills a void in communication theory. Public information also may be seen as nongenetically acquired information (Danchin et al., 2004). Selection acts upon receivers to detect and process the information presented, but there is no intent on the part of the sender to direct the signal to a particular individual. Information in most environments is non-specific and a by-product of other ecological processes (e.g., foraging, predation). Yet public information is important and even lifesaving under certain circumstances.

Thus, the importance of public information drives the selection pressures acting upon receivers for detection and processing. Public information also may represent the platform from which signals evolve (Danchin et al., 2004).

Ostariophysans make excellent test subjects for examining how public information can be used to assess predation risk. Ecologically, Ostariophysans are distributed in almost every part of the globe and hold an important position in most food webs. Ostariophysans comprise taxonomic groups such as minnows and carps, suckers, catfishes, knifefishes, electric eels and milk fishes; combined they make up approximately 74% of all freshwater fishes (Nelson, 1994). Because of their abundance and diversity, the sophistication of chemosensory assessment of predation risk has been studied intensely within this group of fishes.

CHEMICAL ALARM CUES

Humans often are visually biased and tend to discount chemical information. However, chemosensation is likely to be the oldest sensory modality and has a long evolutionary history with water, thereby shaping a variety of responses of aquatic organisms (Wisenden, 2000). Moreover, water is ubiquitous and a universal solvent that can dissolve numerous compounds, giving a large number of potential signals to be detected (Klerekoper, 1969; Hara, 1994). Chemical information also may be more reliable when other sensory modalities are limited, such as at night or in turbid or heavily vegetated waters (Smith, 1992; Wisenden, 2000). However, chemical information can be limited under certain situations, such as in fast-flowing water. Because of the amount of public chemical information that is available, it makes evolutionary sense that aquatic organisms (including Ostariophysans) are selected to use chemical cues in assessment of local ecological conditions, including predation risk.

In a landmark review paper, Lima and Dill (1990) described predation as proceeding through a sequence of steps and calculated the probability that successful predation (i.e., consumption) would occur at each step. Wisenden (2000) condensed the predation event into three stages: (1) detection (pre-encounter) (2) capture and (3) ingestion (post-capture). At each stage, chemical cues may be released that can be used to assess the local predation risk. The chemical cues can be categorised based on the point of the predation event at which each is released (Fig. 10.1). During

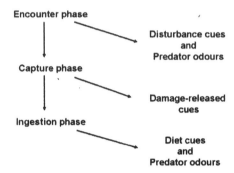

Fig. 10.1 Flow diagram representing the predation event. At each stage of the predation event, chemical alarm cues are released.

the detection/pre-encounter stages, two types of chemical alarm cues may be present. The first of these cues emanates from the predator and are called 'kairomones' (interspecific chemical cues that provide a benefit to the receiver: Brown *et al.*, 1970). The second type of alarm cues released during this stage of the predation event is termed disturbance cues. Disturbance cues are believed to be correlated with low threat levels and released by animals that are startled or stressed most likely via the urine (Wisenden, 2003). During the capture phase, damage-released or injury-released cues are released from a prey animal after mechanical damage to the body (e.g., Smith, 1992). The third category of chemical alarm cues comprises chemicals released after the prey has been digested; these cues are referred to as 'diet cues'. Diet cues refer to any chemical released by the predator that is influenced by the predator's recent diet (e.g., through its bodily secretions and/or cues released from the feces) (Mathis and Smith, 1993b; Brown *et al.*, 1995). Diet cues may bear a similarity to damage-released cues in some respects because some of the presence of same components.

Logically, accurate assessment of predation risk is beneficial because each anti-predator defence has an innate cost to the user (Lima and Bednekoff, 1999). Anti-predator behaviour generally makes use of time and energy that otherwise would have been allocated to foraging and reproduction and is therefore a costly. This trade-off loss is likely to be an important selection pressure driving prey species to develop efficient risk assessment systems (Lima and Dill, 1990). Upon detecting alarm cues, receivers typically respond with an anti-predator response, which includes behavioural and morphological responses as well as shifts in life-history traits. Behavioural responses include dashing, freezing, area avoidance, stronger shoaling, increased use of shelter, decreased foraging activity and decreased movement (Lima and Dill, 1990; Chivers and Smith, 1998; Kats and Dill, 1998; Lima, 1998).

Damage-released Alarm Cues

Damage-released alarm cues have received the most attention in the literature. The work originated in the late 1930s, when Nobel Laureate Karl von Frisch conducted a simple yet elegant experiment with European minnows (*Phoxinus phoxinus*). Von Frisch sat at the edge of a pond with a small feeding table, threw food to the minnows and observed the schools foraging. He then prepared a chemical stimulus by taking a minnow and lacerating the sides several times, rinsing the body and retaining the rinsed water. Von Frisch slowly poured the stimulus into the water and the minnows responded by scattering. Von Frisch concluded that the skin contained a special chemical that induced a fright response in conspecifics. He termed the chemical Schreckstoff (von Frisch, 1938), which translates into 'fright stuff'. This response to shreckstoff is believed to be innate and widespread amongst Ostariophysans (Smith, 1977, 1992).

Soon after von Frisch's initial study, histological analysis of Ostariophysan skin found the presence of special epidermal club cells that were believed to contain schreckstoff/damage-released alarm cue (Fig. 10.2). The cells do not possess any ducts to the surface and are ruptured through mechanical damage as would occur during a predation event. The presence of epidermal club cells and shreckstoff has become a taxonomic character uniting Ostariophysans (Nelson, 1994). Moreover, within the Ostariophysi, species more closely related to each other respond more intensely to each other's damage-released alarm cues than those that are distantly related (Schutz, 1956). The evolution of these specialized cells as well as the function of shreckstoff is unclear and will be discussed later in this chapter.

Fish skin is very fragile and easily ruptured, so even the slightest damage causes the release of alarm cues into the water. The amount of alarm cue required to elicit a fright response is believed to be minimal. Lawrence and Smith (1989) found that 1 cm^2 of fathead minnow (*Pimephales promelas*) skin creates an active space of approximately 58,000 L, which is the equivalent of a sphere with a radius of 3.2 m. Active

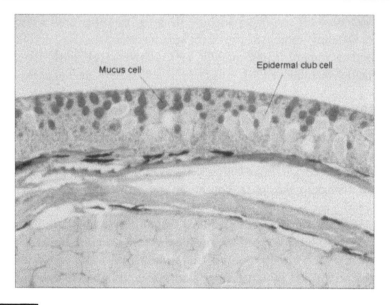

Fig. 10.2 Cross section of common shiner (*Notropis cornutus*) skin showing epidermal club cells.

space is defined as the volume in which the concentration of the stimulus exceeds the detection threshold of the receiver (Lawrence and Smith, 1989). Similarly, Dupuch *et al.* (2004) found that 1 cm^2 of northern redbelly dace skin (*Phoxinus eos*) creates an active space of 110,000 L (equivalent to a sphere with a radius of 4.8 m). Moreover, dace still exhibited fright reactions to conspecific alarm cues after the cues were diluted by a factor of 10,000.

A typical predation event is likely to rupture many more club cells than the experimental injuries in Lawrence and Smith (1989) and studies carried out by Dupuch *et al.* (2004), but the amount released would depend on the size and type of predator as well as the predator's handling time. Conspecifics that detect these alarm cues typically avoid or flee the area, increase shoal cohesion or decrease activity so as to avoid detection, which increases the probability of survival when encountering a predator (Mathis, 1993a; Mirza and Chivers, 2001). In addition to responding behaviourally, Ostariophysans can also alter their morphology over prolonged exposures to alarm cues. Stabell and Lwin (1997) found that Crucian carp (*Carassius carassius*) exhibited deeper body shapes when exposed to carp alarm cues for 45 days, as compared to carp exposed to cues from injured arctic charr (*Salvelinus alpinus*). The increase in body depth is an attempt to exceed the gape-limit of a common predatory northern pike (*Esox lucius*). Deeper-bodied carp survive longer rather than shallow-bodied carp when encountering pike (Nilsson *et al.*, 1995).

The majority of studies examining anti-predator responses upon exposure to Ostariophysan alarm cues are conducted under controlled laboratory conditions. Recent criticisms in literature have encouraged researchers to conduct studies of alarm cues under natural conditions (Magurran *et al.*, 1996; Smith, 1997; Chivers and Smith, 1998). Field studies are particularly important because of the recent suggestion that some prey animals responding strongly to chemical alarm signals in the laboratory may not respond strongly (if at all) when tested under natural conditions (Magurran *et al.*, 1996). Von Frisch (1938) provided the first evidence that Ostariophysans avoided areas labelled with damage-released alarm cues under natural conditions. A similar avoidance response was seen in creek chub (*Semotilus atromaculatus*) in two independent studies (Newsome, 1975; Smith, 1976). Recently, several studies on area avoidance in fathead minnows under natural conditions have been conducted using minnow traps containing small pieces of cellulose sponge saturated with minnow alarm cues. Traps labelled with minnow alarm cue caught fewer fishes than traps labelled with controls (Mathis and Smith, 1992, 1993; Wisenden *et al.* 1995a, b; Chivers *et al.*, 1995). Moreover, minnows would not return to the areas where traps were deployed for 7-8 hours (Wisenden *et al.*, 1995a).

Magurran *et al.* (1996) disagreed with the conclusions based on the results of the trap experiments, stating that minnow traps were not natural. They conducted a study using underwater video cameras and found that European minnows did not avoid areas labelled with alarm cues. The authors argued that responses of Ostariophysans to conspecific alarm cues were simply an artefact of laboratory studies or unnatural field studies. The naturalness of underwater cameras in aquatic systems not withstanding, Wisenden *et al.* (2004) conducted a study using the methods of Magurran *et al.* (i.e., underwater video) in order to observe the responses of cyprinids in three different lakes in Minnesota; they found that cyprinids avoided areas labelled with damage-released alarm cues in each lake. Similarly, Friesen and Chivers (2006) used underwater video cameras to demonstrate strong area avoidance by fathead minnows and finescale dace (*Chrosomus neogaeus*) to areas labelled with fathead minnow alarm cue under natural conditions. Thus, there is strong evidence that Ostariophysans use damage-released alarm cues in the wild to avoid areas labelled as potentially dangerous thereby decreasing the probability of encountering predators.

Damage-released alarm cues serve to warn conspecifics of potential danger, but this does not preclude predators from manipulating this information for their own benefit. Mathis *et al.* (1995) conducted a series of laboratory and field experiments testing the response of predators to Ostariophysan alarm cues using two different predators. In the laboratory, pike oriented more often towards and spent more time in compartments of aquaria containing alarm cue compared to controls. Moreover in field experiments, significantly more predaceous adult diving beetles (*Colymbetes sculptilis*) were attracted to traps labelled with minnow alarm cues than traps labelled with swordtails cues. Similarly, Wisenden and Thiel (2002) found in a field experiment that predators were attracted to and struck sponges saturated with fathead minnow alarm cue significantly more than sponges saturated with cues from convict cichlids (*Archocentrus nigrofasciatus*) or plain distilled water. Although these chemicals function as alarm cues for Ostariophysans, predators can use the same information to locate a potential meal. However, predator attraction to alarm cues may not be universal. Cashner (2004) failed to find attraction of spotted bass (*Micropterus punctulatus*) to damage-released alarm cues from five different sympatric cyprinids.

From an evolutionary perspective it would seem maladaptive that selection would favour an alarm system that attracted predators, but Ostariophysans may actually gain a benefit from attracting additional predators to a predation event. The predator attraction hypothesis states that additional predators attracted to a predation event may attack the initial predator in an attempt to kleptoparasitize the prey. In the ensuing struggle, the prey individual may escape (Smith, 1977, 1992; Mathis *et al.*, 1995). Chivers *et al.* (1996) tested the predator attraction hypothesis by staging encounters between pike and fathead minnows. Once a minnow had been captured, a second pike was introduced into the test arena. The second predator interfered directly or indirectly with the first predator, allowing the minnow to escape.

Disturbance Cues and Predator Kairomones

Disturbance cues have received very little attention in the literature, and have been studied primarily in crayfish and non-Ostariophysan species

(Hazlett 1985, 1989; Wisenden *et al.*, 1995b; Mirza and Chivers, 2002). Some evidence does that suggests that these cues have a nitrogenous based and released via the urine. Kiesecker *et al.* (1999) found that red-legged frog tadpoles (*Rana aurora*) decreased activity when exposed to chemical stimuli from conspecifics that were startled or stressed. A similar decrease in activity was seen when tadpoles were exposed to ammonia at environmentally relevant concentrations. Within Ostariophysans, there is only one published example of a response to disturbance cues. Jordao and Volpato (2000) found that pacus, *Piaractus mesopotamicus*, avoided portions of the tank containing water from conspecifics that had recently viewed a predator. More studies are needed to assess how Ostariophysans respond to cues from stressed conspecifics.

Conversely, much more work has been conducted on responses to predator kairomones. Predator kairomones (commonly referred to as predator odours) can be detected within the predation sequence at both the pre-capture and post-capture stages. The post-capture stage usually also incorporates the odour of a recently consumed prey item and will be discussed separately (see: Diet Cues). Detection of a predator odour prior to direct contact decreases the probability of encountering or being detected by the predator. Through various mechanisms prey can learn the identity of the predator and use this information to assess predation risk (see: Learned Recognition of Novel Stimuli). For example, Chivers and Smith (1994a) found that fathead minnows from a pike free population do not exhibit a fright response to the odour of pike fed a diet of swordtails, while minnows of the same size and age from a pike syntopic population exhibited fright responses to the same pike stimulus. This suggests that responses to predator odours are learned and not innate. Thus, experience is essential to assessing risk accurately during future encounters and may alter or modify the response exhibited by the prey.

Once the identity of the predator has been learned, the odour of the predator alone is sufficient to evoke a fright response regardless of diet. Moreover, based on experience prey can use chemical cues to determine the size of the predator. From an ecological perspective, different sizes of the same predator may represent different levels of risk. Kusch *et al.* (2004) found that fathead minnows could differentiate between pike of two different sizes based on the odour of the predator, pike of each size class were fed a neutral diet of red swordtails (*Xiphophorus helleri*). Fathead minnows responded with a higher intensity fright response to small pike than large pike. The authors argue that smaller pike represented the greater predation threat; hence the higher intensity antipredator response. Similarly, Pettersson *et al.* (2000) found that juvenile crucian carp responded more intensely to the odour of large pike fed swordtails than small pike fed the same diet. Jachner (1997) found that bleak (*Alburnus alburnus*) responded more intensely to the odour of northern pike that had recently consumed bleak (three days before trials) compared to pike that had not been fed for seven days. Prey animals can modify their fright response based on how recent the predator has had a meal.

Diet Cues

One of the most interesting recent advancements in our understanding of the role of chemical alarm cues in predation risk assessment comes from studies that have manipulated the predator's diet. Diet cues are released after the prey has been digested and provide essential information regarding not only the presence of a predator, but also whether the predator is likely to be a current risk (Jachner, 1997). Chivers and Mirza (2001) recently reviewed several papers that assessed the influence of predator diet on the intensity of antipredator responses among aquatic vertebrates. Generally, these studies indicate that the intensity of the response of prey is reduced (or even absent) if the predator is fed a diet that does not contain conspecific prey. The majority of this work has been conducted over the last 15 years.

The specific diet-based chemical cues recognized by prey are as yet unknown. Most likely there exist similarities between diet cues and damage-released alarm cues. In a series of experiments, Mathis and Smith (1993b, c) determined that the presence of epidermal club cells (believed to contain the alarm cue) were necessary to elicit a fright response. In their first study, Mathis and Smith demonstrated that pike-naïve minnows exhibited an anti-predator response to chemical cues of pike that were fed fathead minnows but not to chemical cues of pike that were fed swordtails (*Xiphophorus helleri*). Thus, the response is specific to the presence of conspecifics in the diet. Mathis and Smith (1993c) then elucidated the nature of the cues responsible for the diet effect. They exposed pike-naïve fathead minnows to cues of pike that were fed breeding male fathead minnows, non-breeding fathead minnows or swordtails. Breeding male fathead minnows temporarily lose their epidermal club cells during the breeding season (Smith, 1973). Test minnows showed an anti-predator response to chemical stimuli from pike that had been fed non-breeding minnows (epidermal club cells were present), but not to chemical cues from pike fed a diet of breeding males (epidermal club cells absent) or swordtails. These results demonstrate that the alarm cue contained in the skin of minnows is part of the diet cue to which the minnows respond. Minnow alarm cue, or some active component thereof, survives passage through the gut of the predator and provides the diet-based cue to which minnows respond (Mathis and Smith, 1993b, c; Brown et al., 1995). The cue also would include predator odours that are independent of diet and additional metabolic byproducts of digestion. Brown et al. (1995) found that fathead minnows exhibited a fright response to chemical cues derived from a fecal extract from pike that had been fed minnows versus fed swordtails. Although the exact chemical composition of a diet cue is not known, presence or absence of a source of damage-released alarm cue alters the response of the receiver.

Like damage-released alarm cues, diet cues also can elicit morphological and life-historical responses in Ostariophysans. Brönmark and Miner (1992) examined predator-induced morphological changes in crucian carp exposed to pike that had been feeding on carp. They divided a pond in half with a barrier and placed an equal number of juvenile carp on each side. On one side, they introduced juvenile northern pike and allowed them to interact with the carp for 14 days. They found that carp on the side with the pike had significantly deeper bodies than carp on the non-pike side. The increase in body depth excluded carp from the gape of the pike. Brönmark and Pettersson (1994) repeated the study in the laboratory and found the same morphological changes occurred after 60 days of exposure to pike fed crucian carp. Once the carp were removed from the diet cue exposure, body depth regressed over the next 60 days. Diet cues also can influence life-history traits. Kusch and Chivers (2004) found that fathead minnow eggs exposed to cues from virile crayfish (*Orconectes virilis*) that had been fed minnow eggs hatched sooner than controls. Moreover, newly hatched minnows were also shorter in length than control minnows. Earlier hatching allows minnows to escape predation from egg predators and the smaller size also may help decrease the probability of detection by other predators.

Prey animals should have a selective advantage if they can use information from the last meal the predator consumed to modulate the intensity of their response. By being able to differentiate between predators that consumed different diets, prey individuals will not waste time and energy responding to predators that do not pose an imminent threat.

Moreover, diet cues may facilitate learning (see below), which can translate into increased survival for the prey (Mirza and Chivers, 2003).

LEARNED RECOGNITION OF NOVEL STIMULI

Ostariophysans clearly benefit through direct responses to alarm cues (injury-released, disturbance-released or dietary). However, alarm cues also can be used in indirect assessment of predation risk, such as by playing a role in associative learning of novel dangerous stimuli. Behavioural ecologists have been most interested in using alarm cues to facilitate learning by prey fishes of unfamiliar predators or potentially dangerous habitats, but the learning process works equally well when experimenters use stimuli that are not ecologically relevant.

Releaser-induced Learning

Prey individuals can learn to recognize novel stimuli through releaser-induced recognition learning, i.e., simultaneously pairing of an aversive or 'releasing' stimulus with a neutral stimulus leading to learned aversion to the neutral stimulus (Suboski, 1990). The first demonstration of this ability in fishes was by Göz (1941), who found that blinded European minnows exhibited anti-predator behaviour in response to chemical stimuli from northern pike only after pike attacked minnows in their presence. The introduction of damage-released alarm cues in the presence of pike conditioned the blind minnows to respond to the chemical stimuli from pike in subsequent tests. Similarly, Magurran (1989) found that predator-naïve European minnows raised in the laboratory did not respond to cues from northern pike, but once pike odour was presented simultaneously with minnow alarm cue, the minnows learned the identity of the predator. This learning process is very similar to classical conditioning except for the fact that classical conditioning typically requires repeated conditioning trials, but releaser-induced learning occurs after a single conditioning trial. This rapid learning process may reflect the strength of selective pressures involving recognition of predatory cues. If a prey individual cannot learn the identity of a predator quickly, then the probability of survival during subsequent encounters would diminish rapidly. Learned recognition of a predator is retained for days to weeks after the initial learning event without any reinforcement (Chivers and Smith, 1998). Magurran (1989) demonstrated that European minnows still responded to chemical stimuli from pike two days after being conditioned. Chivers and Smith (1994b) found that fathead minnows retained acquired recognition of visual cues from pike or goldfish (*Carassius auratus*) for 2 months after being conditioned with alarm cues, but the intensity of the response tended to decrease over time, suggesting that some level of reinforcement may be necessary to retain full memory for long periods of time.

The importance of conspecific alarm cues in the general process of learning has been shown in studies where Ostariophysans have learned to identify a novel stimulus that was not ecologically relevant as dangerous when the stimulus was paired with conspecific alarm cues. Studies with zebra danios (*Danio rerio*) and fathead minnows found that releaser-induced learning occurred when conspecific alarm cues were paired with morpholine, a synthetic odorant (Suboski 1990), or a red light (Hall and Suboski, 1995; Yunker et al., 1999). Learning mediated by alarm cues can also be used to learn biologically relevant stimuli other than predators. Chivers and Smith (1995a) trained fathead minnows to recognise potentially dangerous habitats. Juvenile northern pike are sit-and-wait predators that are typically found within shallow weedy areas. Chivers and Smith exposed fathead minnows to a paired stimulus of minnow alarm cue and water from weedy areas of a lake typically occupied by juvenile pike. Upon subsequent exposure to water from the weedy habitat alone, minnows exhibited a fright reaction. The role of chemical alarm cues in learning demonstrates the sophistication of chemical information in assessing predation risk.

Learning via Diet Cues

Diet cues can also be used to acquire recognition of predators. This process is also a form of releaser-induced learning, except that in this case the process paired two types of cues from the same predator: diet cues, which the prey recognized as dangerous, and cues that were independent of diet, which were viewed as novel. Mathis and Smith (1993b) exposed groups of predator-naïve fathead minnows to chemical stimuli from pike fed a diet of either swordtails or minnows, and then subsequently exposed minnows to cues from pike fed a diet of swordtails. Minnows that had been previously exposed to cues from pike fed a diet of minnows responded to subsequent exposure to the pike stimulus with a fright response, demonstrating acquired recognition of the predator.

In many aquatic systems, predators tend to be generalists and prey upon a variety of prey items. Most prey animals are members of prey guilds—animals that share a common habitat and suite of predators. Prey individuals often respond to chemical alarm cues released from other prey guild members (Mathis and Smith, 1993d; Chivers and Smith, 1998), which allows more efficient application of their antipredator responses because more information is available to assess predation risk. In order to acquire recognition of a prey guild member or a new predator in the system via diet cues, there must be a known cue present in the mixed diet stimulus. It is well established that the alarm pheromone of fathead minnows and other Ostariophysan fishes can pass through the digestive system of fishes and can still be recognized (Mathis and Smith, 1993b, c; Brown et al., 1995).

Mirza and Chivers (2001) exposed fathead minnows to chemical cues of a yellow perch (*Perca flavescens*) that were fed a mixed diet of either minnows and brook stickleback (*Culaea inconstans*) or perch that were fed a mixed diet of swordtails and brook stickleback, and then tested to see whether the minnows responded to stickleback alarm cues alone. Minnows previously exposed to perch that had been fed minnows and sticklebacks subsequently exhibited anti-predator behaviour to stickleback cues alone. In contrast, minnows exposed to cues from perch that had been fed sticklebacks and swordtails did not subsequently respond to stickleback cues alone. These results indicate that minnows learn the identity of the stickleback alarm signal based on associating it with conspecific alarm cues in the diet of the perch. Additional studies in different predator/prey systems are needed to determine whether learned recognition of heterospecific alarm cues through a mixed predator diet is widespread phenomenon. Learning additional cues to assess predation risk also can enhance survival of the prey upon encountering a predator. Prey warned by alarm cues may increase their group cohesion, increase their vigilance and maintain a greater distance from the predator. As such, we can expect that warned prey might be less likely to be attacked or less likely to be captured during an attack.

In a follow-up study, Chivers et al. (2002) examined the survivorship of minnows that had been trained to recognize brook stickleback (*Culaea inconstans*) alarm cues. Fathead minnows were exposed to rainbow trout, *Oncorhynchus mykiss*, fed either a mixed diet of minnows and sticklebacks or swordtails and sticklebacks. Subsequently, two sets of staged encounters were conducted using predators with different prey capture efficiencies.

One set of encounters was conducted using yellow perch, an inefficient predator, and the other using northern pike, an efficient predator. Prior to interaction with the predator, minnows were exposed to stickleback alarm cue. Minnows conditioned to the mixed diet of minnow and stickleback were able to evade predators significantly longer than minnows exposed to a mixed diet of swordtails and stickleback indicating they had learned the heterospecific alarm cue (Fig. 10.3).

Social Learning

Another method by which learning could occur is by observing the responses of nearby individuals. Social learning is defined as the acquisition of some biologically relevant information in the absence of direct experience (Mathis *et al.*, 1996). Ostariophysans are very gregarious, and, thus, the opportunity to observe other shoal mates is common. Predator-naïve individuals may be alerted to potential predation threats through spatial associations with experienced conspecifics and/or

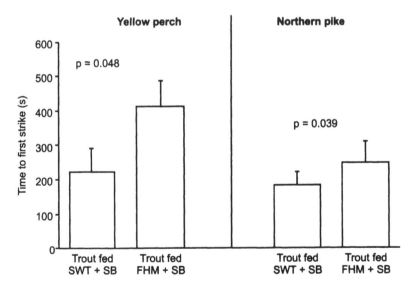

Fig. 10.3 Mean + SE of time to first strike of an inefficient (yellow perch) and an efficient (northern pike) predator towards minnows previously conditioned to rainbow trout fed on a mixed diet of either swordtail (SWT) and brook stickleback (SB) or fathead minnows (FHM) and brook stickleback (SB). Fathead minnows conditioned to a mixed diet containing minnows were able to evade the predator significantly longer (i.e., higher survival) than fathead minnows conditioned to a mixed diet not containing minnows. Figure modifed from Chivers *et al.* (2002).

heterospecifics (Verheijen, 1956; Magurran and Higham, 1988; Krause, 1993; Mathis *et al.*, 1996). Although no direct interactions occur, prey can learn to recognise potential predation threats. Mathis *et al.* (1996) found that pike-naïve fathead minnows could acquire recognition of a predator when they were paired with an experienced tutor that was exhibiting a fright response. Similarly, Chivers and Smith (1995b) found that experienced fathead minnows that had learned risky habitats could socially transmit this learned response to other minnows observing them.

Although evidence of social learning from laboratory studies is compelling, it is important to know whether social learning would occur under natural conditions. Chivers and Smith (1995c) tested whether social learning could occur under natural conditions by stocking a small 0.5 Ha pond containing 20,000 pike-naïve fathead minnows with 10 juvenile northern pike. After 14 days, minnows were collected from the pond and tested for their response to pike odour (pike fed a diet of swordtails). Chivers and Smith found that minnows had learned to identify pike. In a similar experiment, Brown *et al.* (1997) found that after stocking 39 juvenile pike into a 2-Ha pond containing 79,000 pike-naïve fathead minnows, minnows had learned to recognise the odour of pike after only 4 days, but learning to recognize visual cues from pike took 6 days. The initial learning events would have most likely been facilitated by minnow damage-released alarm cues or diet cues. However, the probability that every individual in the pond would have encountered a pike is low and it is, therefore, likely that the knowledge spread via social transmission.

Multiple Cues

A criticism of learned predator-recognition studies is that typically a single predator is used. Single predator studies may lack ecological realism because prey are most likely faced with multiple predators under natural conditions. In these situations, prey must learn to recognise all of the predators that represent a common threat so as to maximize their probability of survival. Learning may be constrained through biological and ecological factors (reviewed in Brown and Chivers, 2005), but ideally, the prey should possess the ability to learn the identity of multiple predators. Prey individuals could learn to recognise predators one at a time (sequentially) or, if multiple predators are present, sequential learning predators may be costly, and thus selection should favour the ability of prey to learn the identity of multiple predators simultaneously.

Darwish *et al.* (2005) tested the ability of glow light tetras (*Hemigrammus erythrozonus*) in order to learn multiple predator odours simultaneously. Predator-naïve glow light tetras were conditioned to conspecific alarm cues paired with a cocktail of odours from two predatory fishes (smallmouth bass, *Micropterus salmoides* and convict cichlids) and one non-predatory fish (goldfish). Tetras were tested 48 h after conditioning for recognition of each of the individual odours within the cocktail, yellow perch odour (a novel predator control), or distilled water. Tetras exhibited a fright response to each of the three odours within the cocktail, but not to yellow perch odour or distilled water. In a second experiment, the authors examined whether tetras conditioned to a cocktail of odours exhibited higher survival when encountering a predator. They conditioned groups of glow light tetras to tetra alarm cues paired with either: a cocktail of predator odours (pumpkinseed sunfish *Lepomis gibbosus*, convict cichlids and rainbow trout), odour of pumpkinseed sunfish alone or swordtail skin extract paired with distilled water and, subsequently, allowed tetras to interact with a live pumpkinseed sunfish. Tetras that had been conditioned to the cocktail of predator odours or sunfish odour only exhibited higher survival than controls (Fig. 10.4).

The survival benefit of simultaneous learning of multiple predators was the same as that of learning the identity of a single predator. Learning multiple predators simultaneously is potentially an ecologically adaptive strategy for prey. Temporal and spatial variability of predators within a community occurs continuously, and, as a result, prey must be able to quickly adjust their behaviour to current levels of risk (Lima and Bednekoff, 1999; Sih, 2000). Moreover, if learning requires reinforcement, potential predators that do not attack will be forgotten in favour of solidifying responses to those predators that are consistently dangerous as antipredator responses are fine-tuned with experience.

Constraints on Learning

The use of alarm cues to facilitate learning appears to be a powerful and important factor contributing to survival of prey animals. In almost every case, learned recognition has resulted from pairing of alarm cues with novel stimuli. However, learning events require reinforcement and appropriate contexts. If the appropriate context is not provided, then learning may be impaired. Recent research with virile crayfish found that if a novel odour is presented before it is paired with the alarm cue during

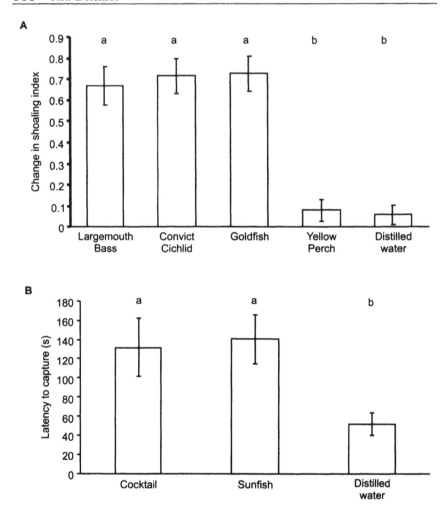

Fig. 10.4 Mean ± SE of A) change in shoaling index of glowlight tetras (previously conditioned to an odour cocktail subsequently exposed to chemical stimuli of either: largemouth bass, convict cichlid, goldfish, yellow perch or distilled water and B) latency to capture by pumpkinseed sunfish of tetras previously conditioned to an odour cocktail, sunfish odour or distilled water. Tetras significantly increased group cohesion to odours present in the cocktail and also exhibited significantly greater survival if conditioned to odours that included the predator (sunfish). Different letters above bars represent significant differences, p < 0.05. Figure modified from Darwish *et al.* (2005).

a learning trial, then releaser-released recognition learning did not occur (Acquistaplace *et al.*, 2003; Hazlett, 2003). This phenomenon is referred to as latent inhibition—a process by which pre-exposure of a stimulus without consequence retards the learning of subsequent conditioned associations with that stimulus.

Although a widespread phenomenon in mammals, it is not known whether latent inhibition occurs in fishes. Ferrari and Chivers (2006) tested whether latent inhibition occurred in fathead minnows. They exposed groups of charr-naive minnows to the odour of brook charr (*Salvelinus fontinalis*) fed a diet of swordtails or distilled water for 60 minutes for a period of 5 days. On the sixth day, minnows were exposed to a paired stimulus of charr odour plus minnow alarm cues or charr odour added to distilled water. Minnows were subsequently tested 24 h later to charr odour only to determine whether learning had occurred. Minnows that had received distilled water for the five days prior to conditioning to a paired stimulus of charr odour and alarm cues decreased activity and increased shoal cohesion when exposed to charr odour on day 7 indicating learning. Minnows that had been exposed to charr odour for 5 days previous to exposure to a paired stimulus of either charr odour and alarm cue or charr odour and distilled water did not change any activity when exposed to charr odour alone on day 7. This result demonstrates latent inhibition on learning of a novel predator.

What is the ecological relevance of latent inhibition? Is it detrimental to fishes or does it provide some benefit? In one sense, latent inhibition may be an adaptive mechanism to avoid learning irrelevant information or false learning of non-dangerous stimuli as being potentially dangerous. Fishes are exposed to a plethora of chemical and visual stimuli in their environments. If they associated everything around them with danger, then excess time would be spent on antipredator responses and less on other important ecological activities such as foraging or reproduction (Lima and Dill, 1990). Conversely, latent inhibition also may be harmful in that if learning of potential threats does not occur, then survival is compromised. However, latent inhibition most likely works in conjunction with other mechanisms to fine tune the learning response of novel stimuli in an adaptive manner.

THREAT SENSITIVITY

Because ecological systems are dynamic, temporal components must be considered; different predators may be active at different times of day or even seasonally (Lima and Bednekoff, 1999; Sih, 2000). Moreover, the activity patterns of the prey must also be considered. Thus, the probability of encountering a predator is variable, and the prey must have flexibility in their antipredator responses to match the current situation. Consequently, there should be strong selection pressure to distinguish between predators that pose a risk and those that do not. The ability of prey species to assess and respond flexibly towards varying degrees of predation threat is described by the threat-sensitive predator-avoidance hypothesis (Helfman, 1989).

The threat-sensitive predator-avoidance hypothesis states that the intensity of the antipredator response performed by prey animals will reflect the level of current predation risk. The hypothesis makes three assumptions: (1) prey face conflicting demands on their time and energy, (2) prey must trade-off antipredator responses versus other activities, and (3) prey will respond in an appropriately graded manner. There have been a number of recent studies that provide evidence for the threat-sensitive predator-avoidance hypothesis in Ostariophysans. Kusch *et al.* (2004) found that fathead minnows from a pike-syntopic population responded more intensely to the odour of small pike than to that of large pike. Moreover, minnows showed a graded increase in fright response to increasing volumes of small-pike odour. Minnows did not waste time and energy responding to large pike, which appeared to represent a lower threat than small pike. As the threat level from the primary predator increased, so did the intensity of the antipredator response (Fig. 10.5). An increase in the intensity of antipredator behaviour also has been demonstrated in response to increasing concentrations of damage-released alarm cues in redbelly dace (Dupuch *et al.*, 2004), goldfish (Zhao and Chivers, 2005) and fathead minnows (Ferrari *et al.*, 2005).

From an ecological perspective, varying concentrations of alarm cues may represent either the distance from the predation event along a concentration gradient or may indicate that time has elapsed since the predation event occurred. The predator may be either a long distance away or else have completely left the area. Ferrari *et al.* (2006a) found that fathead minnows responded more intensely to predators that represented closer proximity than to a larger group of predators. They exposed fathead minnows to the odour of either two northern pike at a concentration of 30 ml pike odour/pike or the odour of twelve pike at a concentration of 5 ml pike odour/pike. Minnows responded with a greater intensity antipredator response to the two-pike-odour stimulus. When they equalised the concentration of pike odour at 5 ml pike odour/pike, minnows exhibited a stronger antipredator response to stimuli from 12 pike versus 2. When the odour per pike was greater, minnows perceived

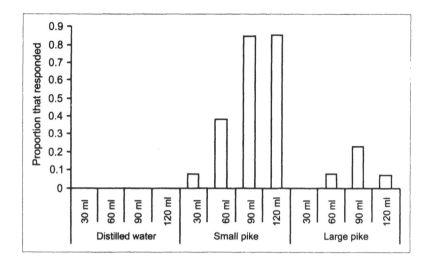

Fig. 10.5 Proportion of fathead minnows that responded with a fright response after exposure to varying concentrations of chemical cues from small or large pike, or to distilled water. Minnows responded more intensely to increasing concentrations of cues from small pike and also responded more intensely to cues from small pike versus large pike. Figure modified from Kusch *et al.* (2004).

the risk to be higher, most likely due to closer proximity, and therefore responded more strongly. However, once the concentration per pike was held to be constant, the larger number of pike represented the higher level of risk.

Threat-sensitive responses also may influence learned recognition of predators. Ferrari *et al.* (2005) found that the intensity of the learned response matches the intensity of the fright response during the conditioning period. Groups of minnows were conditioned with three different concentrations of damage-released alarm cues from conspecifics paired with the odour of brook charr (a novel predator). The intensity of their responses during subsequent exposures to predator odour alone matched that of their responses in medium and high concentration 'training trials' (Fig. 10.6); the learned response reflected the level of threat that minnows had been exposed to during the conditioning period. In a similar study, Ferrari *et al.* (2006b) examined influences on threat-sensitive learning by holding the concentration of damage-released alarm cues constant, but manipulating the concentration of predator odour that minnows were conditioned with. Groups of pike-naïve fathead minnows were conditioned with either a low or high concentration of pike odour paired with minnow alarm cue for 1 h. After 24 h, minnows from each

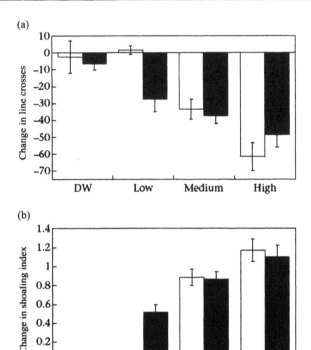

Fig. 10.6 Mean ± SE change in (a) number of line crosses and (b) shoaling index for fathead minnows exposed to different concentrations of minnow alarm cue or distilled water (DW) paired with charr odour during conditioning trials (open bars), or exposed to charr odour alone during recognition trials (closed bars). From Ferrari *et al.*, 2005.

conditioning treatment were exposed to either low or high concentrations of pike odour alone. Regardless of conditioning treatment, if minnows were subsequently exposed to high concentrations of pike odour, they exhibited a higher intensity of fright response than minnows exposed to low concentrations of alarm cues. Minnows matched the intensity of their fright response to reflect the level of current risk irrespective of the concentration of pike odour used for learning.

EVOLUTION OF OSTARIOPHYSAN ALARM CUES

In this chapter, I have described numerous examples of the ways in which Ostariophysan fishes use chemical alarm cues to assess predation risk. From the simple responses described by von Frisch to the complex effects shown in recent studies, it is evident that chemosensory assessment of predation risk via alarm cues is important for Ostariophysans. However, despite these findings, one question that has plagued researchers for decades is 'what is the evolutionary function of chemical alarm cues from the point of view of the sender?' If the schreckstoff system evolved as an alarm system, then the benefits of 'warning' should apply to senders as well as to receivers; this dual benefit has been demonstrated in some alarm systems for terrestrial species (Seyfarth and Cheney, 2003). Although benefits to receivers are clear, the benefits to senders are not. Because alarm cues are released when the prey is captured or consumed by the predator, there seems little opportunity for senders to benefit. As described above, there is some evidence that prey may escape when secondary predators and attracted by the alarm cues and interfere with the primary predator; however, escape occurs relatively infrequently, so selection via this mechanism should be fairly weak. It is possible (and perhaps even likely) selection has acted on receivers to recognize and respond to these cues with appropriate antipredator (in prey) or attraction (in predators) responses in the context of public information (Wisenden and Chivers, 2006).

There are alternative hypotheses regarding the evolution of production of the alarm cue. The system may have evolved to provide indirect benefits to kin, repel predators or provide an antipathogen/ antiparasite function (reviewed in Smith, 1992, 1999) that have not been tested until recently. Hugie (1990) found that fathead minnows from pike-allotopic populations had fewer epidermal club cells than minnows from pike-syntopic populations. This difference in club cells may have been attributed to differences in the levels of predation risk. However, Chivers and colleagues (2007) found that manipulating the level of predation risk did not increase alarm cell numbers in fathead minnows. It appears that predation may not be the selection pressure behind the evolution of Schreckstoff or epidermal club cells.

Wisenden and Smith (1997, 1998) found that the number of epidermal club cells increased when fathead minnows were raised with unfamiliar shoal mates for 16 days. The authors explained the increased number of cells in the context of predation risk, but increased risk in this case could arise from pathogens rather than predators. Most members of any population are most likely exposed to the same suite of pathogens and therefore have built up immunity to these diseases. However, when foreign individuals are introduced into the population, new pathogens may also be introduced. An increase in epidermal club cells could be seen as a preventative response to potential new infections. The schreckstoff alarm system may have evolved to combat infections by pathogens and parasites that burrow through the skin to infect the host. Common pathogens such as water moulds (*Saprolegnia* spp.) grow on the skin of infected individuals. Similarly, common parasites such as trematodes enter their hosts by burrowing through the skin. Chapman and Johnson (1997) examined the histology of epidermal club cells under electron microscopy in channel catfish (*Ictalurus punctatus*) and found that there were several intrusions into the epidermal club cells from microvilli of the epidermal cells, lymphocytes and cells containing intranuclear and intracytoplasmic virus particles. Recently, Chivers and colleagues (unpublished data) found that there might be an antipathogenic/antiparasitic function to schreckstoff. Pathogens and parasites are ubiquitous in aquatic systems and thus could provide the selective pressures necessary to develop essential defences that may be found within epidermal club cells.

CONCLUSIONS AND FUTURE DIRECTIONS

In this chapter, the essential role of Ostariophysan alarm cues in assessing predation risk has been examined. Different categories of alarm cues are released from different points in the predation event. Each category of alarm cue provides important information regarding current levels of predation risk. Earlier studies helped to identify the different types of alarm cues and demonstrate that Ostariophysans responded to these alarm cues with antipredator behaviour. More recent studies have delved more deeply into the intricacies of how chemical information is used in assessing predation risk. The use of chemosensory information in risk assessment is more sophisticated than previously thought. Although predation may not be the selection pressure driving the presence of epidermal club cells and predation, selection has acted upon receivers to utilise alarm cues to assess predation risk. Ostariophysans have served as models for these tests since the first discovery of alarm responses by von Frisch. Since then, the use of alarm cues to assess predation risk has been demonstrated in several non-Ostariophysan fishes, larval amphibians and aquatic invertebrates (reviewed in Chivers and Smith, 1998).

Where do we go from here? To fully understand how chemical alarm cues are used in ecological systems more studies are needed at both the proximate and ultimate levels. The chemical composition of the alarm cue needs to be characterized. Although some researchers have stated that hypoxanthine-3(N)-oxide (H3NO) is the putative Ostariophysan alarm cue (Pfeiffer et al., 1985; Brown et al., 2000), there does not appear to be any known metabolic pathways to producing H3NO in Ostariophysans. However, Brown et al. (2000) also suggested that the primary functional molecular component might be the nitrogen oxide component. In addition to further research on the chemistry of the alarm cue, more information is needed about the chemosensory mechanisms involved in detecting and processing alarm cues. This research should include identification of receptors and neural pathways between the olfactory epithelium and higher brain centres.

At the ultimate level, more information is needed about how the aquatic shapes responses to chemical alarm cues by Ostariophysans. The natural aquatic habitat presents a chemical-rich environment that presents a plethora of information to fishes. From this broth, relevant stimuli must be isolated and synergistic or contradictory stimuli must be integrated. Most studies are conducted in the laboratory under controlled conditions in chemically depauperate environments. More studies need to be conducted under natural conditions or at least more chemically complex captive environments. To understand the ways in which chemical alarm chemicals can mitigate interactions among individuals, further research is needed regarding interactions among conspecifics and among heterospecifics (including fishes and other taxa). Very little research has examined whether chemical alarm signals influences processes at the population or community levels (e.g., population growth rates). Moreover, influences of temporal variation, and diurnal and seasonal cycles should also be examined.

Finally, from a conservation standpoint, it is critical that we determine whether environmental changes affect the alarm cues because they clearly are significantly important to survival of prey. Changes in water chemistry, such as pH, dissolved organic matter, and temperature, could alter the cue or the fish's capacity to receive it. Anthropogenic inputs, such as metals and pesticides, also could also influence chemosensory systems. Understanding these effects on ostariophysian fishes will help environmental regulators and managers provide more meaningful ecological risk assessment. Karl von Frisch started with a simple experiment with schools of European minnows. Little did he know he was opening the door to such an important aspect of fish ecology.

References

Acquistaplace, P., B.A. Hazlett and F. Gherardi. 2003. Unsuccessful predation and learning of predator cues by crayfish. *Journal of Crustacean Biology* 23: 364-370.

Adler, F.R. and C.D. Harvell. 1990. Inducible defenses, phenotypic variability and biotic environments. *Trends in Ecology and Evolution* 5: 407-410.

Bradbury, J.W. and S.L. Vehrencamp. 1998. *Principles of Animal Communication*. Sinauer, Sunderland, Massachusetts.

Brönmark, C. and J.G. Miner. 1992. Predator-induced phenotypical change in body morphology in crucian carp. *Science* 258: 1348-1350.

Brönmark, C. and L.B. Pettersson. 1994. Chemical cues from piscivores induce a change in morphology in crucian carp. *Oikos* 70: 396-402.

Brown, G.E. and D.P. Chivers. 2005. Learning as an adaptive response to predation. In: *Ecology and Evolution of Predator/Prey Interactions*, P. Barbosa and I. Castellanos (eds.). Oxford University Press, Oxford, pp. 34-54.

Brown, W.L., Jr., T. Eisner and R.H. Whittaker. 1970. Allomones and kairomones: Transspecific chemical messengers. *BioScience* 20: 21-22.

Brown, G.E., D.P. Chivers and R.J.F. Smith. 1995. Localized defecation by pike: a response to labelling by cyprinid alarm pheromone? *Behavioural Ecology and Sociobiology* 36: 105-110.

Brown, G.E., D.P. Chivers and R.J.F. Smith. 1997. Differential learning rates of chemical versus visual cues of a northern pike by fathead minnows in a natural habitat. *Environmental Biology of Fishes* 49: 89-96.

Brown, G.E., J.C. Adrian, E. Smyth, H. Leet and S. Brennan. 2000. Ostariophysan alarm pheromones: laboratory and field tests of the functional significance of nitrogen-oxides. *Journal of Chemical Ecology* 26: 139-154.

Cashner, M. 2004. Are spotted basss (*Micropterus punctulatus*) attracted to Schreckstoff? A test of the predator attraction hypothesis. *Copeia* 2004: 592-598.

Chapman, G.B. and E.G. Johnson. 1997. An electron microscope study of intrusions into alarm substance cells of the channel catfish. *Journal of Fish Biology* 51: 503-514.

Chivers, D.P. and R.J.F. Smith. 1994a. The role of experience and chemical alarm signalling in predator recognition by fathead minnows, *Pimephales promelas*. *Journal of Fish Biology* 44: 273-285.

Chivers, D.P. and R.J.F. Smith. 1994b. Fathead minnows, *Pimephales promelas*, acquire predator recognition when alarm substance is associated with the sight of unfamiliar fish. *Animal Behaviour* 48: 597-605.

Chivers, D.P. and R.J.F. Smith. 1995a. Fathead minnows, *Pimephales promelas*, learn to recognize chemical stimuli from high risk habitats by the presence of alarm substance. *Behavioural Ecology* 6: 155-158.

Chivers, D.P. and R.J.F. Smith. 1995b. Chemical recognition of risky habitats is culturally transmitted among fathead minnows, *Pimephales promelas* (Osteichthyes, Cyprinidae). *Ethology* 99: 286-296.

Chivers, D.P. and R.J.F. Smith. 1995c. Free-living fathead minnows rapidly learn to recognize pike as predators. *Journal of Fish Biology* 46: 949-954.

Chivers, D.P. and R.J.F. Smith. 1998. Chemical alarm signalling in aquatic predator-prey systems: a review and prospectus. *Écoscience* 5: 338-352.

Chivers, D.P. and R.S. Mirza. 2001. Predator diet cues and the assessment of predation risk by aquatic vertebrates: a review and prospectus. In: *Chemical Signals in Vertebrates*, A. Marchlewska-Koj, J.J. Lepri and D. Müller-Schwarze (eds.). Plenum Press, New York, Vol. 9, pp. 277-284.

Chivers, D.P., G.E. Brown and R.J.F. Smith. 1996. Evolution of chemical alarm signals: attracting predators benefits alarm signal senders. *American Naturalist* 148: 649-659.

Chivers, D.P., R.S. Mirza and J. Johnston. 2002. Learned recognition of heterospecific alarm cues enhances survival during encounters with predators. *Behaviour* 139: 929-938.

Chivers, D.P., B.D. Wisenden, C.J. Hindman, T.A. Michalak, R.C. Kusch, S.G.W. Kaminskyj, K.L. Jack, M.C.O. Ferrari, R.J. Pollock, C.F. Halbgewachs, M.S. Pollock, S. Alemadi, C.T. James, R.K. Savaloja, C.P. Goater, A. Corwin, R.S. Mirza, J.M. Kiesecker, G.E. Brown, J.C. Adrian Jr., P.H. Krone, A.R. Blaustein and A. Mathis. 2007. Epidermal 'alarm substance' cells of fishes are maintained by non-alarm functions: possible defence against pathogens, parasites and UVB radiation. *Proceedings of the Royal Society of London* B 274: 2611-1619.

Danchin, E., L.-A. Giraldeau, T.J. Valone and R.H. Wagner. 2004. Public information: from nosy neighbors to cultural evolution. *Science* 305: 487-491.

Darwish, T.L., R.S. Mirza, A.O.H.C. Leduc and G.E. Brown. 2005. Acquired recognition of novel predator odour cocktails by juvenile glowlight tetras. *Animal Behaviour* 70: 83-89.

Dupuch, A., P. Magnan and L.M. Dill. 2004. Sensitivity of northern redbelly dace, *Phoxinus eos*, to chemical alarm cues. *Canadian Journal of Zoology* 82: 407-415.

Ferrari, M.C.O. and D.P. Chivers. 2006. The role of latent inhibition in acquired predator recognition by fathead minnows. *Canadian Journal of Zoology* 84: 505-509.

Ferrari, M.C.O., F. Messier and D.P. Chivers. 2006a. The nose knows: minnows determine predator proximity and density through detection of predator odours. *Animal Behaviour* 72: 927-932.

Ferrari, M.C.O., T. Kapitania-Kwok and D.P. Chivers. 2006b. The role of learning in the development of threat-sensitive predator avoidance: the use of predator cue concentration by fathead minnows. *Behavioural Ecology and Sociobiology* 60: 522-527.

Ferrari, M.C.O., J.J. Trowell, G.E. Brown and D.P. Chivers. 2005. The role of leaning in the development of threat-sensitive predator avoidance in fathead minnows. *Animal Behaviour* 70: 777-784.

Friesen, R.G. and D.P. Chivers. 2006. Underwater video reveals strong avoidance of alarm cues by prey fishes. *Ethology* 112: 339-345.

Frisch, K. von. 1938. Zur Psychologie des Fisch-Schwarmes. *Naturwissenschaften* 26: 601-606.

Frisch, K. von. 1941. Über einen Schreckstoff der Fischhaut und seine biologische Bedeutung. *Zeitschrift Vergleichende Physiologie* 29: 46-149.

Göz, H. 1941. Über den Art-und Individualgeruch bei Fischen. *Zeitschrift Vergleichende Physiologie* 29: 1-45.

Hall, D. and M.D. Suboski. 1995. Sensory preconditioning and second-order conditioning of alarm reactions in zebra danio fish (*Brachydanio rerio*). *Journal of Comparative Psychology* 109: 76-84.

Hara, T.J. 1994. The diversity of chemical stimulation in fish olfaction and gustation. *Reviews in Fish Biology and Fisheries* 4: 1-35.

Harvell, C.D. 1990. The ecology and evolution of inducible defenses. *Quarterly Review of Biology* 65: 323-340.

Havel, J.E. 1987. Predator-induced defenses: a review. In: *Predation: Direct and Indirect Impacts on Aquatic Communities*, W.C. Kerfoot and A. Sih (eds.). University Press of New England: Hanover, pp. 263-278.

Hazlett, B.-A. 1985. Disturbance pheromones in the crayfish, *Orconectes virilis*. *Journal of Chemical Ecology* 11: 1695-1711.

Hazlett, B.A. 1989. Additional sources of disturbance pheromone affecting the crayfish *Orconectes virilis*. *Journal of Chemical Ecology* 15: 381-385.

Hazlett, B.A. 2003. Predator recognition and learned irrelevance in the crayfish *Orconectes virilis*. *Ethology* 109: 765-780.

Helfman, G.S. 1989. Threat-sensitive predator avoidance in damselfish-trumpetfish interactions. *Behavioural Ecology and Sociobiology* 24: 47-58.

Hugie, D.M. 1990. *An Intraspecific Approach to the Evolution of Chemical Alarm Signaling in the Ostariophysi*. M.Sc. Thesis. University of Saskatchewan, Saskatoon, Canada.

Jachner, A. 1997. The response of bleak to predator odour of unfed and recently fed pike. *Journal of Fish Biology* 50: 878-886.

Johnstone, R.A. 1997. The evolution of animal signals. In: *Behavioural Ecology: An Evolutionary Approach*, J.R. Krebs and N.B. Davies (eds.). Blackwell Science, London, pp. 155-178.

Jordao, L.C. and G.L. Volpato. 2000. Chemical transfer of warning information in non-injured fish. *Behaviour* 137: 681-690.

Kats, L.B. and L.M. Dill. 1998. The scent of death: Chemosensory assessment of predation risk by prey animals. *Écoscience* 5: 361-394.

Kiesecker, J.M., D.P. Chivers, A. Marco, C. Quilchano, M.T. Anderson and A.R. Blaustein. 1999. Identification of a disturbance signal in larval red-legged frogs (*Rana aurora*). *Animal Behaviour* 57: 1295-1300.

Kleerekoper, H.A. 1969. *Olfaction in Fishes*. Indiana University Press, Bloomington.

Krause, J. 1993. Transmission of fright reaction between different species of fish. *Behaviour* 127: 37-48.

Kusch, R.C. and D.P. Chivers. 2004. The effects of crayfish predation on phenotypic and life history variation in fathead minnows. *Canadian Journal of Zoology* 82: 917-921.

Kusch, R.C., R.S. Mirza and D.P. Chivers. 2004. Making sense of predator scents: investigating the sophistication of predator assessment abilities of fathead minnows. *Behavioural Ecology and Sociobiology* 55: 551-555.

Lawrence, B.J. and R.J.F. Smith. 1989. Behavioural response of solitary fathead minnows, *Pimephales promelas*, to alarm substance. *Journal of Chemical Ecology* 15: 209-219.

Lima, S.L. 1998. Stress and decision-making under the risk of predation: recent developments from behavioral, reproductive and ecological perspectives. *Advances in Study of Behaviour* 27: 215-290.

Lima, S.L. and L.M. Dill. 1990. Behavioural decisions made under the risk of predation: a review and prospectus. *Canadian Journal of Zoology* 68: 619-640.

Lima, S.L. and P.A. Bednekoff. 1999. Temporal variation in danger drives antipredator behavior: the predation risk allocation hypothesis. *American Naturalist* 153: 649-659.

Magurran, A.E. 1989. Acquired recognition of predator odour in the European minnow, *Phoxinus phoxinus*. *Ethology* 82: 216-223.

Magurran, A.E. and A. Higham. 1988. Information transfer across fish shoals under predator threat. *Ethology* 78: 153-158.

Magurran, A.E., P.W. Irving and P.A. Henderson. 1996. Is there a fish alarm pheromone? A wild study and critique. *Proceedings of the Royal Society of London* 263: 1551-1556.

Mathis, A. and R.J.F. Smith. 1992. Avoidance of area marked with a chemical alarm substance by fathead minnows (*Pimephales promelas*) in a natural habitat. *Canadian Journal of Zoology* 70: 1473-1476.

Mathis, A. and R.J.F. Smith. 1993a. Chemical alarm signals increase the survival time of fathead minnows (*Pimephales promelas*) during encounters with northern pike (*Esox lucius*). *Behavioural Ecology* 4: 260-265.

Mathis, A. and R.J.F. Smith. 1993b. Fathead minnows, *Pimephales promelas*, learn to recognize northern pike, *Esox lucius*, as predators on the basis of chemical stimuli from minnows in the pike's diet. *Animal Behaviour* 46: 645-656.

Mathis, A. and R.J.F. Smith. 1993c. Chemical labelling of northern pike (*Esox lucius*) by the alarm pheromone of fathead minnows (*Pimephales promelas*). *Journal of Chemical Ecology* 19: 1967-1979.

Mathis, A. and R.J.F. Smith. 1993d. Intraspecific and Cross-Superorder responses to chemical alarm signals by brook stickleback. *Ecology* 74: 2395-2404.

Mathis, A., D.P. Chivers and R.J.F. Smith. 1995. Chemical alarm signals: predator-deterrents or predator attractants? *American Naturalist* 146: 994-1005.

Mathis, A., D.P. Chivers and R.J.F. Smith. 1996. Cultural transmission of predator recognition in fishes: intraspecific and interspecific learning. *Animal Behaviour* 51: 185-201.

Mirza, R.S. and D.P. Chivers. 2001a. Do chemical alarm signals enhance survival of aquatic vertebrates? An analysis of the current research paradigm. In: *Chemical Signals in Vertebrates*, A. Marchlewska-Koj, J.J. Lepri and D. Müller-Schwarze (eds.). Plenum Press, New York, Vol. 9, pp. 19-26.

Mirza, R.S. and D.P. Chivers. 2001b. Learned recognition of heterospecific alarm signals: The importance of a mixed predator diet. *Ethology* 107: 1007-1018.

Mirza, R.S. and D.P. Chivers. 2002. Behavioural responses to conspecific disturbance chemicals enhance survival of juvenile brook charr, *Salvelinus fontinalis*, during encounters with predators. *Behaviour* 139: 1099-1109.

Mirza, R.S. and D.P. Chivers. 2003. Fathead minnows learn to recognize heterospecific alarm cues they detect in the diet of a known predator. *Behaviour* 140: 1359-1370.

Nelson, J.S. 1994. *Fishes of the World*. Third Edition. John Wiley and Sons, New York.

Newsome, G.E. 1975. *A Study of Prey Preference and Selection by Creek Chub*, Semotilus atromaculatus, *in the Mink River*, Manitoba. PhD. Thesis. University of Manitoba, Winnipeg, Canada.

Nilsson, P.A., C. Brönmark and L.C. Pettersson. 1995. Benefits of a predator-induced morphology in crucian carp. *Oecologia* 104: 291-296.

Pfeiffer, W. 1977. The distribution of fright reaction and alarm substance cells in fishes. *Copeia* 1977: 653-665.

Pfeiffer, W., G. Reiglbauer, G. Meir and B. Scheibler. 1985. Effect of hypoxanthine-3(N)-oxide and hypoxanthine-1(N)-oxide on central nervous excitation of the black tetra *Gymnocorymbus ternetzi* (Characidae, Ostariophysi, Pisces) indicated by dorsal light response. *Journal of Chemical Ecology* 11: 507-524.

Schutz, F. 1956. Vergleichende Untersuchungen über die Schrekreaktion bei Fischen und deren Verbreitung. *Zeitschrift Vergleichende Physiologie* 38: 84-135.

Seyfarth, R.M. and D.L. Cheney. 2003. Signalers and receivers in animal communication. *Annual Reviews in Psychology* 54: 145-175.

Sih, A. 1987. Predator and prey lifestyles: an evolutionary and ecological overview. In: *Predation: Direct and Indirect Impacts on Aquatic Communities*, W.C. Kerfootand and A. Sih (eds.). University Press of New England, Hanover, pp. 203-224.

Sih, A. 2000. New insights on how temporal variation in predation risk shapes prey behavior. *Trends in Ecology and Evolution* 15: 3-4.

Smith, R.J.F. 1973. Testosterone eliminates alarm substance in male fathead minnows. *Canadian Journal of Zoology* 51: 875-876.

Smith, R.J.F. 1976. Male fathead minnows (*Pimephales promelas* Rafinesque) retain their fright reaction to alarm substance during the breeding season. *Canadian Journal of Zoology* 54: 2230-2231.

Smith, R.J.F. 1977. Chemical communication as adaptation: Alarm substance of fish. In: *Chemical Signals in Vertebrates*, D. Müller-Schwarze and M.M. Mozell (eds.). Plenum, New York, pp. 303-320.

Smith, R.J.F. 1992. Alarm signals in fishes. *Reviews in Fish Biology and Fisheries* 2: 33-63.

Smith, R.J.F. 1997. Does one result trump all others? A response to Magurran, Irving and Henderson. *Proceedings of the Royal Society of London* B264: 445-450.

Stabell, O.B. and M.S. Lwin. 1997. Predator-induced phenotypic changes in crucian carp are caused by chemical signals from conspecifics. *Environmental Biology of Fishes*, pp. 139-144.

Suboski, M.D. 1990. Releaser-induced recognition learning. *Psychological Reviews* 97: 271-284.

Tollrian, R. and C.D. Harvell. 1999. The evolution of inducible defenses: current ideas. In: *The Ecology and Evolution of Inducible Defenses*, R. Tollrian and C.D. Harvell (eds.). Princeton University Press, Princeton, New Jersey, pp. 303-321.

Verheijen, F.J. 1956. Transmission of a fright reaction amongst a school of fish and the underlying sensory mechanisms. *Experientia* 12: 202-204.

Wisenden, B.D. 2000. Scents of danger: the evolution of olfactory ornamentation in chemically-mediated predator-prey interactions. In: *Animal Signals: Signalling and Signal Design in Animal Communication*, Y. Espmark, T. Amundsen and G. Rosenqvist (eds.). Tapir Academic Press, Trondheim, Norway, pp. 365-386.

Wisenden, B.D. 2003. Chemically mediated strategies to counter predation. In: *Sensory Processing in the Aquatic Environment*, S.P. Collin and N.J. Marshall (eds.). Springer-Verlag, New York, pp. 236-251.

Wisenden, B.D. and R.J.F. Smith. 1997. The effect of physical condition and shoal-mate familiarity on proliferation of alarm substance cells in the epidermis of fathead minnows. *Journal of Fish Biology* 50: 799-808.

Wisenden, B.D. and R.J.F. Smith. 1998. A re-evaluation of the effect of shoalmate familiarity on the proliferation of alarm substance cells in fathead minnows. *Journal of Fish Biology* 53: 841-846.

Wisenden, B.D. and T.A. Thiel. 2002. Field verification of predator attraction to minnow alarm substance. *Journal of Chemical Ecology* 28: 433-438.

Wisenden, B.D. and D.P. Chivers. 2006. The role of public chemical information in antipredator behaviour. In: *Communication in Fishes*, F. Ladich, S.P. Collins, P. Moller and B.G. Kapoor (eds.). Science Publishers, Enfield, NH, USA, Vol. 1, pp. 259-278.

Wisenden, B.D., D.P. Chivers, G.E. Brown and R.J.F. Smith. 1995a. The role of experience in risk assessment: Avoidance of areas chemically labelled with fathead minnow alarm pheromone by conspecifics and heterospecifics. *Écoscience* 2: 116-122.

Wisenden, B.D., D.P. Chivers and R.J.F. Smith. 1995b. Early warning of risk in the predation sequence: A disturbance pheromone in Iowa darters (*Etheostoma exile*). *Journal of Chemical Ecology* 21: 1469-1480.

Wisenden, B.D., K.A. Vollbrecht and J.L. Brown. 2004. Is there a fish alarm cue? Affirming evidence from a wild study. *Animal Behaviour* 67: 59-67.

Yunker, W.K., D.E. Wein and B.D. Wisenden. 1999. Conditioned alarm behavior in fathead minnows (*Pimephales promelas*) resulting from association of chemical alarm pheromone with a nonbiological visual stimulus. *Journal of Chemical Ecology* 25: 2677-2686.

Zhao, X. and D.P. Chivers. 2005. Response of juvenile goldfish (*Carassius auratus*) to chemical alarm cues: Relationship between response intensity, response duration and the level of predation risk. In: *Chemical Signals in Vertebrates*, R.T. Mason, M. LeMaster and D. Müller-Schwarze (eds.). Plenum Press, New York, Vol. 10, pp. 334-341.

Alarm Responses as a Defense: Chemical Alarm Cues in Nonostariophysan Fishes

Alicia Mathis

INTRODUCTION

Response to Alarm Cues as a Defense

A predation event consists of an escalation of steps that proceeds from encounter to detection, identification, approach, subjugation and consumption of prey (Endler, 1991). At each step of the interaction, prey can evolve antipredator responses that increase their probability of survival. Because the probability of successful capture increases at each succeeding step in the predation sequence (Lima and Dill, 1990), there is strong selection favoring prey that perform antipredator responses early in the encounter.

In fishes, one method of early detection of predators is through chemical alarm cues. Following the terminology of previous studies

Author's address: Department of Biology, Missouri State University, Springfield, Missouri, USA.

E-mail: aliciamathis@missouristate.edu

(Smith, 1986), 'alarm' cues in this chapter will refer to chemicals released from injured fishes ('senders') that lead to antipredator ('fright') responses by nearby conspecifics ('receivers'). Although, in some species, distressed fishes can produce 'disturbance' cues without actual injury (Wisenden, 2003), this chapter will address only injury-released cues.

Over the decades, periodic reviews of fish alarm cues have been published, and they provide an excellent history of progress in this field of study. In 1977, extensive reviews by Pfeiffer (discussed in detail below) and Smith firmly established the early view that alarm chemicals were virtually ubiquitous for fishes in the superorder Ostariophysi (minnows, suckers, catfishes), but were absent or rare in nonostariophysans. Smith's next review, a decade later (Smith, 1986), included an 'analogous chemical alarm system' in darters, and his 1992 review added gobies and sculpins to the list. Finally, a 1998 overview of alarm signals in vertebrate and invertebrate aquatic species also listed cichlids, poeciliids, salmonids and sticklebacks (Chivers and Smith, 1998).

1938-1977

Wolfgang Pfeiffer did English-speaking researchers a substantial service by reviewing the research conducted prior to 1977 on fish alarm chemicals (Pfeiffer, 1977). Many of these papers were published in German, including seminal works by von Frisch (1938; 1941a, b), Schutz (1956) and Pfeiffer (1963). The nonostariophysan taxa included in Pfeiffer's review (1977) included species in three classes (Cephalaspidomorpha, Chondricthyes, Osteichthyes); 10 superorders and 21 orders of bony fishes were included (Table 11.1).

Behavioral responses to conspecific extracts were included for 94 nonostariophysan species in Pfeiffer's review (1977), and most (90) did not yield fright responses based on the methodology in the studies. However, most of the early studies used behavioral assays that were subjective (e.g., von Frisch's qualitative scale of - = no reaction to +++ = intense reaction), and so they may have missed subtle responses; responses by ostariophysan fishes often are more dramatic than the responses by some nonostariophysan species (pers. obs.). More detailed studies of some of the same species that were reported in Pfeiffer's review have yielded dramatically different results. I strongly recommend that the

Table 11.1 Taxa of nonostariophysan fishes included in Pfeiffer's (1977) review of studies of fish alarm cues; some taxonomic names have been updated to reflect recent changes. For each order, there is an indication in parentheses as to whether at least one species was tested for behavioral responses or for the presence of epidermal Alarm Substance Cells (ASCs); minus signs mean that all tests were negative, and plus signs mean that at least one species tested positive. With one exception (Cyprinodontiformes), all behavioral tests were negative. With one exception, a salmonid in the Family Glaxiidae with questionable results, all histological studies were negative for the presence of ASCs. To see which species were tested, see Pfeiffer (1977). Families for which post-1977 tests have been reported are indicated by an asterisk (*); these families are covered in detail later in this chapter.

Class Cephalaspidomorpha
      Order Petromyzoniformes (– ASCs)
          Family Petromyzontidae: Lampreys
Class Chondrichthyes
      Order Batidiomorpha (– Alarm behavior, – ASCs)
          Family Torpedinidae: Electric rays
          Family Dasyatidae: Stingrays
      Order Chimaeriformes (– ASCs)
          Family Chimaeridae: Chimaeras
Class Osteichthyes
    Superorder Dipnoi
      Order Ceratodontiformes (– ASCs)
          Family Ceratodontidae: Australian lungfish
      Order Lepidosirenidae (– ASCs)
          Family Lepidosirenidae: South American lungfish
          Family Protopteridae: African lungfish
    Superorder Crossopterygii
      Order Porolepiformes (– ASCs)
          Family Coelacanthidae: Coelacanths
    Superorder Palaeonisciformes
      Order Polypteriformes (– ASCs)
          *Family Polypteridae: Bichirs
      Order Acipenseriformes (– Alarm behavior, – ASCs)
          Family Acipenseridae: Sturgeons
    Superorder Elopomorpha
      Order Anguilliformes (– Alarm behavior, – ASCs)
          Family Congridae: Conger eels
          Family Opichthidae: Snake eels
    Superorder Clupeomorpha (– Alarm behavior, – ASCs)
      Order Clupeoformes
          Family Clupeidae: Herrings, sardines
          Family Engraulidae: Anchovies

(Table 11.1 contd.)

(Table 11.1 contd.)

Superorder Osteoglossomorpha
    Order **Osteoglossiformes** (– Alarm behavior, – ASCs)
        Family Osteoglossidae: Arowana
        Family Notopteridae: Knifefishes
        Family Mormyridae: Elephantfishes
        Family Gymnarchidae: Electric knifefishes
Superorder **Protacanthopterygii**: (– Alarm behavior, –/? ASCs)
    Order **Salmoniformes: Salmon, Trout**
        Family Salmonidae: Salmon, trout
        Family Esociformes: Pikes, Mudminnows
        Family Omeriformes: Smelts
        Family Galaxiidae: Galaxids
Superorder **Cyclosquamata** (– ASCs)
    Order **Aulopiformes**
        Family Synodontidae: Lizardfishes
Superorder **Paracanthopterygii**
    Order **Gobiesociformes** (–Alarm behavior, – ASCs)
        Family Gobiesocidae: Klingfishes, singleslits
    Order **Gadiformes** (– Alarm behavior, – ASCs)
        Family Gadidae: Cods
        Family Lotidae: Hakes, Burbots
    Order **Ophidiiformes** (– ASCs)
        Family Ophidiidae: Brotulas, congriperlas, cusk-eels
        Family Carapidae: Peralfishes
Superorder **Acanthopterygii**
    Order **Atheriniformes** (–/+ Alarm behavior, - ASCs)
        Family Exocoetidae: Flying fishes
        Family Belonidae: Needle fishes
        Family Atherinidae: Silversides
    Order **Cyprinodontiformes** (–/+ Alarm behavior, - ASCs)
        *Family Fungilidaeidae: Killifishes
        *Family Poeciliidae: Guppies, swordtails, mosquitofish
    Order **Beryciformes** (– ASCs)
        Family Holocentridae: Soldierfishes, Squirrelfishes
    Order **Gasterosteiformes** (– Alarm behavior, – ASCs)
        *Family Gasterosteidae: Sticklebacks
        Family Aulorhynchidae: Tubesnouts
        Family Syngnathidae: Pipefishes, Seahorses
    Order **Scorpaeniformes** (– Alarm behavior, – ASCs)
        Family Scorpaenidae: Scorpionfishes, rockfishes
        Family Triglidae: Searobins, gunards
        Family Hexagrammidae: Greenlings
        Family Anoplopomatidae: Sablefishes

(Table 11.1 contd.)

(Table 11.1 contd.)

        Family Agonidae: Poachers, alligatorfish
        Family Cyclopteridae: Lumpfishes
        *Family Cottidae: Sculpins
Order Perciformes (–Alarm behavior, –ASCs)
        Family Serranidae: Groupers, sea basses
        *Family Centrarchidae: Sunfishes, basses
        Family Priacanthidae: Bigeyes, catalufas
        Family Apogonidae: Cardinalfishes
        *Family Percidae: Perches, darters
        Family Echeneidae: Remoras, sharksuckers
        Family Carangidae: Jacks, Pompanos
        Family Maenidae (= Centracanthidae): Picarels
        Family Lutjanidae: Snappers
        Family Sparidae: Porgies
        Family Sciaenidae: Drums, croakers
        Family Mullidae: Goat fishes
        Family Kyphosidae: Sea chubs, pilotfishes
        Family Chaetodontidae: Butterflyfishes
        *Family Cichlidae: Cichlids
        *Family Pomacentridae: Damselfishes
        Family Mugilidae: Mullets
        Family Labridae: Parrotfishes, Rainbowfishes, wrasses
        Family Bathymasteridae: Ronquils
        Family Trachinidae: Weeverfishes, weevers
        Family Dactyloscopidae: Sand Stargazers
        Family Blennidae: Blennies
        Family Tripterygiidae: Threadfin blennies
        Family Clinidae: Clinids, Kelpfishes, Scaled blennies
        Family Chaenopsidae: Pike-, Tube-, flagblennies
        Family Stichaeidae: Pricklebacks, shannies
        Family Ammodytidae: Sand lances
        Family Callionymidae: Dragonets
        *Family Gobiidae: Gobies
        Family Acanthuridae: Surgeonfishes, tangs
        Family Scombridae: Mackerels, tunas, bonitos
        Family Anabantidae: Climbing Perches
        Family Zoarcidae: Eelpots
Order Pleuronectiformes (– Alarm behavior, – ASCs)
        Family Bothidae: Lefteye flounders
        Family Pleuronectidae: Righteye flounders
        Family Soleidae: Soles
Order Tetraodontiformes (– Alarm behavior, – ASCs)
        Family Blastidae: Triggerfishes
        Family Tetraodontidae: Puffers

results for all of the species in Pfeiffer's review be verified using the following criteria:

(1) **Replication must be adequate.** Some early studies made inferences about the presence of alarm chemicals following only a single exposure. Most recent studies of alarm behavior have used 10-20 replicates per treatment.

(2) **Response variables should be appropriate for individual species.** Recent studies have included a variety of responses that must be carefully quantified, including shoaling intensity (Mathis and Smith, 1993), use of shelter (Mathis and Smith, 1993), activity levels (Mirza and Chivers, 2001a), avoidance of areas marked with the stimulus (Friesen and Chivers, 2006), use of substrate (Garcia et al., 1992), foraging behavior (Leduc et al., 2003) and opercular movements (Gibson and Mathis, 2006).

(3) **Experiments should include exposure to appropriate controls that include chemical blanks and heterospecific extracts.** Use of chemical blanks will allow the researcher to distinguish between response to alarm chemicals and a general disturbance response. Response to extracts from heterospecific extracts will allow researchers to distinguish between specific responses to alarm chemicals and responses to blood or other tissues associated with damage to skin. The choice of heterospecific extract can be complex. Ideally, it is important to choose a species that is not closely related to the focal species because a positive response could result if a chemical homology exists between the two species. However, it is also important to choose a heterospecific that does not co-exist with the focal species and does not have close relatives that co-exist with the focal species. Some species have been shown to learn to respond to extracts from co-existing species that are members of the same prey guild (e.g., Mathis and Smith, 1993), so responses to extracts from syntopic species could confuse chemical homologies with learned responses. For example, studies in Canada have tended to use members of tropical families such as poeciliids (swordtails and guppies) as heterospecific control stimuli because there are no native members of this family in the study area (e.g., Berejikian et al., 1999; Mirza and Chivers, 2001b). However, in the Midwestern/southern United States, researchers have avoided using poeciliids as a control because the mosquitofish (*Gambusia affinis*), an introduced member of this family, is widespread in the region (e.g., Commens and Mathis, 1999).

Pfeiffer's review included anatomical studies of 160 nonostariophysan species that lack ostariophysan-like Alarm Substance Cells and no nonostariophysan species that possess the cells (although one was listed as questionable) (Table 11.1). Discussion of the skin glands in nonostariophysan species will be covered later in this chapter.

TAXONOMIC OVERVIEW OF ALARM RESPONSES

Superorder Protacanthopterygii, Family Salmonidae: Trout, Charr and Salmon

The family Salmonidae probably contains the most economically important species that have been tested for alarm chemicals. Adult salmonids are relatively large bodied and thus are less vulnerable to gape-limited predators than juveniles. Therefore, studies typically have focused on fright responses by juveniles (Table 11.2).

Fright responses of juvenile salmonids to conspecific skin extracts have been reported for three genera in this family, including brook charr (*Salvelinus fontinalis*), rainbow trout (*Oncorhynchus mykiss*), Chinook salmon (*O. tshawytscha*), and brown trout (*Salmo trutta*) (Table 11.2). The form of the behavioral response to the alarm cues under laboratory conditions resembles that of other species and includes decreased activity and feeding behavior (Brown and Smith, 1998; Scholz et al., 2000; Mirza and Chivers, 2001a, b; Scott et al., 2003; Leduc et al., 2004b), decreased aggression (Leduc et al., 2004b), and increased shelter use (Brown and Smith, 1998). In field studies with brook charr, both avoidance of areas marked with the alarm cues (Mirza and Chivers, 2001b) and decreased foraging and aggressive behavior (Leduc et al., 2004b) were observed.

An interesting effect of body size on response to conspecific cues was noted by Mirza and Chivers (2002) for brook charr. Small individuals responded most strongly to skin extracts from small charr and large individuals responded most strongly to extracts from large charr. These results suggest that either the alarm chemicals are not identical for the two size classes or that other chemicals associated with the alarm cues allow fish to distinguish between cues from different size classes (Mirza and Chivers, 2002).

Juvenile rainbow trout exhibited fright responses when exposed to conspecifics that were infected with skin-penetrating parasites (cercariae of the trematode *Diplostomum* sp.) (Poulin et al., 1999). The authors suggested that skin damage by the trematodes resulted in release of alarm

Table 11.2 Behavioral responses to chemical stimuli from injured conspecifics by members of the Family Salmonidae (Superorder Protacanthopterygii) published since 1977. Responses to cues from related (within-family) heterospecifics and other stimuli (extracts from extra-familial fishes, water, etc.) also are reported. "None" means that the stimulus was tested, but there was no fright response. "Not tested" means that the study did not include exposure to the cues listed in the column heading.

Response to conspecific cues	Response to cues from related heterospecifics	Response to other stimuli	Reference
Brook charr: *Salvelinus fontinalis*			
1. Associative learning of predator odor; Higher survival of trained fish (juveniles)	1. Not tested	1. Swordtails (*Xiphophorus helleri*): None	1. Mirza and Chivers (2000)
2. Lab: Decreased activity, foraging; increased survival to unfamiliar predators (juveniles). **Field** (trap): Avoidance	2. Not tested	2. Swordtails (*Xiphophorus helleri*): None	2. Mirza and Chivers (2001b)
3. **Field** (video): No response in weakly acidic stream; response (decreased foraging, aggression) in pH-neutral stream	3. Not tested	3. Not tested	3. Leduc et al. (2004b)
4. Decreased activity, decreased foraging behavior	4. Rainbow trout (*Oncorhynchus mykiss*), brown trout (*Salmo trutta*): Fright response, but weaker than to conspecific cues	4. Swordtails (*Xiphophorus helleri*): None	4. Mirza and Chivers (2001a)
5. Size effect: Small individuals respond strongest to cues from small individuals; large individuals respond strongest to cues from large individuals	5. Not tested	5. Swordtails (*Xiphophorus helleri*): None	5. Mirza and Chivers (2002)
Rainbow trout: *Oncorhynchus mykiss*			
1. Decreased activity, foraging behavior (juveniles)	1. Brown trout (*S. trutta*); Brook charr (*S. fontinalis*): Fright response, but weaker than to conspecifics	1. Swordtails (*X. helleri*): None	1. Mirza and Chivers (2001a)
2. Freezing, darting in response to conspecifics that were parasitized by skin parasites (juveniles)	2. Not tested	2. Noninjured, nonparasitized conspecifics: None	2. Poulin et al. (1999)
3. Associative learning: hatchery-reared recognize pike scent (decreased foraging and area use; increased shelter use and freezing (juveniles))	3. Not tested	3. Water: None	3. Brown and Smith (1998)

(*Table 11.2 contd.*)

(Table 11.2 contd.)

4. Not tested	4. Not tested	4. Nitrous oxides: None (juveniles)	4. Brown et al. (2003)
5. Decreased activity, decreased foraging, increased survival with unknown predator — even at concentrations below behavioral threshold (juveniles)	5. Not tested	5. Water: None	5. Mirza and Chivers (2003)
6. Decreased activity, decreased foraging; increased shelter use; increased plasma cortisol. Water-borne (not food-borne) Cadmium eliminated behavioral and plasma cortisol response. Cd accumulated in olfactory rosette, olfactory nerve and olfactory bulb (juveniles)	6. Not tested	6. Not tested	6. Scott et al. (2003)
7. Lab: Reduced activity, avoidance, increased shelter use. **Field** (snorkeling): Decreased foraging, aggression. Response eliminated under weakly acidic conditions (juveniles)	7. Not tested	7. Water: None	7. Leduc et al. (2004)
8. Predator-recognition learning eliminated under weakly acidic conditions	8. Not tested	8. Water: None	8. Leduc et al. (2004)
9. Plasma cortisol levels increase 12 h after exposure to extracts	9. Not tested	9. Not tested	9. Toa et al. (2005)
10. Frequency of exposure influenced intensity of response (high freq = low response) (juveniles)	10. Brook charr (*S. fontinalis*): Fright response, but weaker than to conspecific cues	10. Not tested	10. Mirza et al. (2006)
Chinook salmon: *Oncorhynchus tshawytscha*			
1. **Lab**: Associative learning of predator (cutthroat trout) odor; **Field**: Predator-trained had higher survival (juveniles)	1. Not tested	1. Swordtails (*X. helleri*): None	1. Berejikian et al. (1999)
2. Decreased activity and foraging; Response lowers in the presence of diazinon (insecticide) (juveniles)	2. Not Tested	2. Not tested	2. Scholz et al. (2000)
Brown trout: *Salmo trutta*			
1. Decreased Activity, decreased foraging behavior (juveniles)	1. Rainbow trout (*O. mykiss*), brook charr (*S. fontinalis*): weaker than to conspecific	1. Swordtails (*X. helleri*): None	1. Mirza and Chivers (2001a)
Atlantic salmon: *Salmo salar*			
1a. **Field** (snorkeling, video): Decreased activity, decreased foraging, increased time on substrate, avoidance b. Response eliminated in acidic streams	1. Not tested	1. Water	1. Leduc et al. (2006)

chemicals into the water because the trout did not respond to the scent of the cercariae alone. This response could be adaptive if it results in avoidance of parasites, but this hypothesis remains to be tested.

Increased survival for fishes 'warned' by the alarm chemical has been observed in staged encounters with predators for brook charr (Mirza and Chivers, 2001b) and rainbow trout (Mirza and Chivers, 2003). The latter study was particularly interesting because increased survival was noted even at a concentration of alarm pheromone that was below the threshold for behavioral responses in other trials. This finding introduces a cautionary note into interpretation of apparent nonresponses to alarm cues; lack of an overt behavioral response does not necessarily mean that the alarm cue was not detected.

Alarm chemicals also can influence the survival of salmonids indirectly. Rainbow trout (Brown and Smith, 1998), Chinook salmon (Berejikian et al., 1999) and brook charr (Mirza and Chivers, 2000) can learn to associate the odor of unfamiliar predators with the alarm cue, and show fright responses to those predators during future encounters. This associative learning was demonstrated to increase survival for both brook charr in laboratory trials (Mirza and Chivers, 2000) and Chinook salmon in natural habitats (Berejikian et al., 1999). Predator-recognition training may have applications for salmonids and other species for increasing post-release survival of hatchery-reared fishes.

Although physiological responses to alarm cues have been studied for only a few species, preliminary indications are that some aspects may be comparable across a wide range of species. In rainbow trout, plasma cortisol levels were seen to increase 12 hours after exposure to conspecific extracts (Scott et al., 2003; Toa et al., 2005), indicating that a physiological stress response is part of the alarm response. A similar response was reported for pearl dace, Semotilus margarita (Ostariophysi, Cyprinidae), which showed an increase in levels of plasma cortisol and glucose 15-hours post-exposure (Rehnberg and Smith, 1987).

Environmental pollutants may inhibit the ability of salmonids to detect or recognise salmonid alarm cues. Environmentally relevant weakly acidic conditions reduced behavioral responses to alarm chemicals by both brook charr and rainbow trout (Leduc et al., 2004b), and nullified associative learning of unfamiliar predators by rainbow trout (Leduc et al., 2004a). Exposure to water-borne Cadmium (Scott et al., 2003) and the insecticide Diazinon (Scholz et al., 2000) reduced or eliminated the behavioral response to conspecific extracts, and Cadmium also eliminated the physiological stress response (Scott *et al.*, 2003).

Although there have been only a few studies of cross-species reactions within the Salmonidae, the data thus far indicate that there may be some degree of homology for alarm chemicals within this group. Brown trout respond to skin extracts from rainbow and brown trout (Mirza and Chivers, 2001a), rainbow trout respond to extracts from brook charr (Mirza and Chivers, 2001a; Mirza *et al.*, 2006) and brown trout (Mirza and Chivers, 2001a), and brown trout respond to extracts from brook charr and rainbow trout (Mirza and Chivers, 2001a). In all the cases, the form of the fright response to heterospecific cues was similar to the response to conspecific cues (decreased activity and feeding behavior), but the intensity of the response is weaker. For fishes to distinguish between conspecific and heterospecific extracts could require other chemicals in addition to the alarm chemicals, or there may be some chemical differences between the alarm cues of different species (Mirza and Chivers, 2001a).

Like most nonostariophysan species, initial reports of tests for alarm cues in salmonids were negative (Pfeiffer, 1977). As reported by Pfeiffer, the species tested included the common whitefish (*Coregonus lavaretus*), Danube salmon (*Hucho hucho*), Coho salmon (*Oncorhynchus kisutch*), Chinook salmon (*O. tshawytscha*), rainbow trout (*Salmo gairdneri*), Arctic charr (*Salvelinus alpinus*), brook charr (*S. fontinalis*) and grayling (*Thymallus thymallus*). The positive results of re-tests of three of these species (Coho salmon, Chinook salmon, and brook charr) using more sensitive assays than the original tests underline the need for additional experiments in this group. For the species in the more recent studies, the lack of fright responses following exposure to control stimuli (water and extracts from injured heterospecific fishes: Table 11.2) rule out disturbance and general cues associated with injured fishes as explanations for the observed fright responses to conspecific stimuli.

Superorder Acanthopterygii

Order Cyprinodontiformes: Livebearers, Killifishes and Topminnows

As early as 1956, there was a suggestion of a possible response to conspecific skin extracts by a member of the families Poeciliidae (*Gambusia affinis*, mosquitofish) and Fundulidae (*Fundulus olivaceus*, blackspotted topminnow), but some of the results were contradictory (reviewed by Pfeiffer, 1977). Reed (1969) also observed a weak fright response for mosquitofish; his study used von Frisch's qualitative scale for assessing fright reactions and so may have missed subtle responses. The response to conspecific extracts by mosquitofish was confirmed by Garcia *et al.* (1992) (Table 11.3), who showed that both large and small fish responded to conspecific extracts—but not to extracts from *Betta splendens*—by moving to the tank bottom. Another poeciliid, the least killifish (*Heterandria formosa*), also was tested in early studies with contradictory results (Pfeiffer, 1977); additional tests are needed to clarify the response for this species.

Both male and female red swordtails, *Xiphophorus helleri*, decreased their activity in response to conspecific extracts but not to extracts of fathead minnows, *Pimephales promelas* (Mirza and Chivers, 2001c). Interestingly, the details of the response differed between the sexes. While females moved to the bottom of the tank and sought shelter, males tended to move to the top of the tank. The authors attributed this difference in behavior to differences in body shape; males are more streamlined and so are more agile swimmers and less likely to be captured during attack. Few studies have assessed whether males and females differ in their responses to alarm cues (see Matity *et al.*, 1994 for an example in ostariophysan). Because female livebearers carry young, the differences in body shapes of males and females are common, and so examination of behavioral differences between the sexes may be a fruitful area of research. The study by Mirza *et al.* (2001a) included a control for general disturbance (water) and chemicals associated with general damage to fish skin (fathead minnows). Pfieffer (1977) reported negative results for tests of two other members of this genus—*X. maculates* and *X. variatus*—but these results need to be verified.

The most comprehensive study of response to alarm cues by livebearers was Brown and Godin's (1999) study of Trinidadian guppies (*Poecilia reticulata*). In a laboratory experiment, guppies increased their shoaling, dashing, and freezing behaviors in response to conspecific extracts (Nordell, 1998), but not to extracts of swordtails (see below). Brown and Godin (1999) confirmed the laboratory results in two field experiments. One experiment showed that fewer guppies were caught in traps marked with conspecific extracts, and the second showed that fewer guppies inspected a model that was paired with conspecific extract. An additional twist showed that guppies from a high-predation population had

Table 11.3 Behavioral responses to chemical stimuli from injured conspecifics by members of two families (Poecoliidae and Fundulidae) in the Order Cyprinodontiformes (Superorder Acanthopterygii) published since 1977. Responses to cues from related (within-family) heterospecifics and other stimuli (extracts from extra-familial fishes or water) also are reported. "None" means that the stimulus was tested, but there was no fright response. "Not tested" means that the study did not include exposure to the cues listed in the column heading.

Response to conspecific cues	Response to cues from related heterospecifics	Response to other stimuli	Reference
Family Poeciliidae			
Red swordtails: Xiphophorus helleri			
1. Both sexes decreased activity; males occupied the top of the tank, females occupied the bottom of the tank and sought shelter	1. Guppy (Poecilia reticulate): Decreased activity, but weaker response than to conspecific; no difference between sexes	1a. Water: Noneb. Fathead minnows (Pimephales promelas): None	1. Mirza et al. (2001)
Trinidadian guppies: Poecilia reticulata			
1. Lab: Increased shoaling, dashing, freezing; decreased area use; **Field:** Trap avoidance, predator model avoidance	1. Lab: Swordtails (Xiphophorus helleri)	1. **Field**—Water: None	1. Brown and Godin (1999)
2. Increased shoaling	2. Not tested	2. Water: None	2. Nordell (1998)
Mosquitofish: Gambusia affinis			
1. Move to bottom of tank; increased sensitivity to fright stimuli after resuming "normal" activity (small individuals only; qualitative assessment)	1. Not tested	1a. Water: None 1b. Siamese fighting fish (Betta splendens): None	García et al. (1992)
2. Decreased changes in direction	2. Not tested	2. Water: None	2. Cashner (2001)
Family Fundulidae			
Banded killifish: Fundulus diaphanous			
1. Increased shoaling	1. Not tested	1. Water: None	1. Hoare et al. (2004)
Blackspotted topminnow: F. olivaceus			
1. None	1. Not tested	1. Water: None	1. Cashner (2001)

weaker (though significant) responses to extracts from guppies from their own population than to extracts from their own population, possibly indicating a weaker concentration of the alarm chemical in the low-predation fishes (Brown and Godin, 1999).

The question of whether the alarm chemical is homologous within some or all of the live-bearing fishes has not been well studied. The results, thus far, are not consistent. Swordtails responded to extracts from guppies, although the response was weaker than to conspecific extracts (Mirza and Chivers, 2001c). In contrast, guppies showed no significant responses to swordtail extracts (Brown and Godin, 1999). Clearly, more research on cross-species responses is needed in this group. At the most basic level, within-genus comparisons are needed. Interpretation of cross-species responses may be more complex because of population differences in the alarm cue (e.g., Brown and Godin, 1999).

Two members of the family Fundulidae also have tested positive for fright responses to chemical alarm cues from conspecifics. Using von Frisch's qualitative criteria, Reed (1969) reported a possible response (3/11 trials with conspecific extracts versus 0/11 control responses) by the blackspotted topminnow, *Fundulus olivaceus*. However, in a study that examined several response variables associated with fright, Cashner (2001) saw no significant change in behavior for blackspotted topminnows exposed to conspecific extracts. A second species, the banded killifish, *F. diaphanous*, was the subject of an experiment that demonstrated that individuals occupy larger groups upon detection of the conspecific alarm cue, and that this decision can be balanced with the demands of feeding (Hoare *et al.*, 2004). Both Reed (1969) and Hoare *et al.* (2004) observed no fright responses following exposure to a blank water control, eliminating disturbance as a cause for the fright responses observed in the alarm cue treatments.

The family Poeciliidae contains several of the nonostariophysan species in the review by Pfeiffer (1977) that had some early indication of alarm responses. Pfeiffer reported contradictory results both for guppies, *P. reticulata*, and the least killifish, *Heterandria formosa*. Brown and Godin's (1999) detailed study confirmed the presence of the alarm response for guppies, but follow-up study is needed for the least killifish. Weakly positive responses were reported in initial tests of mosquitofish, *G. affinis* (3/6 trials for tests with conspecific extracts versus 0/8 trials for the controls) (Reed, 1969), and this positive response was confirmed by Garcia *et al.* (1992). For two species of *Xiphophorus*, the southern platyfish, *X. maculates*, and the variable platyfish, *X. variatus*, negative results were reported by Pfeiffer (1977). Given the positive tests for another member of this genus (see above), additional tests of these species of *Xiphophorus* should be performed. For recent studies of swordtails, guppies, and mosquitofish, focal fishes did not exhibit fright responses to blank (water) controls and heterospecific skin extracts (Table 11.3), eliminating general disturbance or general injury-based chemicals as a cause for the fright responses to conspecific alarm cues.

Order Gasterosteiformes: Sticklebacks

Surprisingly, although the three-spine stickleback, *Gasterosteus aculeatus*, has been one of the best-studied species in all of behavioral/evolutionary ecology, it has received little attention with respect to alarm cues. This lack of attention is probably primarily due to initial indications of negative results by Schutz in the 1950s (Pfeiffer, 1977). However, a more recent test using a more sensitive assay indicated that individuals respond to conspecific cues with a fright response that included increased use of cover and increased time near the substrate (Brown and Godin, 1997). Three-spine sticklebacks in that study also performed fright responses following exposure to extracts from four-spine sticklebacks, *Apeltes quadracus*. However, interpretation of this response is complicated because the two species are sympatric. Response to the heterospecific cue could indicate either a chemical homology or a learned response to a cue from a co-occurring species (e.g., Chivers *et al.*, 1995b). Brown and Godin's study (1997) also included exposure to skin extracts from an allopatric species (swordtails, *X. helleri*), with negative results, so the positive responses to the two stickleback stimuli was not due to a general disturbance response or to general injury-based cues.

The best-studied species of stickleback with respect to chemical alarm cues is the brook stickleback, *Culaea inconstans* (Table 11.4). In laboratory trials, brook sticklebacks increased shoaling in response to chemical alarm cues (Mathis and Smith, 1993). This response was confirmed in field trap experiments: if a shoal of sticklebacks already were present in the traps, free-ranging sticklebacks were attracted to the trap even though it was marked with the alarm cue (Wisenden *et al.*, 2003). However, if the alarm-marked trap was empty, free-ranging sticklebacks avoided the trap.

Table 11.4 Behavioral responses to chemical stimuli from injured conspecifics by members of the Family Gasterosteidae (Superorder Acanthopterygii, Order Gasterosteiformes) published since 1977. Responses to cues from related (within-family) heterospecifics and other stimuli (extracts from extra-familial fishes or water) also are reported. "None" means that the stimulus was tested, but there was no fright response. "Not tested" means that the study did not include exposure to the cues listed in the column heading.

Response to conspecific cues	Response to cues from related heterospecifics	Response to other stimuli	Reference
Brook stickleback: _Culaea inconstans_			
1. Field (trap): Avoidance; Lab: Increased shoaling	1. Not tested	1a. Allopatric swordtails (_Xiphophorus helleri_): None b. Sympatric fathead minnows (_Pimephales promelas_)—**Field (trap)**: Avoidance; Lab: Decreased activity	1. Mathis and Smith (1993)
2. **Field** (trap): Avoidance; size effect = experience effect?	2. Not tested	2. Water: None	2. Chivers and Smith (1994)
3. **Field** (video): Avoidance	3. Not tested	3a. Water: None b. Allopatric swordtails (_X. helleri_)—**Field** (video): Avoidance c. Sympatric fathead minnows (_Pimephales promelas_)—**Field** (video): Avoidance	3. Friesen and Chivers (2006)

(Table 11.4 contd.)

(Table 11.4 contd.)

4. Field (traps): Attraction to shoals; avoidance w/o shoals	4. Not tested	4a. Water: None b. Sympatric fathead minnows (*P. promelas*)—Field (traps): Attraction to shoals; Avoidance w/o shoals	4. Wisenden *et al.* (2003)
5. Prefer to shoal with fathead minnows (*P. promelas*) over conspecifics	5. Prefer to shoal with Fathead minnows (*P. promelas*) over conspecifics	5a. Water: None b. Sympatric fathead minnows (*P. promelas*): Prefer to shoal with fathead minnows over conspecifics	5. Mathis and Chivers (2003)
6. Increased shoaling; movement toward substrate; predator recognition learning	6. Not tested	6a. Water: None b. Sympatric fathead minnows (*P. promelas*): predator recognition learning	6. Chivers *et al.* (1995)
Three-spine stickleback: *Gasterosteus aculeatus* 1. Increased cover use; increased time near bottom	1. Sympatric four-spine sticklebacks *Apeltes quadracus*: Increased shelter use; increased time near bottom	1. Allopatric swordtails (*X. helleri*): none	1. Brown and Godin (1997)

Although shoaling with conspecifics clearly is preferable to remaining solitary when predation risk is high (Mathis and Smith, 1993; Wisenden *et al.*, 2003), the shoaling response becomes more complex in the presence of multiple species of the same prey guild. If a solitary stickleback has the option of joining a group of conspecifics or a group of fathead minnows (*Pimephales promelas*), it chooses to join the minnow shoal (Mathis and Chivers, 2003; Fig. 11.1). This result is counter to the prediction of the 'oddity hypothesis': prey should group with similar individuals because predators focus on individuals in a group that stand out (Godin, 1997). This unexpected result may have occurred because some predators prefer minnows over sticklebacks, although they will readily consume sticklebacks when no other choice is available (Mathis and Chivers, 2003).

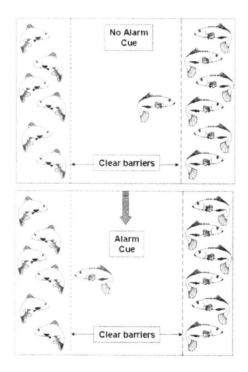

Fig. 11.1 Shoaling preferences of solitary brook stickleback, *Culaea inconstans*. In the absence of alarm cues, sticklebacks prefer to associate with conspecifics (upper panel). In the presence of alarm cues, sticklebacks prefer to associate with fathead minnows, *Pimephales promelas*. Some predators preferentially consume minnows over sticklebacks, so shoaling with minnows when predation risk is high may increase the probability of survival for sticklebacks.

Avoidance of alarm-marked traps (without shoals) by brook sticklebacks in the field also was reported by Chivers and Smith (1994). Henderson *et al.* (1997) criticized trap experiments because fishes are unnaturally confined; they advocated the use of video cameras to track behavior of free-ranging fishes. Friesen and Chivers (2006) recently confirmed avoidance of alarm cues by brook sticklebacks in the open water through use of underwater video.

Several studies have shown that brook sticklebacks also respond to alarm cues from fathead minnows. In the laboratory, responses include decreased activity (Mathis and Smith, 1993), increased shoaling (Chivers *et al.*, 1995a), movement toward the substrate (Chivers *et al.*, 1995a) and preference for shoaling with minnows over conspecifics (Mathis and Chivers, 2003). Field trials demonstrated attraction to alarm-marked traps containing shoals (Wisenden *et al.*, 2003), avoidance of traps without shoals (Mathis and Smith, 1993; Chivers and Smith, 1994; Wisenden *et al.*, 1994, 1995, 2003) or area avoidance by free-ranging minnows (Friesen and Chivers, 2006). The authors generally interpret these responses as selection for response to alarm cues from members of a common prey guild rather than indication of a chemical homology between the two groups. In at least one experiment (Chivers and Smith, 1994), sticklebacks that responded to the minnow alarm cues were significantly larger, and therefore older, than sticklebacks in control traps. This size effect may indicate that the sticklebacks' response to the minnow cue requires experience. Studies of brook sticklebacks also have included negative responses to extracts from allopatric controls (Mathis and Smith, 1993; Brown and Godin, 1997; Friesen and Chivers, 2006) and blank (water) controls (Chivers and Smith, 1994; Mathis and Chivers, 2003; Wisenden *et al.*, 2003; Friesen and Chivers, 2006).

Brook and three-spine sticklebacks are the only two species in the family Gasterosteidae that have been tested for the presence of chemical alarm cues. In comparison to most other taxa, this family is relatively small (seven species in five genera), and so a detailed study involving responses to conspecifics and all possible combinations of heterospecific sticklebacks is a realistic goal that should be pursued.

Order Scorpaeniformes, Family Cottidae: Sculpins

Only a few species of sculpins have been tested for potential responses to alarm chemicals (Table 11.5). Sculpins testing positive for responses to

Table 11.5 Behavioral responses to chemical stimuli from injured conspecifics by members of the Family Cottidae (Superorder Acanthopterygii, Order Scorpaeniformes) published since 1977. Responses to cues from related (within-family) heterospecifics and other stimuli (extracts from extra-familial fishes or water) also are reported. "None" means that the stimulus was tested, but there was no fright response. "Not tested" means that the study did not include exposure to the cues listed in the column heading.

Response to conspecific cues	Response to cues from related heterospecifics	Response to other stimuli	Reference
Slimy sculpin: *Cottus cognatus*			
1. Decreased activity, increased shelter use	1. Not tested	1. Swordtails (*Xiphophorus helleri*): None	1. Bryer et al. (2001)
Reticulate sculpin: *Cottus perplexus*			
1a. Decreased activity	1. Not tested	1a. Sympatric speckled dace (*Rhinichthyes osculus*): None	1. Chivers et al. (2000)
1b. Hunger eliminated alarm response		1b. Swordtails (*X. helleri*): None	
Tidepool sculpin: *Oligocottus maculosus*			
1. Decreased activity, decreased foraging, increased substrate use, increased shelter use	1. Not tested	1. Guppies (*Poecilia reticulata*): None	1. Hugie et al. (1991)
2. Response depends on substrate color: Strong response only when on matching (gray sand) background (not when on white background)	2. Not tested	2. Not tested	2. Houtman and Dill (1994)

alarm chemicals include two freshwater species of the genus *Cottus*, the reticulate sculpin (*C. perplexus*) and the slimy sculpin (*C. cognatus*). Both of these freshwater species showed decreased activity in response to skin extracts from conspecifics (Chivers *et al.*, 2000; Bryer *et al.*, 2001); increased shelter use in response to conspecific extracts also was reported for slimy sculpins (Chivers *et al.*, 2000). In both cases, individuals did not respond to damage-released chemicals from poeciliids (Table 11.5), suggesting at least some degree of specificity for the response. Neither species was exposed to chemicals from congeners. Kinziger *et al.* (2005) found that these two species belong to different monophyletic clades within the genus, so the response is not limited to a single clade.

Overall, few marine species have been tested for the presence of alarm chemicals. As most of the 70 genera of cottids are marine fishes, members of this family are excellent candidates for exploration of alarm responses in marine fishes. One intertidal species, the tidepool sculpin (*Ologocottus maculosus*), has been shown to respond to alarm cues from conspecifics (but not poeciliids) with decreased activity and increased use of both shelter and the substrate (Hugie *et al.*, 1991). The alarm response appears to be somewhat plastic, depending on local conditions. Individuals decreased activity only when on matching backgrounds, supporting the hypothesis that the function of decreased activity is to decrease the risk of detection by visually oriented predators (Houtman and Dill, 1994). Plasticity of response also was reported for reticulate sculpins, which failed to respond to conspecific alarm cues after two days of food deprivation (Chivers *et al.*, 2000). These latter data suggest that there is a foraging cost to decreased activity, and that sculpins can alter their behavior depending upon perceived costs and benefits at a given time.

Pfeiffer (1977) reported negative results for two additional species, *Cottus gobio* and *Myoxocephalus scorpius*. These results should be verified through additional studies.

Order Perciformes

Family Centrarchidae: Sunfishes and Basses

Two species of sunfishes (green and pumpkinseed) have been studied fairly intensively (Table 11.6): the green sunfish (*Lepomis cyanellus*) and the pumpkinseed sunfish (*Lepomis gibbosus*). In laboratory studies, responses to chemical cues from injured conspecifics include freezing, decreased activity, and increased time spent with the dorsal fin rays erect (Brown and

Table 11.6   Behavioral responses to chemical stimuli from injured conspecifics by members of the Family Centrarchidae (Superorder Acanthopterygii, Order Perciformes) published since 1977. Responses to cues from related (within-family) heterospecifics and other stimuli (extracts from extra-familial fishes, water, or other stimuli) also are listed. "None" means that the stimulus was tested, but there was no fright response. "Not tested" means that the study did not include exposure to the cues listed in the column heading.

Response to conspecific cues	Response to cues from related heterospecifics	Response to other stimuli	Reference
Green sunfish: Lepomis cyanellus			
1. Freezing, increased time with dorsal fin rays erect (juveniles)	1. Not tested	1. Swordtails (*Xiphophorus helleri*): None	1. Brown and Brennan (2000)
2. Ontogentic shift—Small: Decreased activity, increased time with fin erect; Large: Foraging response (juveniles)	2. Not tested	2a. Swordtails (*Xiphophorus helleri*): None 2b. Finescale dace (*Phoxinus neogeaus*): Response negatively correlated with size	2. Golub and Brown (2003)

(Table 11.6 contd.)

(Table 11.6 contd.)

Pumpkinseed sunfish: *Lepomis gibbosus*			
1. Ontogenetic shift: Young juveniles: Decreased activity; increased shoaling; Older juveniles: Foraging. Stronger response at higher concentrations	1. Not tested	1. Water: None	1. Marcus and Brown (2003)
2. Decreased activity, increased time with fin erect erection; declined under weakly acidic condition (juveniles)	2. Green sunfish (*Lepomis cyanellus*): Response declined under weakly acidic conditions	2. Hypozanthine-3N-oxide (ostariophysan putative alarm chemical): Response eliminated under weakly acidic condition	2. Leduc et al. (2003)
3. Field (snorkeling): Avoidance (all size classes)	3. Largemouth bass (*Micropterus salmoides*): Field (snorkeling): Avoidance; effect of habitat complexity and size	3. Swordtails (*X. helleri*): None	3. Golub et al. (2005)
Bluegill sunfish: *L. macrochirus*			
1. None (juveniles)	1. Not tested	1. Water: None	1. Cashner (2001)
Largemouth bass: *Micropterus salmoides*			
1. Not tested	1. Green sunfish (*Lepomis cyanellus*): Decreased activity, area use (juveniles)	1. Not tested	1. Brown et al. (2002)

Brennan, 2000; Leduc et al., 2003). The intensity of responses increased with higher concentrations of the alarm stimulus (Marcus and Brown, 2003). Golub et al. (2005) observed responses by free-ranging pumpkinseed sunfish, and noted the avoidance responses under natural conditions. These fright responses are not a result of general disturbance (i.e., no fright response to a blank control: Marcus and Brown, 2003) or to chemicals that would be released when the skin of any fish is damaged (i.e., no fright response to heterospecific control: Brown and Brennan, 2000; Golub and Brown, 2003; Golub et al., 2005). A negative result was reported for bluegill sunfish (*Lepomis macrochirus*), which showed no significant change in behavior following exposure to extracts from conspecifics (Cashner, 2001).

There has been a limited amount of investigation into whether centrarchid alarm chemicals might be homologous. Juveniles of both largemouth bass, *Micropterus salmoides* (Brown et al., 2002) and pumpkinseed sunfish (Leduc et al., 2003) responded to chemical stimuli from injured green sunfish with a fright response. However, more data are needed to determine definitively whether these responses indicate a chemical homology within the family or whether the responses are because juveniles of these species are members of the same prey guild. Both largemouth bass (Brown et al., 2001b) and green sunfish (Golub and Brown, 2003) responded to alarm chemicals from ostariophysan fishes, presumably because these chemicals are good indicators of predation risk and not due to a homology.

Antipredator responses of centrarchids may be influenced negatively by weakly acidic conditions. At a pH of about 6.0, the response of pumpkinseed sunfish to conspecific and heterospecific (green sunfish) alarm chemicals declined (Leduc et al., 2003). In addition, the response of pumpkinseed sunfish to the putative ostariophysan alarm chemical (Hypozanthine-3N-Oxide) was eliminated under acidic conditions (Leduc et al., 2003). More study is needed to determine whether other pollutants can negatively affect antipredator responses of centrarchids.

Adult centrarchids grow to fairly large body sizes, and so there is considerable difference in predation risk between adults and juveniles. Consequently, it is not surprising that there are ontogenetic differences in responses of adults and juveniles to alarm cues. Ontogenetic shifts have been reported for both green (Golub and Brown, 2003) and pumpkinseed (Leduc et al., 2003) sunfishes, with young (small) individuals responding to conspecific extracts with a fright response and older (larger) individuals showing foraging responses. Similar ontogentic shifts have been observed for green sunfish exposed to alarm cues from a sympatric ostariophysan, the finescale dace (*Phoxinus neogaeus*) (Golub and Brown, 2003).

It is notable that one of the species for which Pfeiffer (1977) reported negative results was the pumpkinseed sunfish. The contradictory results described above for this species emphasises the need for re-tests of many species that were initially tested using only qualitative criteria. In addition to the pumpkinseed sunfish, Pfeiffer (1977) also reported early negative results for the black-banded sunfish, *Mesogonistius chaetodon*. In light of recent positive results with other sunfish, this species should be re-tested using more sensitive assays.

Family Percidae: Perch-like Fishes (including darters)

Three subfamilies within the family Percidae have been tested for responses to alarm cues since Pfeiffer's review in 1977 (Table 11.7). Most of these studies are of darters (subfamily Etheostominae), which are so named because they generally move via short hops ('darts') along the substrate. Three genera of darters have been reported to give fright responses following exposure to conspecific alarm cues: *Ammocrypta* (naked sand darter, *A. beani*), *Etheostoma* (rainbow darter, *E. caeruleum;* Iowa darter, *E. exile*; Johnny darter, *E. nigrum*; orangethroat darter, *E. spectabile*; gulf darter, *E. swaini*; redfin darter, *E. whipplei*), and *Percina* (blackbanded darter, *P. nigrofasciata*) (Table 11.7). Activity of darters is relatively easy to quantify by counting movements, and the alarm response of all darters that have been studied has included reduced activity. In addition, avoidance (McPherson *et al.*, 2004), decreased opercular movements (Gibson and Mathis, 2006), use of an 'alert' posture (arched back, dorsal fins erect and head up: Smith, 1979), increased shelter use (Smith, 1982) and cryptic coloration (Smith, 1982) also have been observed. These responses were not a result of general disturbance or exposure to general injury related cues because tests of all species have included negative responses to water or injury released cues from heterospecific fishes (Table 11.7). The attack of a natural predator (northern pike, *Esox lucius*) releases sufficient alarm chemicals to be detected by nearby conspecifics (Smith, 1979).

Two species of darters have tested negative for responses to conspecific alarm cues: the fantail darter, *E. flabellare* (Commen-Carson and Mathis,

Table 11.7 Behavioral responses to chemical stimuli from injured conspecifics by three subfamilies in the Family Percidae (Superorder Acanthopterygii, Order Perciformes) published since 1977. Responses to cues from related (within-family) heterospecifics and other stimuli (extracts from extra-familial fishes or water) also are reported. "None" means that the stimulus was tested, but there was no fright response. "Not tested" means that the study did not include exposure to the cues listed in the column heading.

Response to conspecific cues	Response to cues from related heterospecifics	Response to other stimuli	Reference
Subfamily Etheostomatinae			
Naked sand darter: Ammocrypta beani			
1. Decreased activity	1a. Blackbanded darter (Percina Nigrofasciata): Decreased activity b. Gulf darter (Etheostoma swaini): None	1. Not tested	1. Smith (1982)
Rainbow darter: Etheostoma caeruleum			
1. Decreased activity	1a. Yoke darters (E. juliae): Decreased activity	1a. Water: None b. Bumblebee gobies (Brachygobius sabanus): None	1. Commens and Mathis (1999)
2. Not tested	2a. Redfin darters (E. whipplei): Decreased activity b. Fantail darters (E. flabellare): None	2a. Water: None b. Bumblebee gobies (Brachygobius sabanus): None	2. Commens and Mathis (2007)
3. Decreased opercular movements	3a. Yoke darters (E. juliae): Decreased opercular movements b. Banded darter (E. zonale): Decreased opercular movements c. Greenside darters (E. blennioides): None	3a. Bumblebee gobies (Brachygobius doriae): None b. Sympatric fathead minnows (Pimephales promelas): Decreased activity	3. Gibson and Mathis (2006)

(Table 11.7 contd.)

(Table 11.7 contd.)

Iowa darter: E. exile			
1. Decreased activity; arched back, dorsal fins erect and head up	1. Not tested	1. Water	1. Smith (1979)
2. Decreased activity; alarm response eliminated when hungry	2. Not tested	2. Not tested	2. Smith (1981)
3. Field: Avoidance (trap); Alarm response eliminated in metal-contaminated lakes	3. Not tested	3. Swordtails (*Xiphophorus helleri*): None	3. McPherson et al. (2004)
Fantail darter: E. flabellare			
1. None	1a. *E. caeruleum*: None 1b. *E. whipplei*: None	1. *Brachygobius sabanus*: Bumblebee gobies	1. Commens and Mathis (2007)
Johnny darter: E. nigrum			
1. Decreased activity; arched back; dorsal fins erect; head up	1. Not tested	1. Not tested	1. Smith (1979)
2. Decreased activity	2a. Orangethroat darter (*E. spectabile*): None b. Logperch (*P. caprodes*): None c. Blackside darter (*P. maculata*): None d. Slenderhead darter (*P. phoxocephala*): Decreased activity	2. Not tested	2. Haney et al. (2001)
Orangethroat darter: E. spectabile			
1. Decreased activity by lab-reared individuals	1. Not tested	1. Not tested	1. Vokoun and Noltie (2002)
2. Decreased activity	2a. Johnny darters (*E. nigrum*): Decreased activity b. Logperch (*P. caprodes*): Decreased activity	2. Not tested	2. Haney et al. (2001)

(Table 11.7 contd.)

(Table 11.7 contd.)

Species / Defense	Predator response	Reference
Gulf darter: E. swaini 1. Decreased activity, darting, cryptic coloration, increased shelter use 1a. Blackbanded darter (P. nigrofasciata): Decreased activity b. Ammocrypta beani: Decreased activity c. Slenderhead darters (P. phoxocephala): Decreased activity d. Blackside darters (P. maculata): None	1. Not tested	1. Smith (1982)
Redfin darter: E. whipplei 1. Decreased activity 1a. E. caeruleum: Decreased activity b. E. flabellare: Decreased activity	1. Brachygobius sabanus: Bumblebee gobies: None	1. Commens and Mathis (2007)
Logperch: Percina caprodes 1. None 1a. Johnny darters (E. nigrum): None b. Orangethroat darters (E. spectabile): None c. Blackside darters (P. maculata): None d. Slenderhead darters (P. phoxocephala): None	1. Not tested	1. Haney et al. (2001)
Blackbanded darter: P. nigrofasciata 1. Decreased activity 1a. Naked sand darter (A. beani): Decreased activity b. Gulf darter (E. swaini): Decreased activity …	None	1. Smith (1982)

(Table 11.7 contd.)

(Table 11.7 contd.)

Subfamily Percinae **Eurasian Ruffe: Gymnocephalus cernuus** 1. Decreased activity; decreased foraging; avoidance	1. Yellow perch (Perca flavescens): None	1. Goldfish (Garassius auratus): None	1. Maniak et al. (2000)
Yellow perch: Perca flavescens (juveniles) 1. Increased shoaling; increased use of substrate; increased shelter use; freezing	1. Iowa darters (Etheostoma exile): Increased shelter use; freezing	1a. Swordtails (Xiphophorus helleri): None b. Sympatric spottail shiners (Notropis hudsonius): Increased shelter use; freezing	1. Mirza et al. (2003)
2. Field (trap): Avoidance	2. Not tested	2. Water: None	2. Wisenden et al. (2004)
3a. Juveniles: Decreased activity, increased time with spines erect; decreased foraging b. Adults: Increased activity, increased foraging, decreased time with spines erect c. Neither age class responded differently to cues from juveniles nor adults	3. Not tested	3. Not tested	3. Harvey and Brown (2004)
Subfamily Luciopercinae **Walleye: Sander (Stizostedion) (vitreum)** 1. Decreased activity; associative learning of novel predators (juveniles)	1. Not tested	1. Water: None	1. Wisenden et al. (2004)

2007) and the logperch, *P. caprodes* (Haney *et al.*, 2001). Failure to respond to cues from injured conspecifics might be due to the potential for 'accidental' injury to the skin resulting from scraping against rocks: fantail darters are prone to hiding in rock crevices and logperch turn over rocks to find prey (Pflieger, 1997). However, in both of the above articles, the authors suggested that additional study is needed before inferences about the absence of alarm chemicals in these species can be made with confidence. In a study by Commens and Mathis (in press), the fantail darters showed low overall activity (even under control conditions), making reductions in activity difficult to detect; the authors recommended that future studies should quantify shelter use because this species may be more likely than other darters to hide among crevices. In contrast, the logperch were more active than the other species in a study by Haney *et al.* (2001); the authors suggested that future studies should measure the length of movements rather than the number of movements.

Like most of species, most studies of darter alarm cues have been conducted with wild-caught focal individuals, making it impossible to know whether observed fright responses are the result of learning or of innate responses. The one exception is Vokoun and Noltie's demonstration (2002) of reduced activity of lab-reared orangethroat darters following exposure to conspecific alarm cues. Interestingly, the fright responses only occurred when the skin extracts were taken from wild-caught darters. The authors suggested three hypotheses to account for the lack of response to cues from lab-reared darters: (1) Lab-reared darters were younger than their wild-caught counterparts; alarm cue production may not occur until later in life or may occur in very low concentrations in young darters. (2) The darter alarm cue may contain chemicals that are derived from their natural diet. (3) The close relatedness/familiarity of the lab-reared donors and receivers may have resulted in habituation to alarm cues from those individuals; this hypothesis assumes that there are individual/kin differences in the make-up of the alarm cue.

The first step in determining whether alarm chemicals are homologous among taxa is testing for cross-species responses. For the Percidae, cross-species tests have included both within and between genus comparisons. Within the genus *Etheostoma*, tests for cross-species responses to alarm cues have been reported for numerous species (Table 11.7). The frequency of the positive cross-reactions suggests that there may be some degree of homology for alarm chemicals within the genus *Etheostoma*, but the picture is not at all clear.

Rainbow darters exhibited fright responses following exposure to alarm chemicals from yoke darters, redfin darters, and banded darters, but did not respond to extracts from fantail darters or greenside darters (Commens and Mathis, 1999; Commens-Carson and Mathis, 2007; Gibson and Mathis, 2006). Redfin darters responded to alarm cues from rainbow darters, but also did not respond to extracts from fantail darters. Fantail darters, which did not respond to conspecific extracts (see above), also did not respond to extracts from rainbow darters or redfin darters (Commens and Mathis, in press). The negative results for responses to cues from fantail and greenside darters deserve further study. Two hypotheses could explain the lack of response. First, fantail and greenside darters produce alarm chemicals, but they are different from the alarm chemicals of the focal species. The authors (Commens-Carson and Mathis, 2007; Gibson and Mathis, 2006) suggested that habitat differences might make injury-released cues from these species poor indicators of predation risk for other species of darters, and so the alarm chemicals could be ignored by heterospecifics. Second, these species may not produce alarm chemicals at all. Fantail darters do not respond to chemicals from injured conspecifics (see discussion above) and responses of greenside darters to conspecific cues have not been tested.

Another pair of *Etheostoma* species that has been tested for cross-species responses to alarm cues is Johnny and orangethroat darters, which occur together in a stream in north-central Missouri (Haney et al., 2001). The pattern of responses is difficult to interpret with respect to the implications for the hypothesis of homology. Although orangethroat darters exhibited a weak response to extracts from Johnny darters, Johnny darters did not respond to cues from orangethroat darters. Unlike the case of fantail darters, the explanation for negative results cannot be that the cue is not produced; both species respond to alarm cues from conspecifics. The data, therefore, suggest that either an additional species-identifying chemical is necessary to elicit the alarm response or the alarm cues are not homologous. The possibility that orangethroat darters learned to respond to alarm cues from the sympatric Johnny darters cannot be ruled out.

Within the genus *Percina*, only one cross-species reaction has been tested. Logperch did not respond to extracts from slenderhead darters (Haney et al., 2001). Slenderhead darters have not been tested for responses to conspecific extracts, so it is not known whether they produce chemical alarm cues.

Several cross-genus tests have been conducted within the subfamily Etheostomatinae. Smith (1982) tested a trio of species that co-occurred in a river in southern Louisiana. The naked sand darter (*Ammocrypta*) and the blackbanded darter (*Percina*) exhibited fright responses following exposure to each other's alarm cues, as did blackbanded darters and gulf darters (*Etheostoma*). Although these data are consistent with the hypothesis of homology, they also could be explained as a learned response because the three species are sympatric. Naked sand darters failed to respond to extracts from gulf darters (*Etheostoma*), but the sample size was relatively small (n = 7), the stimulus solution was relatively weak (Smith, 1982), and the trend was in the predicted direction.

The study by Haney et al. (2001) also included cross-genus exposures. Both Johnny (*Etheostoma*) and orangethroat (*Etheostoma*) darters exhibited fright reactions following exposure to alarm cues from slenderhead darters (*Percina*). It is possible that this response could have been learned, because the three species co-occur. In contrast, neither Johnny nor orangethroat darters responded to alarm cues from blackside darters (*Percina*). Haney et al. (2001) suggested that the lack of response might be due to habitat differences between blackside darters and the other two species. Whether blackside darters respond to extracts from conspecifics is not known. Logperch (*Percina*) did not respond to extracts from Johnny or orangethroat darters, and individuals in these two species did not respond to extracts from logperch. The failure of other species to respond to extracts from logperch supports the hypothesis that logperches do not produce chemical alarm cues (but see discussion above). Alternatively, injury released cues from logperch may not be a good indicator of danger for Johnny and orangethroat darters.

Two species in the percid Subfamily Percinae have been tested for response to conspecific alarm cues, both with positive results. The fright response of the Eurasian ruffe (*Gymnocephalus cernuus*) included avoidance and decreased activity and foraging (Maniak et al., 2000). For the yellow perch (*Perca flavescens*), the fright response consisted of freezing, increased shoaling, increased use of the substrate, increased shelter use and increased time with the spines erect (Mirza et al., 2003; Harvey and Brown, 2004). Yellow perch showed an ontogenetic switch from feeding on invertebrates to piscivory, so cues from injured conspecifics might indicate the presence of wounded prey. Consistent with the shift in foraging, perch also showed an ontogenetic shift in response to conspecific alarm cues; young-of-the-year responded to conspecific cues with a fright response and adults responded with a feeding response (Harvey and Brown, 2004). Individuals from neither age class discriminated between conspecific cues from juveniles and adults, indicating that there may be no substantive difference between cues from individuals in different age classes.

Eurasian ruffe did not respond to alarm cues from yellow perch, suggesting either a lack of homology between the two species or the absence of additional identifying cues (Maniak et al., 2000). The response of yellow perch to cues from ruffe has not been tested, but perch do exhibit fright responses following exposure to alarm cues from another percid, the Iowa darter (*Etheostoma exile*). This latter response could either indicate some degree of homology between perch and darter alarm cues or could be the result of a learned response by members of the same prey guild.

One species in the percid Subfamily Luciopercinae has been tested for responses to conspecific cues. Juvenile walleye (*Sander vitreum*) responded to conspecific alarm cues with decreased activity (Wisenden et al., 2004). This response was not a result of general disturbance because tests with a blank (water) control were negative. There were no tests of walleye responses to heterospecific stimuli. Like some species in other taxa, walleye can use conspecific alarm cues as an unconditioned cue in associative learning of chemical cues from unfamiliar predators (Wisenden et al., 2004).

Two species of percids have been shown to respond to alarm cues from unrelated conspecifics. Rainbow darters showed increased opercular activity following exposure to alarm cues from fathead minnows, *Pimephales promelas* (Gibson and Mathis, 2006), and yellow perch showed increased shelter use and freezing following exposure to alarm cues from spot tail shiners, *Notropis hudsonus* (Mirza et al., 2003); both donor species belong to the superorder Ostariophysi. In both the cases, the authors attributed these fright responses to learned responses to alarm chemicals from a member of the same prey guild.

The only member of the family Percidae reported to have been tested in the review by Pfeiffer (1977), with negative results, is the European perch, *Perca fluviatilis*. There is a particular need for verification of this result in light of the recent positive response by another member of this genus.

Family Cichlidae: Cichlids

The first indication of an alarm response by cichlids to chemical cues from injured conspecifics was in a paper by Noble (1939), who stated that juvenile African mouthbrooders, *Tilapia macrocephala*, had the same reaction to chemicals from injured conspecifics as that of minnows (escape); he did not include any data to support this statement. Five decades later, Jaiswal and Waghray (1990) also reported avoidance behavior by young of another member of the cichlid family, the Mozambique tilapia, *Oreochromis mossambicus*. Their study did not include a control for disturbance or a control for general cues (e.g., blood) from any injured fish.

The only other member of the Family Cichlidae to be tested for alarm responses to conspecific cues is the convict cichlid, *Archocentrus* (= *Cichlosoma*) *nigrofasciatum*; this species has been studied intensively (Table 11.8). Behavioral responses to cues from injured conspecifics are variable and include avoidance (Alemadi and Wisenden, 2002), decreased general activity (Brown *et al.*, 2004, 2006a, b; Pollock *et al.*, 2006a, b), decreased aggression (Wisenden and Sargent, 1997; Brown *et al.*, 2004; Kim *et al.*, 2004; Foam *et al.*, 2005a), decreased foraging (Brown *et al.*, 2004, 2006a; Foam *et al.*, 2005a; Pollock *et al.*, 2006), increased shelter use (Wisenden and Sargent, 1997), increased time near the bottom of tank (Wisenden and Sargent, 1997), and increased use of a vigilance ('heads-up') posture (Foam *et al.*, 2005a). In addition, individuals in groups increase shoaling intensity (Alemadi and Wisenden, 2002; Brown *et al.*, 2004), and the make-up shoals includes individuals that are more similar to each other following exposure to conspecific alarm cues (Kim *et al.*, 2004) ('odd' prey individuals in groups tend to be favored by predators (Godin, 1997)). These fright responses are not a result of general disturbance (i.e., no fright response to a blank control: (Alemadi and Wisenden, 2002; Kim *et al.*, 2004; Roh *et al.*, 2004; Pollock *et al.*, 2006; Brown *et al.*, 2006a, b) or to chemicals that would be released when the skin of any fish is damaged (i.e., no fright response to heterospecific control): (Wisenden and Sargent, 1997; Brown *et al.*, 2004; Foam *et al.*, 2005a, b; Pollock *et al.*, 2006).

Decreased foraging in the presence of increased predation risk, such as indicated by the presence of alarm chemicals, is common in fishes (see other taxa reviewed in this chapter). Although decreased *foraging* should have energetic consequences, very few studies have shown long-term

Table 11.8 Behavioral responses to chemical stimuli from injured conspecifics by members of the Family Cichlidae (Superorder Acanthopterygii, Order Perciformes) published since 1977. Responses to cues from related (within-family) heterospecifics and other stimuli (extracts from extra-familial fishes or water) also are reported. "None" means that the stimulus was tested, but there was no fright response.

Response to conspecific cues	Response to cues from related heterospecifics	Response to other stimuli	Reference
Convict cichlids:			
Archocentrus (= Cichlosoma) nigrofasciatum:			
1. Increased time near tank bottom, increased shelter use, decreased aggression (juveniles)	1. Not tested	1. Mosquitofish (*Gambusia affinis*): None	1. Wisenden and Sargent (1997)
2. Area avoidance, increased schooling (juveniles; during and after period of parental care)	2. Not tested	2. Water: None	2. Alemadi and Wisenden (2002)
3. Decreased foraging, Decreased aggression, use of heads-up (vigilant) foraging posture; background level of risk influenced level of response (juveniles)	3. Not tested	3. Swordtails (*Xiphophorus helleri*): None	3. Foam et al. (2005a)
4. Decreased area use, decreased general activity, decreased foraging; increased level of response with increased concentration of cue; background level of risk influenced level of response (juveniles)	4. Not tested	4. Water: None	4. Brown et al. (2006a)

(Table 11.8 contd.)

(Table 11.8 contd.)

5. Background level of risk influenced level of response (juvenile)	5. Not tested	5. Swordtails (*X. helleri*): None	5. Foam *et al.* (2005b)
6. Decreased general activity, decreased area use, increased shoaling, decreased aggression; focal individuals (juveniles) responded differently to cues based on the donor's diet and body condition influenced production of alarm cues, but not maturity	6. Not tested	6. Swordtails (*X. helleri*): None	6. Brown *et al.* (2004)
7. Focal individuals (juveniles) responded differently to cues based on the donor's body condition	7. Not tested	7. Water: None	7. Roh *et al.* (2004)
8. Decreased aggression, increased similarity of shoal-mates, change in number of foragers; nature of response depends on foraging patch size (juveniles)	8. Not tested	8. Water: None	8. Kim *et al.* (2004)
9. Decreased general activity, decreased foraging; group size influenced responses (juveniles)	9. Not tested	9. Water: None	9. Brown *et al.* (2006b)
10. Decreased general activity, decreased foraging, decreased growth (juveniles)	10. Not tested	10a. Water: None; b. Swordtails (*X. helleri*): Increased foraging	10. Pollock *et al.* (2006)
Mozambique tilapia: *Oreochromis mossambicus*			
1. Avoidance (juveniles)	1. Not tested	1. Not tested: (pre-versus post-stimulus only)	1. Jaiswal and Waghray (1990)

effects from prolonged exposure to alarm chemicals. Pollock *et al.* (2006) exposed pairs of convict cichlids to conspecific alarm cues or to a blank control daily for a period of 41 days. Both males and females exposed to the alarm cues showed significantly lower growth than individuals in the control condition, including percentage change in body length, body depth and mass.

Another factor that influences foraging success for many species is the presence of competitors. Convict cichlids will readily establish feeding territories in the laboratory, and so provide a good model for examining the combined influence of competition and predation risk on foraging success. Kim *et al.* (2004) found that the relationship between the two variables can be complex. When predation risk is low, dominant individuals exclude some individuals from small feeding patches and growth rates are highly variable; decreased aggression following exposure to alarm cues allowed more individuals to co-occur in the patch, decreasing the variability in individual growth rates. In contrast, in large patches, overall aggression was low regardless of the level of predation risk, but the number of individuals foraging declined following exposure to the alarm stimuli leading to increased variability of individual growth rates. Interestingly, the overall effect of increased predation risk eliminated the effect of patch size on variability of growth rates.

Very little is known about the ontogeny of responses to nonostariophysan alarm cues. Convict cichlids provide an interesting case study because young cichlids rely on parental care for protection until about the age of one month (Wisenden, 1995). Very young cichlids might be expected not to develop responses to cues from injured conspecifics until after they are independent of parental protection. However, Alemadi and Wisenden (2002) found that very young convict cichlids (within the size range where parental care occurs) responded to extracts from injured conspecifics with the same degree of effectiveness as older individuals. Skin extracts from both older juveniles and adults elicit fright responses from conspecifics (Brown *et al.*, 2004).

Is production of the cichlid alarm chemical energetically expensive? Direct determination of the cost of alarm chemical production is difficult, especially when the source of production of the chemical and the identity of the chemical(s) are not known. However, Brown *et al.* (2004) provided two lines of evidence that support the hypothesis that convict cichlid alarm cues are expensive to produce. First, alarm cues from individuals that were fed a high quality diet (brine shrimp) elicited stronger responses than cues from individuals fed low quality diets (tubifex worms). Second, alarm cues from individuals with higher body condition scores elicited stronger responses than cues from individuals with lower body condition scores. Roh et al. (2004) also found a response of donor body condition on strength of alarm responses by conspecific cichlids. In addition, they found a lower concentration threshold for behavioral responses when the alarm cues were produced by donors in better body condition.

Recent theoretical work in antipredator behavior has suggested that the background level of predation risk should influence the intensity of responses to current levels of risk (Lima and Bednekoff, 1999). For convict cichlids, the intensity of antipredator responses to conspecific cues is lower when individuals had been exposed to frequent episodes of high predation risk than when exposed to predation risk infrequently (Foam et al., 2005b). However, the threshold concentration that produced a behavioral response was lower for fish exposed to frequent episodes of high risk (Brown et al., 2006a). Clearly, antipredator responses to alarm cues by cichlids (and other fishes) are not all-or-nothing, but can be modified based on a number of factors, including the background level of predation risk.

Group size can influence the way individual convict cichlids respond to different concentrations of the alarm cue (Brown et al., 2006b). Single individuals have been shown to be hypersensitive, responding strongly to a threshold concentration of alarm cues that have a dilution factor as low as 12.5% of the 'normal' experimental concentration. Individuals in small groups ($n = 3$) have a threshold response that is slightly higher (25% of normal strength). In contrast, for individuals in large groups ($n = 6$), alarm responses were graded with respect to concentration, declining gradually with decreasing concentrations. The effect of group size was eliminated at a concentration of about 50% of the normal strength. One cautionary inference from this experiment is that a lack of visible response does not necessarily mean that the alarm cue is not detected. Foam et al. (2005a) also found that convict cichlids switched to a low risk food patch after exposure to sub-threshold concentrations of the alarm cue, indicating that the cue had been detected even though there were none of the typical overt antipredator behavioral responses.

The only cichlid reported to have been tested for responses to conspecific cues in Pfeiffer's review (1977) was the South American angelfish, *Pterophyllum eimekei*. Given the consistent responses of convict cichlids and the demonstrated utility of these fishes in tests of numerous conceptual hypotheses about antipredator behavior, there is a definite need for testing additional species of cichlids for responses to alarm chemicals, including verification of the negative results for the angelfish. Additional tests also will help to determine whether there exist any chemical alarm homologies within the family; to date no studies have tested for cross-species reactions within the Cichlidae (Table 11.8).

Family Gobiidae: Gobies

Gobies belong to a speciose family with around 2000 species in over 200 genera (Nelson, 1994). The distribution is mostly marine and tropical, but some species do occur in fresh or brackish water. Coral reefs are the most common habitat, and most species are benthic, using a modified pelvic fin as a suction device for adhering to the substrate. In most other taxa, studies of alarm chemicals have focused on freshwater species. However, for marine species, chemical communication appears to be particularly important for benthic species (e.g., McClintock and Baker, 1997; Zimmer and Butman, 2000), so the Gobiidae should be an excellent taxon for future investigation into the role of alarm cues in marine systems. In addition, because some gobies have become invasive (e.g., *Gobius* in the Great Lakes of the USA), a more thorough understanding of basic goby biology may be particularly important for conservation purposes.

Since Pfeiffer's review (1977), only three species of gobies have been studied for detection of alarm chemicals. Positive responses have been noted in species belonging to two subfamilies (Gobionellinae; Gobiinae), indicating that chemical alarm cues are not limited to a single narrow taxon. Tests of all three species have included negative results for heterospecific controls, indicating that any observed fright responses were not elicited by general stimuli associated with injured fishes.

Like most members of the subfamily Gobionellinae, bumblebee gobies (genus *Brachygobius*) occur in fresh/brackish water. One of these species, *B. sabanus*, shows the typical response of decreased activity and also color change, reducing the contrast between the black and yellow bands when exposed to conspecific alarm cues (Smith and Lawrence, 1992). Individuals of this species do not respond to alarm cues from the starry goby (*Asterropteryx semipunctatus*), a member of the Subfamily Gobiinae; this lack of response suggests that there is no simple homology or convergence between the two taxa (Hugie *et al.*, 1991). However, to complicate the picture, starry gobies do respond to alarm cues from bumblebee gobies! There are at least two possible explanations for this inconsistency: (1) There is enough chemical similarity between the alarm cues of bumblebee and starry gobies for the starry gobies to recognize the chemical as an indication of danger; and (2) Starry gobies have learned that alarm cues from bumblebee gobies (or from other species whose alarm cue is homologous with that of bumblebee gobies) are good indicators of danger. In either case, it is unclear as to why the cross-reaction is unidirectional. Bumblebee gobies also have been used as 'control' sources of alarm cues for tests of fishes in more distant taxa (darters, see Table 11.7); none of these species responded to goby alarm cues.

In contrast, the other gobionellid that has been tested, *Gnatholepsis anjerensis* shows no change in behavior following exposure either to conspecific cues or to cues from starry gobies, *A. semipunctatus* (Smith, 1989). In addition, starry gobies also do not respond significantly to skin extracts from *G. anjerensis* (Smith, 1989). Unlike the bumblebee gobies, *G. anjerensis* occupies marine reefs, but whether this difference in habitat accounts for the lack of response is not known.

Starry gobies (*A. semipunctatus*; Subfamily Gobiinae) show one of the most interesting behavioral responses to conspecific alarm cues: bobbing. According to Smith and Smith (1989), bobbing occurred when 'the anterior end of the goby rose slowly off of the bottom, by extension of the pelvic fins and, sometimes, beating of the pectoral fins. Bobs were given either singly or in bouts of 2 to about 10, with an interbob interval of about 1 to 3 s.' Starry gobies bob in response to visual and chemical cues from predators (Smith, 1989; Smith and Smith, 1989) and to alarm chemicals from conspecifics (Smith, 1989) and bumblebee gobies, *B. sabanus* (Hugie *et al.*, 1991). Bobbing also acts as a visual alarm signal to conspecifics; gobies begin bobbing following exposure to other bobbing gobies (Smith and Smith, 1989). Although bobbing certainly has an alarm function, the evolutionary forces driving the evolution of this behavior are not known. Smith and Smith (1989) suggested that one possibility is that bobbing might serve as a pursuit deterrent signal by informing the predator that it has been detected. 'Warned' prey may be more difficult to catch than unaware prey.

Chemical alarm cues can serve as unconditioned stimuli in associative learning of unfamiliar predators. In a similar process as that of the well-known example of Pavlov's dog, individuals that showed no initial response to novel predators were exposed to predatory cues plus alarm cues; fright responses to the novel predator alone were then shown during subsequent exposures. This alarm-based predator-recognition learning has been demonstrated in many ostariophysan fishes (see Brown, 2003; Kelley and Magurran, 2003), but has not been studied very well in other taxa of fishes. In gobiids, only starry gobies have been tested, and they too have been shown to be able to learn to associate neutral stimuli with danger through this alarm-based process (Larson and McCormick, 2005). Like the ostariophysan species, learning was associated with a single exposure to the alarm chemical paired with the neutral stimulus.

In recent years, animal care issues have included concern about the potential effects of anesthesia on subsequent behavior. In laboratory tests, Losey and Hugie (1994) found that prior exposure to two fish anesthetics, quinaldine and phenoxyethanol, reduced fright responses of starry gobies to alarm cues from conspecifics. In contrast, prior exposure to MS-222 did not affect the alarm response. Clearly, more study is needed, but researchers should refrain from using quinaldine and phenoxyethanol if at all possible.

Pfeiffer (1977) listed four additional species (Subfamily Gobiinae) that had been tested for response to conspecific extracts, and all tests yielded negative results: *Gobius niger, G. paganellus, Pomatoschistus microps*, and *P. minutus*. In addition to verification of the negative results for these species, tests of other representatives of this subfamily should yield important information.

Miscellaneous Perciform Families: Eleotridae (Bullies), Pomacentridae (Damselfishes), Centropomidae (Glassfishes) and Belontiidae (Gouramis)

Single species in four other families of perciform fishes have been tested for responses to conspecific skin extracts since Pfeiffer's (1977) review (Table 11.10). Positive responses were reported for two of these species, the common bully (*Bogiomorphus cotidianus*: family Eleotridae) and the ambon damselfish (*Pomacentrus amboinensis*: family Pomacentridae) (Kristensen and Closs, 2004; Larson, 2003 cited in McCormick and Holmes, 2006). The two species with negative responses were Indian fishes that are members of the families Centropomidae (the Asian glassfish, *Chana nama*) and Belontiidae (the banded gourami, *Colisa fasciata*) (Ahsan and Prasad, 1982). The only cross-species test was a negative result for the Asian glassfish exposed to extracts from an ostariophysan fish, the Indian catla carp (*Catla catla*) (Ahsan and Prasad, 1982). Only one of these families (Pomacentridae) was included in Pfeiffer's review (1977), with negative responses for the blue-green chromis (*Chromis chromis*) and the brown damselfish (*Pomacentrus rectifraenum*).

THE SOURCE OF THE ALARM CUE

The Integument: Epidermal Sacciform Cells

Description and Evidence for Alarm Function

For nonostariophysans, the anatomical source of the alarm chemical has not been identified definitively for species in any taxon. This situation contrasts with that for ostariophysans, where several lines of evidence support the hypothesis that epidermal club cells are responsible for production of the alarm cues. This evidence includes a lack of response to conspecific extracts in species that lack club cells and the elimination of a response in other species that coincides with a seasonal loss of club cells (Pfeiffer, 1977; Smith, 1977, 1986). Ostariophysan club cells ('alarm substance cells') have a distinct appearance; they are ductless, located in the middle of the epidermis, have centrally located nuclei and a negative reaction to certain histological stains (Smith, 1977). Club cells are rare outside of ostariophysans (Table 11.1), and the one nonostariophysan species with club cells that has been tested did not exhibit an alarm response following exposure to cues from injured conspecifics (reed fish, *Calamoichthys calabaricus*, superorder Palaeonisciformes, Order Polypteriformes) (Whitear, 1981; Hugie and Smith, 1987).

Is the skin the source of the nonostariophysan cues? Certainly, almost all of the researchers studying these cues have proceeded on the assumption that this hypothesis is true, but the evidence is mostly indirect. In many studies, alarm extracts have been produced by making shallow cuts in the skin of freshly killed donor fishes and placing the carcass in water (e.g., Smith, 1979; Smith et al., 1982; Mathis and Smith, 1993; Wisenden and Sargent, 1997; Chivers et al., 2000). Although the depth of the cuts cannot be verified, it seems unlikely that they would have released much muscle material because to do so the cuts would have had to go through the scales which are located in the dermis. Macerated skin fillets (either using blenders or mortars and pestles) have been used to produce alarm extracts in other studies (e.g., Houtman and Dill, 1994; Chivers and

Smith, 1994; Brown and Brennan, 2000; Mirza *et al.*, 2001; Kristensen and Closs, 2004; Foam *et al.*, 2005a; Leduc *et al.*, 2006), but most of these would have included some muscle tissue because it is virtually impossible to remove the skin of small fishes without removing some underlying muscle tissue (pers. obs.). It is important to note, however, that studies of most taxa have ruled out chemical cues that are not species-specific, such as general blood and muscle proteins, by showing negative responses to control stimuli from injured heterospecifics. Smith (1979) experimentally ruled out muscle tissue as the source of the cue for Iowa darters (*Etheostoma exile*) (see for review Zaccone *et al.*, 2001).

Although most nonostariophysans do not have club cells, another type of epidermal cell, the sacciform cell (Fig. 11.2), is common. Unlike club cells, there is considerable taxonomic diversity in the histochemistry and morphology of sacciform cells (Mittal *et al.*, 1981). In general, sacciform cells tend to have a peripheral nucleus and a large, membrane-bound lumen; there may or may not be a duct leading to the surface of the skin (Mittal *et al.*, 1981). The chemical nature of the secretions is not known, but some studies have reported the presence of an acidophilic, proteinaceous substance (Pickering and Fletcher, 1987; Lopez-Doriga and Martinez, 1993), and immunohistochemistry has resulted in positive reactions of sacciform cells to serotonin and the peptides bombesin and caerulein (Fasulo *et al.*, 1993; Zaccone *et al.*, 1986, 1999).

One source of evidence for the alarm function of *club cells* in ostariophysans is that fishes in species that lack the cells also lack the alarm response (Smith, 1977). Similar correlations should be examined for

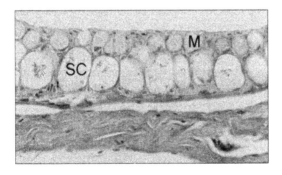

Fig. 11.2 Cross section of the skin of a fantail darter, *Etheostoma flabellare*. Note the large clear sacciform cells (S) in the lower layer of the epidermis (E); the mucous cells (M) are closer to the surface. D = dermis. The section was stained with eosin and Harris hematoxylin stain. Bar = 50 μm.

nonostariophysan species with and without sacciform cells. It is notable that some species that possess sacciform *cells* apparently do not have alarm responses to conspecific extracts (e.g., eyebar gobies, *Gnatholepis anjerensis*: Smith, 1989; Smith and Lawrence, 1992; fantail darters: Commens-Carson and Mathis, 2007), suggesting that nonalarm functions of sacciform cells exist for at least some species. The presence of club cells when the fright response is absent also occurs in some ostariophysan species (Pfeiffer, 1977).

Another line of evidence for the alarm function of club cells in ostariophysan fishes is that skin extracts from areas of the body that lack the cells do not elicit a fright response from conspecific receivers. For example, the barbels of some ostariophysans (carp, some catfishes) lack club cells, and conspecifics do not perform fright responses following exposure to barbel extracts (Smith, 1977). Detailed histological examination of the body surfaces of nonostariophysan fishes might identify areas that lack sacciform cells and thus would provide extracts to test this hypothesis. For example, the tips of pelvic fin knobs of the tessellated darter, *Etheostoma olmstedi*, appear to lack sacciform cells (Bart and Page, 1991) even though the cells are present on other parts of the body (Smith, 1986). The absence of sacciform cells on fin knobs, which develop on breeding males of egg-clustering darters (Page, 1985) may be due to 'false alarms' that might occur due to abrasion of the knobs that occurs during spawning (e.g., Lindquist *et al.*, 1981). Interestingly, Bart and Page (1991) discounted this hypothesis because fin knobs of fantail darters, *E. flabellare*, retain a dense concentration of sacciform cells. However, we now know that fantail darters do not respond to conspecific extracts containing sacciform cells (Commens-Carson and Mathis, 2007).

Taxonomic Overview

Although I will not attempt an exhaustive review of skin morphology in nonostariophysans, I will mention a few studies that describe sacciform cells in major taxa where alarm cues have been described.

Two species of sculpins (Order Scorpaeniformes) have been shown to possess large sacciform cells in the epidermis. In the tidepool sculpin, *Oligocottus maculosus*, at least some of the cells have ducts leading to the surface of the skin (Hugie *et al.*, 1991) such as in the case of the puffer fish *Takifugu pardalis* (Zaccone *et al.*, 2001). Similar cells were seen in slimy sculpins, *Cottus cognatus* (Bryer *et al.*, 2001). In both species, the cells take up a large volume of the epidermis. Another member of the Order

Scorpaeniformes that has sacciform cells is the estuarine species, the bullhead (*Agonus cataphractus*) (Whitear and Mittal, 1986). Therefore, the presence of sacciform cells in this order occurs in species in a wide range of habitats, including shallow ocean tidepools, deeper water estuaries, and freshwater streams.

In salmonids, epidermal sacciform cells have been described for three genera, including brown trout (*Salmo trutta*) (Pickering and Fletcher, 1987; Lopez-Doriga and Martinez, 1993), Arctic charr (*Salvelinus alpinus*) (Pickering and Fletcher, 1987; Witkowski *et al.*, 2004), and Coho salmon (*Oncorhynchus kisutch*) (Fast *et al.*, 2002). However, sacciform cells were not observed in the skin of Atlantic salmon (*Salmo salar*) and rainbow trout (*Oncorhynchus mykiss*) (Fast *et al.*, 2002). As there is considerable variation in the presence and density of sacciform cells seasonally and between individuals in salmonids (Blackstock and Pickering, 1980), additional study is needed.

Smith (1979) reported that one member of the Cichlidae (*Cichlasoma fasciata*) lacked 'darter-like club cells' (Smith's original term for sacciform cells), but this result was based on a sample size of one. However, sacciform cells have been reported to be present in another species of this family, the jewel cichlid, *Hemichromis bimaculatus* (Quilhac and Sire, 1999).

Some species of gobies (Gobiidae) possess epidermal sacciform glands, including *Asteropteryx semipunctatus* (Smith, 1989), *Gnatholepis anjerensis* (Smith, 1989), and *Gobiodon histrio* (Munday *et al.*, 2003). The presence of these glands may not be ubiquitous within the family; *Paragobiodon* sp. apparently lack sacciform glands (Munday *et al.*, 2003). Secretions from skin glands of at least some gobies are thought to function as predator deterrents, but there is substantial variation among species in the level of toxicity (Muday *et al.*, 2003). It would be enlightening to learn whether the alarm response is absent in species that lack the glands and whether the intensity of alarm responses is correlated with the intensity of toxicity.

With the family Poeciliidae, epidermal sacciform cells have been described for mosquitofish, *Gambusia affinis* (Bryant, 1987).

Sacciform cells have been reported to be present in numerous species of darters (Percidae). In the genus *Etheostoma*, positive results for sacciform cells were reported for seven species by Smith (1979) (*E. blennoides, E. caeruleum, E. exile, E. flabellare, E. kennicotti, E. nigrum,* and *E. squamiceps*), and for *E. swaini* (Smith, 1982), *E. olmstedi* (Smith, 1986), and *E. whipplei* (Commens-Carson and Mathis, 2007). Sacciform cells also have been reported in the epidermis of individuals of four species of *Percina* (Smith, 1979: *P. caprodes, P. phoxocephala,* and *P. sciera*; Smith, 1982: *P. nigrofasciata*) and one species of *Ammocrypta* (Smith, 1982: *A. beani*).

Sacciform *cells* also are present in at least some nondarter percids, including two species of *Sander* (= *Stizostedion*), the sauger (*S. canadense*) and the walleye (*S. vitreum*) (Smith, 1979; Wisenden, 2003). The yellow perch (*Perca flavescens*) also has sacciform cells, but they are small and less numerous than in other percid species (Smith, 1979).

Two species in the darter family have been tested that did not show an alarm response following exposure to conspecific extracts (fantail darters, *Etheostoma flabellare*: Commens-Carson and Mathis, 2007; *Percina caprodes*, logperch: Haney *et al.*, 2001). Both of these species do have epidermal sacciform cells (fantail darters: Smith, 1979; Commens-Carson and Mathis, 2007; Fig. 11.2; logperch: Smith, 1979). A number of hypotheses might explain this incongruency between the presence of sacciform cells and the apparent lack of an alarm response. (1) The conclusion that there is no alarm response could be incorrect; researchers who reported the negative results recommended further study of these species to confirm this result (Commens-Carson and Mathis, 2007; Haney *et al.*, 2001). (2) The sacciform cells may not be the source of the alarm cue. (3) Sacciform cells may produce alarm cues in some species, but fantail darters and logperch may have lost either production of the alarm chemical or the response to chemical because they have behaviors that may result in incidental skin damage or 'false alarms'. Both fantail darters and logperch have feeding behaviors in rocky stream substrates that may result in abrasions (Pflieger, 1997). (4) There may be seasonal variation in production of sacciform cells, as seen in some species of ostariophysan fishes (Smith, 1982). However, sacciform cells are present during both the breeding and nonbreeding seasons for fantail darters (Commens-Carson and Mathis, 2007); so, this hypothesis seems unlikely. The phenomenon of the lack of an alarm response in species that possess alarm (club) cells also was seen in some ostariophysan species (Smith, 1977).

The Chemical Nature of the Alarm Cue

Although some of the chemicals that occur within sacciform cells have been identified (e.g., Mittal *et al.*, 1981; Zaccone, 1981; Fasulo *et al.*, 1993), the chemical(s) that is responsible for eliciting the alarm response has not been identified. Because cross-taxa exposures are sometimes negative, either the alarm cue is different among taxa or an additional species-identifying chemical is necessary to elicit the alarm reaction. Freezing does not deactivate the alarm cue of ostariophysan fishes, and the same appears to be true for at least some taxa of nonostariophysans that have been tested, including salmonids (Leduc et al., 2006), cichlids (Brown et al., 2006a), sticklebacks (Chivers and Smith, 1994), cyprinodonts (Mirza et al., 2001b), centrarchids (Brown and Brennan, 2000), and two perciform fishes, walleye (*Sander vitreum*: Wisenden et al., 2004) and common bullies (*Gobiomorphus cotidianus*: Kristensen and Closs, 2004). An experimental study of tidepool sculpins (*Oligocottus maculosus*) demonstrated that responses to cryofrozen and fresh extracts did not differ; other studies of sculpins used fresh extracts. Smith (1992) stated that, in his personal observations, extracts from gobies and darters lose activity when frozen; however, published tests have only used fresh extracts (gobies: Smith, 1989; Smith et al., 1991; Smith and Lawrence, 1992; Larson and McCormick, 2005; darters: Smith, 1979, 1981, 1982; Commens and Mathis, 1999; Haney et al., 2001; Vokoun and Noltie, 2002, in press; Gibson and Mathis, 2006). Tests of single species from three other families of perciform fishes have only used fresh extracts: Centropomidae and Belontiidae (Ahsan and Prasad, 1982) and Eleotridae (Maniak et al., 2000).

Brown et al. (2003) found that nitrogen oxides may be involved in the eliciting an alarm response from ostariophysan fishes, but exposure to this chemical did not lead to an alarm response in convict cichlids, *Cichlasoma nigrofasciatum*, or rainbow trout, *Oncorhynchus mykiss*. The chemical nature of the alarm cue in nonostariophysans should receive extensive attention in future studies.

DISCUSSION

Evolution of Nonostariophysan Alarm System

Chemical alarm cues occur within at least two superorders (Protacanthopterygii and Acanthopterygii) outside of the Ostariophysi. Most species that have been tested belong to the superorder Acanthopterygii, including cichlids, cyprinidonts (guppies and topminnows), centrarchids (basses and sunfishes), percids (darters and perches), gobies, sticklebacks, and sculpins. Clearly, a phylogenetic perspective would provide great insight the evolution of the chemical alarm system of nonostariophysan fishes. Because numerous cross-taxa exposures have yielded negative data, it appears that the alarm response and/or the alarm chemicals have evolved independently numerous times even within the superorder Acanthoperygii. For example, darters do not respond to alarm cues from gobies (Commens and Mathis, 1999), gobies do not respond to alarm cues from guppies (Smith and Lawrence, 1992), and sticklebacks to not respond to skin from swordtails (Mathis and Smith, 1993) (see also Tables 11.2–11.10). Even within a family, there may be substantial variation in response to cross-taxa exposures (e.g., Percidae, Gobiidae, Tables 11.7 and 11.9).

To date, there has been little effort to place alarm responses of nonostariophysan fishes within a phylogenetic framework. Two deficiencies limit our current ability to apply a phylogenetic approach. First, only a few species have been tested within any one family, making it difficult to search for general patterns. Testing a broader array of species is essential. Second, detailed phylogenies are either not available or are not well supported for specific groups of interest. There is tremendous potential for new research in fish systematics to provide numerous opportunities to test phylogenetic hypotheses about the evolution of alarm systems.

This review has focused on the antipredator function of the alarm response (i.e., selection on receivers). Receivers that respond with antipredator behavior to cues from injured conspecifics experience increased survival (Mirza and Chivers, 2001b) and so are favored by selection. However, another interesting line of research is the evolution of production of the alarm chemical from the point of view of the sender. Selection on senders and receivers is not necessarily linked. A number of hypotheses have been proposed to explain evolution of alarm cues in ostariophysan fishes (e.g., Smith, 1986, 1991), and these also would apply to nonostariophysans (see below).

As explained in detail by Smith (1986, 1992), these hypotheses fall into the following general categories: (1) Alarm chemicals benefit signalers by being noxious or distasteful and therefore act as predator deterrents. Although this function does not appear to apply to ostariophysan fishes (Bernstein and Smith, 1984; Mathis et al., 1995), it may play a role for at least some species of nonostariophysans (e.g., gobies: Schubert et al., 2003). (2) Warning nearby conspecifics of a predator's presence could provide protection to the injured individual by initiating group defenses (mobbing, schooling) or distracting the predator. (3) The alarm cue may increase the survival of nearby offspring or potential mates, which would benefit the sender through direct selection. (4) The alarm cue may increase the survival of nearby relatives, which would benefit the sender through indirect (kin) selection. (5) The alarm cue may attract secondary predators that disrupt the predation event, allowing the prey a chance to escape (Mathis et al., 1995; Chivers et al., 1996). (6) There may be certain nonalarm functions that benefit the sender, such as promotion of healing of injured tissues or antipathogenic functions (Chivers et al., 2007). Immunoreactive secretions have been identified in the sacciform cells of some species (e.g., Zaccone et al., 1986). In addition, sacciform cells increase during infestation by some ectoparasites, suggesting a possible antiparasitic function (Pickering and Fletcher, 1987; see also Munday et al., 2003).

The Antipredator Function of the Alarm Response

Fishes typically respond to alarm chemicals in a way that is generally similar to their responses to predators (Smith, 1989; Chivers et al., 1995; Brown and Godin, 1999; Bryer et al., 2001; Kristensen and Closs, 2004; Gibson and Mathis, 2006). The most common response was reduced activity (including freezing, decreased area use, and decreased number of directional changes), which was reported in almost all major taxa that were tested (Tables 11.2–11.3, 11.5–11.10). Decreased activity is an effective antipredator response for prey with visually-oriented predators because it reduces the probability of detection (Azevedo-Ramos et al., 1992; Skelly, 1994). In some species of gobies and darters, decreased activity is accompanied by color changes that appear to enhance crypticity.

Associated with decreased general activity (movements) are decreases in the occurrence of other behaviors that are important in terms of fitness. Because it is usually impossible to perform antipredator behavior simultaneously with other fitness-enhancing behaviors, this area of research often is couched in terms of 'trade-offs' (Lima and Dill, 1990). One of the more common trade-offs is between response to alarm cues and decreased foraging behavior (centrarchids, cichlids, percids, salmonids, sculpins). In one study (cichlids, Pollock et al., 2006), repeated exposure to alarm cues resulted in decreased growth; potential long-term effects should be the subject of more study. Another line of evidence that responses to alarm cues represent a trade-off between increased probability of survival and decreased foraging success is that the response to alarm cues can be condition-dependent, with hungry individuals tending to ignore the alarm cues (darters: Smith, 1981; sculpins: Chivers et al., 2000). Response to alarm cues also has been shown to lead to decreases in aggression associated with interference competition (cichlids: Table 11.8). Trade-offs between response to alarm chemicals and reproductive behaviors have been reported for ostariophysans (Jones and Paszkowski, 1997), but have not yet been studied for nonostariophysans. Studies of trade-offs offer a fertile area for future research with nonostariophysan fishes.

Another behavior that often is associated with decreased activity is movement to the substrate. This behavior may function in decreasing the probability of detection for cryptic prey or may make it less likely that a free-swimming predator would intersect the path of the prey. Nonostariophysan fishes that tend to utilise substrate areas more often when under high risk (alarm) conditions include mosquitofish (Table 11.3), sculpins (Table 11.5), perches (Table 11.7), and cichlids (Table 11.8). Vertical area use by swordtails is particularly interesting. Like the other species, female swordtails tended to move toward the bottom of testing tanks; males, however, swim near the tank's surface (Mirza et al., 2001). This difference may be due to morphological difference between the sexes (males are more streamlined and, thus, better swimmers). Species in other taxa should be carefully examined for potential differences in behavior between the sexes.

Another common response to alarm cues reported in studies of nonostariophysan fishes is avoidance behavior (salmonids, cyprinodonts, sticklebacks, sunfishes, darters, cichlids; Tables 11.2–11.4, 11.6–11.8). These studies are particularly exciting because most of them have occurred under natural conditions. Field techniques have included use of traps labeled with alarm/control cues (guppies, sticklebacks, darters; Tables 11.3, 11.4, 11.7), observation while snorkeling (bass, Table 11.6), observation from the bank (guppies, Table 11.3), and underwater videotaping (salmonids, sticklebacks; Tables 11.2, 11.4). In the past, there was some concern expressed as to whether responses observed under laboratory conditions were artifacts of oversimplified and enclosed environments (Magurran et al., 1996), but these field experiments definitively show that fishes respond to alarm cues under natural conditions.

A common *antipredator* behavior observed in response to nonostariophysan alarm cues is an increase in shoaling/schooling intensity (sunfishes, perch, sticklebacks, cyprinodonts). Shoaling is well known to increase prey survival through processes such as predator confusion, dilution effects, or selfish herd effects (Hamilton, 1971; Pitcher, 1986). To maximise the potential for predator confusion, group mates should be as similar as possible, and one study has shown that shoal composition changes following exposure to alarm chemicals so that the similarity of shoal-mates is increased (cichlids: Kim *et al.*, 2004). In contrast, sticklebacks prefer to shoal with fathead minnows over conspecifics under high risk alarm conditions (Mathis and Chivers, 2003). This unexpected result apparently occurs because some predators prefer minnow as prey over sticklebacks, and so sticklebacks increase their probability of survival by associating with minnows (Mathis and Chivers, 2003). Large groups can offer increased benefits over small groups, particularly with respect to dilution and confusion effects. Group size influenced response to alarm cues for cichlids, with individuals responding to alarm cues at lower concentrations than small groups and large groups exhibiting a graded response to concentrations of alarm cues (Brown *et al.*, 2006b).

Dashing, a rapid burst of apparently disoriented swimming (Lawrence and Smith, 1989), has been observed in a few species of nonostariophysan fishes (salmonids, cyprinodonts, eleotrids; Tables 11.2, 11.3, 11.10). This behavior is typically followed by freezing or low activity. Smith (1989) suggested that dashing was an escape behavior.

Hiding has obvious antipredator benefits in terms of avoiding detection by predators and, in some cases, allowing the prey fish to retreat to inaccessible areas. Increases in shelter use following exposure to conspecific alarm cues have been reported for salmonids, sticklebacks, sculpins, percids, and cichlids (Tables 11.2, 11.4, 11.5, 11.7, 11.8).

Changes in posture have been reported for a few species. Sunfishes and some darters respond to alarm cues by keeping their dorsal spines erect, presumably as a defensive action. Both cichlids (Table 11.8) and gobies (Table 11.9) adopt a 'heads-up' posture that may result in increased efficiency of vigilance.

Some species of gobies respond to both alarm cues and lizardfish predators by bobbing, slowly raising the anterior part of the body off of the substrate apparently by extensions of the fins (Smith, 1989). Lizardfish are cryptic ambush predators, and bobbing may be a pursuit-deterrent signal that communicates to the predator that it has been detected (Smith and Smith, 1989). Bobbing also serves as a visual alarm signal between individuals, both within and between species (Smith and Smith, 1989).

Table 11.9 Behavioral responses to chemical stimuli from injured conspecifics by members of the Family Gobiidae (Superorder Acanthopterygii, Order Perciformes) published since 1977. Responses to cues from related (within-family) heterospecifics and other stimuli (extracts from extra-familial fishes or water) also are reported. "None" means that the stimulus was tested, but there was no fright response. "Not tested" means that the study did not include exposure to the cues listed in the column heading.

Response to conspecific cues	Response to cues from related heterospecifics	Response to other stimuli	Reference
Bumblebee goby: _Brachygobius sabanus_			
1. Decreased activity; head-up (vigilant) posture; color change	1. Not tested	1. Guppies (_Poecilia reticulata_): None	1. Smith and Lawrence (1992)
1. Not tested	2. Starry gobies (_Asterropteryx semipunctatus_): Foraging response	2. Not tested	2. Smith et al. (1991)
Starry goby:			
Asterropteryx semipunctatus			
1. Decreased activity, bobbing	1. Eyebar goby (_Gnatholepis anjerensis_): None	1a. Water: None b. Cave shortfin molly (_Poecilia mexicana_): None	1. Smith (1989)
2. Not tested	2. Bumblebee gobies _Brachygobius sabanus_: Decreased activity, bobbing	2. Not tested	2. Smith et al. (1991)
3. Reduced intensity of response following exposure to 2 anesthetics (quinaldine and phenoxyethanol) (but not after exposure to MS-222)	3. Not tested	3. Not tested	Losey and Hugie (1994)
4. Predator-recognition learning	4. Not tested	4. Swordtails (_Xiphophorus helleri_): None	4. Larson and McCormick (2005)
Eyebar goby: _Gnatholepis anjerensis_			
1. None	1. Starry gobies (_A. semipunctatus_)	1. Not tested	1. Smith (1989)

Table 11.10 Behavioral responses to chemical stimuli from injured conspecifics by four families in the Order Perciformes (Superorder Acanthopterygii) published since 1977. Responses to cues from related (within-family) heterospecifics and other stimuli (extracts from extra-familial fishes or water) also are reported. "None" means that the stimulus was tested, but there was no fright response. "Not tested" means that the study did not include exposure to the cues listed in the column heading.

Response to conspecific cues	Response to cues from related heterospecifics	Response to other stimuli	Reference
Family Centropomidae			
Asian glassfish: *Chanda nama*			
1. None	1. Not tested	1. Indian catla carp (*Catla catla*): None	1. Ahsan and Prasad (1982)
Family Belontiidae			
Banded gourami: *Colisa fasciata*			
1. None	1. Asian glassfish (*Chanda nama*): None	1. Not tested	1. Ahsan and Prasad (1982)
Family Eleotridae			
Common bully: *Gobiomorphus cotidianus*			
1. Dashing followed by inactivity on the substrate	1. Not tested	1. Water	1. Kristensen and Closs (2004)
Family Pomacentridae			
Ambon damselfish:			
Pomacentrus amboinensis			
1. "Antipredator behaviours"	1. ?	1. ?	1. Larson (2003) as cited in McCormick and Holmes (2006)

Physiological responses have not been well studied and can provide important insights into the mechanistic issues associated with alarm responses. Increases in plasma cortisol levels were reported for rainbow trout (Scott et al., 2003; Toa et al., 2005). Another common physiological response to stress is increased oxygen consumption (Sapolsky, 2002). Increased opercular activity in response to conspecific alarm cues was seen by rainbow darters (Gibson and Mathis, in press). The authors suggested two possible functions for such movements. Rapid opercular movements could increase oxygen consumption as part of the fight-or-flight response (Rottmann et al., 1992), or they could result in increased water flow toward the olfactory epithelia, increasing the efficiency of chemoreception.

Condition-dependent responses to alarm cues may be common and should help to shed light on the cost and benefits associated with *antipredator* behavior. For example, different responses are seen by hungry and satiated darters and sculpins and by male and female swordtails (see above). Another difference among individuals that can influence payoffs for responses to alarm cues is body size. In brook sticklebacks (*Culaea inconstans*), small individuals were not as likely to avoid conspecific alarm cues as large individuals, potentially indicating that experience may influence alarm responses for this species (Chivers and Smith, 1994). Ontogenetic effects may be particularly dramatic for species where adults grow to a relatively large body size, and so will experience dramatic differences in predation pressure from juveniles. For example, in both green (*Lepomis cyanellus*) and pumpkinseed (*L. gibbosus*) sunfishes, only younger juveniles responded to alarm cues with antipredator behavior; older juveniles responded with foraging responses (Golub and Brown, 2003; Marcus and Brown, 2003). The same difference was seen between juveniles and adults of yellow perch (*Perca flavescens*) (Harvey and Brown, 2004). The effect of size for brook charr (*Salvelinus fontinalis*) is quite different from that of sunfishes. Small charr responded most strongly to the alarm cues of small conspecifics and large charr responded most strongly to the cues of large conspecifics (Mirza and Chivers, 2002). This result suggests that there is an ontogenetic change in either the alarm cue itself or in some accompanying chemical for this species (Mirza and Chivers, 2002). In contrast, juvenile and adult yellow perch did not show different response to alarm cues from juveniles and adults (Harvey and Brown, 2004).

Alarm cues have been useful as tools in studies of predator-recognition learning (reviewed by Brown, 2003; Kelly and Magurran, 2003). These studies have used associative learning protocols to demonstrate that prey fishes can learn to recognise unfamiliar predators through pairing of predatory cues with conspecific (and sometimes heterospecific) alarm cues. Alarm cue-based predator-recognition learning has been documented for a number of nonostariophysan species: percids (walleye, *Sander vitreum*: Wisenden *et al.*, 2001); salmonids (brook charr, *Salvelinus fontinalis*: Mirza and Chivers, 2000; rainbow trout, *Oncorhynchus mykiss*: Mirza and Chivers, 2003; Chinook salmon, *O. tshawytscha*: Berejikian *et al.*, 1999); sticklebacks (brook sticklebacks, *Culaea inconstans*: Chivers *et al.*, 1995) and gobiids (starry gobies, *Asterropteryx semipunctatus*, Larson and McCormick, 2005). For at least one species, learning can occur even at concentrations below the response threshold required to elicit alarm behavior (Mirza and Chivers, 2003). Two studies (Berejikian *et al.*, 1999; Mirza and Chivers, 2000) showed that predator-recognition training in the laboratory could lead to enhanced post-release survival.

Use of alarm cues to alter levels of predation risk can be beneficial from a practical standpoint because use of chemical stimuli does not require the researcher to maintain live predators, and it allows for relatively easy control of stimulus concentrations and introductions. For example, Foam *et al.* (2005a, b) and Mirza *et al.* (2006) used chemical alarm cues to vary the frequency of exposure to predation risk in tests of a model concerning how the background level of risk influences antipredator and foraging behaviors. In another example, Houtman and Dill (1994) used alarm cues to determine how background matching influenced antipredatory freezing behavior.

Environmental Pollutants and the Alarm Response

Fish conservation biologists should be particularly aware of potential effects of environmental pollutants on antipredatory behavior, including fish responses to chemical alarm cues. Antipredator behavior (and subsequent survival) could be affected if the contaminants interfere with the chemical nature of the alarm cue or with the prey fish's ability to respond to the chemical (i.e., through damage to the olfactory epithelium). Weakly acidic conditions has been shown to negatively affect responses to alarm chemicals by pumpkinseed sunfish, *Lepomis gibbosus* (Leduc *et al.*, 2003), brook charr, *Salvelinus fontinalis* (Leduc *et al.*, 2004b),

Atlantic salmon, *Salmo salar* (Leduc *et al.*, 2006), and rainbow trout, *Oncorhynchus mykiss* (Leduc *et al.*, 2004a). Effects of low pH on rainbow trout also included elimination of alarm cue-based predator-recognition learning (Leduc *et al.*, 2004). Iowa darters (*Eteostoma exile*) that inhabit metal contaminated lakes failed to respond to alarm cues (McPherson *et al.*, 2003), and waterborne cadmium eliminates the alarm response of rainbow trout (*Oncorhynchus mykiss*) (Scott *et al.*, 2003). Another pollutant that has been shown to negatively affect the alarm response is the insecticide diazionon (Scholz *et al.*, 2000). Clearly, study of the effects of environmental contaminants on alarm behavior is a vital area of future research.

Acknowledgments

My interest in nonostariophysan alarm cues was sparked when I was a post-doctoral researcher in Jan Smith's laboratory at the University of Saskatchewan. We discovered the alarm response of brook sticklebacks by accident while we were doing a field study of ostariophysan alarm responses. I am pleased to dedicate this chapter to Jan's memory. D. Cardorette-Hall provided the histological section in Figure 11.1 Nathan Windel commented on parts of the manuscript and Daphne Smith assisted with editing. I thank Shirley and Dannie Broadhead for their support.

References

Ahsan, S.N. and M. Prasad. 1982. Occurrence of fright reaction in Indian fishes. *Biological Bulletin India* 4: 41-47.

Alemadi, S.D. and B.D. Wisenden. 2002. Antipredator response to injury-released chemical alarm cues by convict cichlid young before and after independence from parental protection. *Behaviour* 139: 603-611.

Azevedo-Ramos, C., M. van Sluys, J.-M. Hero and W.E. Magnusson. 1992. Influence of tadpole movement on predation by odonate naiads. *Journal of Herpetology* 26: 335-338.

Bart, H.L., Jr. and L.M. Page. 1991. Morphology and adaptive significance of fin knobs in egg-clustering darters. *Copeia* 1991: 80-86.

Berejikian, B.A., R.J.F. Smith, E.P. Tezak, S.L. Schroder and C.M. Knudsen. 1999. Chemical alarm signals and complex hatchery rearing habitats affect antipredator behavior and survival of chinook salmon (*Oncorhynchus tshawytscha*) juveniles. *Canadian Journal of Aquatic Sciences and Fisheries* 56: 830-838.

Bernstein, J.W. and R.J.F. Smith. 1984. Alarm substance cells in fathead minnows do not affect the feeding preference of rainbow trout. *Environmental Biology of Fishes* 9: 307-311.

Blackstock, N. and A.D. Pickering. 1980. Acidophilic granular cells in the epidermis of the brown trout, *Salmo trutta*. *Cell and Tissue Research* 210: 359-369.

Brown, G.E. 2003. Learning about danger: chemical alarm cues and local risk assessment in prey fishes. *Fish and Fisheries* 4: 227-234.

Brown, G.E. and J.-G. J. Godin. 1997. Anti-predator responses to conspecific and heterospecific skin extracts by three-spine sticklebacks: alarm pheromones revisited. *Behaviour* 134: 1123-1134.

Brown, G.E. and R.J.F. Smith. 1998. Acquired predator recognition in juvenile rainbow trout (*Oncorhynchus mykiss*): conditioning hatchery-reared fish to recognize chemical cues of a predator. *Canadian Journal of Fisheries and Aquatic Sciences* 55: 611-617.

Brown, G.E. and J.-G. J. Godin. 1999. Chemical alarm signals in wild Trinidadian guppies (*Poecilia reticulata*). *Canadian Journal of Zoology* 77: 562-570.

Brown, G.E. and S. Brennan. 2000. Chemical alarm signals in juvenile green sunfish (*Lepomis cyanellus*, Centrarchidae). *Copeia* 2000: 1079-1082.

Brown, G.E., J.C. Adrian, Jr., T. Patton and D.P. Chivers. 2001a. Fathead minnows learn to recognize predator odour when exposed to concentrations of artificial alarm pheromones below their behavioural response thresholds. *Canadian Journal of Zoology* 79: 2239-2245.

Brown, G.E., V.J. LeBlanc and L.E. Porter. 2001b. Ontogenetic changes in the response of largemouth bass (*Micropterus salmoides*, Centrarchidae, Perciformes) to heterospecific alarm pheromones. *Ethology* 107: 401-414.

Brown, G.E., D.L. Gershaneck, D.L. Plata and J.L. Golub. 2002. Ontogenetic changes in response to heterospecific alarm cues by juvenile largemouth bass are phenotypically plastic. *Behaviour* 139: 913-927.

Brown, G.E., P.E. Foam, H.E. Cowell, P. Guevara-Fiore and D.P. Chivers. 2004. Production of chemical alarm cues in the convict cichlids: the effects of diet, body condition and ontogeny. *Annales Zoologicae i Fennicae* 41: 487-499.

Brown, G.E., A.C. Rive, M.C.O. Ferrari and D.P. Chivers. 2006a. The dynamic nature of anti-predator behaviour: prey fish integrate threat-sensitive anti-predator responses within background levels of predation risk. *Behavioural Ecology and Sociobiology* 61: 9-16.

Brown, G.E., T. Bongiorno, D.M. DiCapua, L.I. Ivan and E. Roh. 2006b. Effects of group size on the threat-sensitive response to varying concentrations of chemical alarm cues by juvenile convict cichlids. *Canadian Journal of Zoology* 84: 1-8.

Bryant, P.B. 1987. *A Study of the Alarm System in Selected Fishes of Northern Mississippi*. M.S. Thesis. University of Mississippi, Oxford.

Bryer, P.J., R.S. Mirza and D.P. Chivers. 2001. Chemosensory assessment of predation risk by slimy sculpins (*Cottus cognatus*): Responses to alarm, disturbance, and predator cues. *Journal of Chemical Ecology* 27: 533-546.

Cashner, M.F. 2001. *An Investigation of Chemical Alarm Systems of a Coevolved Assemblage of Freshwater Fishes in the Southeastern United States*. M.S. Thesis. University of Southern Mississippi, Hattiesburg.

Chivers, D.P. and R.J.F. Smith. 1994. Intra- and interspecific avoidance of areas marked with skin extract from brook sticklebacks (*Culaea inconstans*) in a natural habitat. *Journal of Chemical Ecology* 20: 1517-1524.

Chivers, D.P. and R.J.F. Smith. 1998. Chemical alarm signalling in aquatic predator-prey systems: A review and prospectus. *Écoscience* 5: 338-353.

Chivers, D.P., G.E. Brown and R.J.F. Smith. 1995a. Acquired recognition of chemical stimuli from pike by brook sticklebacks, *Culaea inconstans* (Osteichthyes, Gasterosteidae). *Ethology* 99: 234-242.

Chivers, D.P., B.D. Wisenden and R.J.F. Smith. 1995b. The role of experience in the response of fathead minnows (*Pimephales promelas*) to skin extract of Iowa darters (*Etheostoma exile*). *Behaviour* 132: 665-674.

Chivers, D.P., G.E. Brown and R.J.F. Smith. 1996. The evolution of chemical alarm signals: attracting predators benefits alarm signal senders. *American Naturalist* 148: 649-659.

Chivers, D.P., M.H. Puttlitz and A.R. Blaustein. 2000. Chemical alarm signaling by reticulate sculpins, *Cottus perplexus*. *Environmental Biology of Fishes* 57: 347-352.

Chivers, D.P., B.D. Wisenden, C.J. Hindman, T.A. Michalak, R.C. Kusch, S.G.W. Kaminskyj, K.L. Jack, M.C.O. Ferrari, R.J. Pollock, C.F. Halbgewachs, M.S. Pollock, S. Alemadi, C.T. James, R.K. Savaloja, C.P. Goater, A. Corwin, R.S. Mirza, J.M. Kiesecker, G.E. Brown, J.C. Adrian, Jr., P.H. Krone, A.R. Blaustein, and A. Mathis. 2007. Epidermal 'alarm substance' cells of fishes are maintained by non-alarm functions: possible defence against pathogens, parasites and UVB radiation. *Proceedings of the Royal Society B* 274: 2611-2619.

Commens, A. and A. Mathis. 1999. Alarm pheromones of rainbow darters (*Etheostoma caeruleum*): responses to skin extracts of conspecifics and congeners (*Etheostoma juliae*). *Journal of Fish Biology* 55: 1359-1362.

Commens-Carson, A. and A. Mathis. 2007. Response of three species of darters of the genus *Etheostoma* to chemical alarm cues from conspecifics and congeners. *Copeia* 2007: 838-843.

Endler, J.A. 1991. Interactions between predator and prey. In: *Behavioural Ecology: An Evolutionary Approach*, J.R. Krebs and N.B. Davies (eds.). Blackwell Scientific Publications, Oxford, pp. 169-196.

Fast, M.D., D.E. Sims, J.F. Burka, A. Mustafa and N.W. Ross. 2002. Skin morphology and humoral non-specific defence parameters of mucus and plasma in rainbow trout, coho and Atlantic salmon. *Comparative Biochemistry and Physiology* A132: 645-657.

Fasulo, S., G. Tagliafierro, A. Contini, L. Ainis, M.B. Ricca, N. Yanaihara and G. Zaccone. 1993. Ectopic expression of bioactive peptides and serotonin in the sacciform gland cells of teleost skin. *Archives of Histology and Cytology* 56: 117-125.

Foam, P.E., M.C. Harvey, R.S. Mirza and G.E. Brown. 2005a. Heads up: juvenile convict cichlids switch to threat-sensitive foraging tactics based on chemosensory information. *Animal Behaviour* 70: 601-607.

Foam, P.E., R.S. Mirza, D.P. Chivers and G.E. Brown. 2005b. Juvenile convict cichlids (*Archocentrus nigrofasciatus*) allocate foraging and antipredator behaviour in response to temporal variation in predation risk. *Behaviour* 142: 129-144.

Friesen, R.G. and D.P. Chivers. 2006. Underwater video reveals strong avoidance of chemical alarm cues by prey fishes. *Ethology* 112: 339-345.

Frisch, K. von. 1938. Zur psychologie des fisch-schwarmes. *Naturwissenschaften* 26: 601-606.

Frisch, K. von. 1941a. Die bedeutung des geruchssinnes im lepen der fische. *Naturwissenschaften* 29: 321-333.

Frisch, K. von. 1941b. Über einen schreckstoff der fischhaut und seine biologische bedeutung. *Zeitschrift Vergleichende Physiologie* 29: 46-145.

García, C., E. Rolán-Alvarez and L. Sánchez, L. 1992. Alarm reaction and alert state in *Gambusia affinis* (Pisces, Poeciliidae) in response to chemical stimuli from injured conspecifics. *Journal of Ethology* 10: 41-46.

Gibson, A. and A. Mathis. 2006. Opercular beat rate for rainbow darters, *Etheostoma caeruleum*, exposed to chemical stimuli from conspecific and heterospecific fishes. *Journal of Fish Biology* 69: 224-232.

Godin, J.-G. J. 1997. Evading predators. In: *Behavioral Ecology of Teleost Fishes*, J.-G. Godin (ed.). Oxford University Press, Oxford, pp. 191-236.

Golub, J.L. and G.E. Brown. 2003. Are all signals the same? Ontogenetic change in the response to conspecific and heterospecific chemical alarm signals by juvenile green sunfish (*Lepomis cyanellus*). *Behavioural Ecology and Sociobiology* 54: 113-118.

Golub, J.L., V. Vermette and G.E. Brown. 2005. Response to conspecific and heterospecific alarm cues by pumpkinseeds in simple and complex habitats: field verification of an ontogenetic shift. *Journal of Fish Biology* 66: 1073-1081.

Hamilton, W.D. 1971. Geometry for the selfish herd. *Journal of Theoretical Biology* 31: 293-311.

Haney, D.C., J.C. Vokoun and N.B. Noltie. 2001. Alarm pheromone recognition in a Missouri darter assemblage. *Journal of Fish Biology* 59: 810-817.

Harvey, M.C. and G.E. Brown. 2004. Dine or dash? Ontogenetic shift in the response of yellow perch to conspecific alarm cues. *Environmental Biology of Fishes* 70: 345-352.

Henderson, P.A., P.W. Irving and A.E. Magurran. 1997. Fish pheromones and evolutionary enigmas: A reply to Smith. *Proceedings of the Royal Society of London* B264: 451-453.

Hoare, D.J., J.D. Couzin, J.-G. J. Godin and J. Krause. 2004. Context-dependent group size choice in fish. *Animal Behaviour* 67: 155-164.

Houtman, R. and L.M. Dill. 1994. The influence of substrate color on the alarm response of tidepool sculpins (*Oligocottus maculosus*; Pisces, Cottidae). *Ethology* 96: 147-154.

Hugie, D.M. and R.J.F. Smith. 1987. Epidermal club cells are not linked with an alarm response in reedfish, *Erpetoichthys* (= *Calamoichthys*) *calabaricus*. *Canadian Journal of Zoology* 65: 2057-2061.

Hugie, D.M., P.L. Thuringer and R.J.F. Smith. 1991. The response of the tidepool sculpin (*Oligocottus maculosus*) to chemical stimuli from injured conspecifics, alarm signalling in the Cottidae (Pisces). *Ethology* 89: 322-334.

Iger, Y. and M. Abraham. 1990. The process of skin healing in experimentally wounded carp. *Journal of Fish Biology* 36: 421-437.

Jones, H.M. and C.A. Paszkowski. 1997. Effects of exposure to predatory cues on territorial behaviour of male fathead minnows. *Environmental Biology of Fishes* 49: 97-109.

Kelley, J.L. and A.E. Magurran. 2003. Learned predator recognition and antipredator responses in fishes. *Fish and Fisheries* 4: 216-226.

Kim, J.-W., G.E. Brown and J.W.A. Grant. 2004. Interactions between patch size and predation risk affect competitive aggression and size variation in juvenile convict cichlids. *Animal Behaviour* 68: 1181-1187.

Kinziger, A.P., R.M. Wood and D.A. Neely. 2005. Molecular systematics of the genus *Cottus* (Scorpaeniformes: Cottidae). *Copeia* 2005: 303-311.

Kristensen, E.A. and G.P. Closs. 2004. Anti-predator response of naive and experienced common bully to chemical alarm cues. *Journal of Fish Biology* 64: 643-652.

Larson, J.K. and M.I. McCormick. 2005. The role of chemical alarm signals in facilitating learned recognition of novel chemical cues in a coral reef fish. *Animal Behaviour* 69: 51-57.

Leduc, A.O.H.C., M.K. Noseworthy, J.C. Adrian, Jr. and G.E. Brown. 2003. Detection of conspecific and heterospecific alarm signals by juvenile pumpkinseed under weak acidic conditions. *Journal of Fish Biology* 63: 1331-1336.

Leduc, A.O., M.C.O. Ferrari, J.M. Kelly and G.E. Brown. 2004a. Learning to recognize novel predators under weakly acidic conditions: The effect of reduced pH on acquired predator recognition by juvenile rainbow trout. *Chemoecology* 14: 107-112.

Leduc, A.O., J.M. Kelly and G.E. Brown. 2004b. Detection of conspecific alarm cues by juvenile salmonids under neutral and weakly acidic conditions: laboratory and field tests. *Oecologia* 139: 318-324.

Leduc, A.O.H.C., E. Roh, M.C. Harvey and G.E. Brown. 2006. Impaired detection of chemical alarm cues by juvenile wild Atlantic salmon (*Salmo salar*) in a weakly acidic environment. *Canadian Journal of Fisheries and Aquatic Sciences* 63: 2356-2363.

Lima, S.L. and L.M. Dill. 1990. Behavioral decisions made under the risk of predation: a review and prospectus. *Canadian Journal of Zoology* 68: 619-640.

Lima, S.L. and P.A. Bednekoff. 1999. Temporal variation in danger drives antipredator behavior: the predation risk allocation hypothesis. *American Naturalist* 153: 649-659.

Lindquist, D.G., J.R. Shute and P.W. Shute. 1981. Spawning and nesting behaivor of the Wacamaw darter, *Etheostoma perlongum*. *Environmental Biology of Fishes* 6: 177-191.

Lopez-Doriga, M.V. and J.L. Martinez. 1993. Fine structure of sacciform cells in the epidermis of the brown trout, *Salmo trutta*. *Journal of Zoology* 230: 425-432.

Losey, G.S., Jr. and D.M. Hugie. 1994. Prior anesthesia impairs a chemically mediated fright response in a gobiid fish. *Journal of Chemical Ecology* 20: 1877-1883.

Magurran, A.E., P.W. Irving and P.A. Henderson. 1996. Is there a fish alarm pheromone? A wild study and critique. *Proceedings of the Royal Society of London* B263: 1551-1556.

Maniak, P.J., R.D. Lossing and P.W. Sorensen. 2000. Injured Eurasian ruffe, *Gymnocephalus cernuus*, release an alarm pheromone that could be used to control their dispersal. *Journal of the Great Lakes Research* 26: 183-195.

Marcus, J.P. and G.E. Brown. 2003. Response of pumpkinseed sunfish to conspecific chemical alarm cues: an interaction between ontogeny and stimulus concentration. *Canadian Journal of Zoology* 81: 1671-1677.

Mathis, A. and R.J.F. Smith. 1993. Intraspecific and cross superorder responses to a chemical alarm signal by brook stickleback. *Ecology* 74: 2395-2404.

Mathis, A. and D.P. Chivers. 2003. Overriding the oddity effect in mixed-species aggregations: group choice by armored and nonarmored prey. *Behavioural Ecology* 14: 334-339.

Mathis, A., D.P. Chivers and R.J.F. Smith. 1995. Chemical alarm signals: Predator deterrents or predator attractants? *American Naturalist* 145: 994-1005.

Matity, J.G., D.P. Chivers and R.J.F. Smith. 1994. Population and sex differences in antipredator responses of breeding fathead minnows (*Pimephales promelas*) to chemical stimuli from garter snakes (*Thamnophis radix* and *T. sirtalis*). *Journal of Chemical Ecology* 20: 2111-2121.

McClintock, J.B. and B.J. Baker. 1997. A review of the chemical ecology of Antarctic marine invertebrates. *American Zoologist* 37: 329-342.

McCormick, M.I. and T.H. Holmes. 2006. Prey experience of predation influences mortality rates at settlement in a coral reef fish, *Pomacentrus amboinensis*. *Journal of Fish Biology* 68: 969-974.

McPherson, T.D., R.S. Mirza and G.G. Pyle. 2004. Responses of wild fishes to alarm chemical in pristine and metal-contaminated lakes. *Canadian Journal of Zoology* 82: 694-700.

Mirza, R.S. and D.P. Chivers. 2000. Predator-recognition training enhances survivial of brook trout: Evidence from laboratory and field-enclosure studies. *Canadian Journal of Zoology* 78: 2198-2208.

Mirza, R. and D.P. Chivers. 2001a. Are chemical alarm cues conserved within salmonid fishes? *Journal of Chemical Ecology* 27: 1641-1655.

Mirza, R.S. and D.P. Chivers. 2001b. Chemical alarm signals enhance survival of brook charr (*Salvelinus fontinalis*) during encounters with predatory chain pickerel (*Esox niger*). *Ethology* 107: 989-1005.

Mirza, R.S. and D.P. Chivers. 2001c. Do juvenile yellow perch use diet cues to assess the level of threat posed by intraspecific predators? *Behaviour* 138: 1249-1258.

Mirza, R.S. and D.P. Chivers. 2002. Brook char (*Salvelinus fontinalis*) can differentiate chemical alarm cues produced by different age/size classes of conspecifics. *Journal of Chemical Ecology* 28: 555-564.

Mirza, R.S. and D.P. Chivers. 2003. Response of juvenile rainbow trout to varying concentrations of chemical alarm cue: response thresholds and survival during encounters with predators. *Canadian Journal of Zoology* 81: 88-96.

Mirza, R.S., S.A. Fisher and D.P. Chivers. 2003. Assessment of predation risk by juvenile yellow perch, *Perca flavescens*: responses to alarm cues from conspecifics and prey guild members. *Environmental Biology of Fishes* 66: 321-327.

Mirza, R., A. Mathis and D.P. Chivers. 2006. Does temporal variation in predation risk influence the intensity of antipredator responses? A test of the risk allocation hypothesis. *Ethology* 112: 44-51.

Mittal, A.K., M. Whitear and A.M. Bullock. 1981. Sacciform cells in the skin of teleost fish. *Zeitschrift für Mikroskopische-Anatomische Forschung* 95: 559-585.

Munday, P.L., M. Schubert, J.A. Baggio, G.P. Jones, M.J. Calley and A.S. Grutter. 2003. Skin toxins and external parasitism of coral-dwelling gobies. *Journal of Fish Biology* 62: 976-981.

Nelson, J.S. 1994. *Fishes of the World*. John Wiley & Sons, New York. Third Edition.

Noble, G.K. 1939. The experimental animal from the naturalist's point of view. *American Naturalist* 73: 113-126.

Nordell, S.E. 1998. The response of female guppies, *Poecilia reticulata*, to chemical stimuli from injured conspecifics. *Environmental Biology of Fishes* 51: 331-338.

Page, L.M. 1985. Evolution of reproductive behaviors in percid fishes. *Illinois Natural History Survey Bulletin* 33: 275-295.

Pfeiffer, W. 1963. Vergleichende untersuchungen uber die schreckreaktion und den schreckstoff der Ostariophysen. *Zeitschrift Vergleichende Physiologie* 47: 111-147.

Pfeiffer, W. 1977. The distribution of fright reaction and alarm substance cells in fishes. *Copeia* 1977: 653-665.

Pflieger, W.L. 1997. *The Fishes of Missouri*. Missouri Department of Conservation, Jefferson City.

Pickering, A.D. and J.M. Fletcher. 1987. Sacciform cells in the epidermis of the brown trout, *Salmo trutta*, and the Arctic char, *Savelinus alpinus*. *Cell and Tissue Research* 247: 259-265.

Pitcher, T.J. 1986. Functions of shoaling behaviour in teleosts. In: *The Behaviour of Teleost Fishes*, T.J. Pitcher (ed.). Croom Helm, London.

Pollock, M.S., X. Zhao, G.E. Brown, R.C. Kusch, R.J. Pollock and D.P. Chivers. 2006. The response of convict cichlids to chemical alarm cues: an integrated study of behaviour, growth, and reproduction. *Annales Zoologicae i Fennicae* 42: 385-395.

Poulin, R., D.J. Marcogliese and J. D. McLaughlin. 1999. Skin-penetrating parasites and the release of alarm substances in juvenile rainbow trout. *Journal of Fish Biology* 55: 47-53.

Quilhac, A. and J.-Y. Sire. 1999. Spreading, proliferation, and differentiation of the epidermis after wounding a cichlid fish, *Hemichromis bimaculatus*. *Anatomical Record* 254: 435-451.

Reed, J.R. 1969. Alarm substances and fright reaction in some fishes from the southeastern United States. *Transactions of American Fisheries Society* 98: 664-668.

Rehnberg, B.G. and R.J.F. Smith. 1987. The reaction of pearl dace (Pisces, Cyprinidae) to alarm substance: Time-course of behavior, brain amines, and stress physiology. *Canadian Journal of Zoology* 65: 2916-2921.

Roh, E., R.S. Mirza and G.E. Brown. 2004. Quality or quantity? The role of donor condition in the production of chemical alarm cues in juvenile convict cichlids. *Behaviour* 141: 1235-1248.

Rottmann, R.W., R. Francis-Floyd and R. Durborrow. 1992. The role of stress in fish disease. *Southern Regional Aquatic Center Publication* 474: 1-3.

Sapolsky, R.M. 2002. Endocrinology of the stress response. In: *Behavioral Endocrinology*, J.B. Becker, S.M. Breedlove, D. Crews and M. McCarthy (eds.). MIT, Cambridge, Massachusetts, pp. 409-450.

Scholz, N.L., N.K. Truelove, B.L. French, B.A. Berejikian, T.P. Quinn, E. Casillas and T.K. Collier. 2000. Diazinon disrupts antipredator and homing behaviors in chinook salmon (*Oncorhynchus tshawytscha*). *Canadian Journal of Fisheries and Aquatic Sciences* 57: 1911-1918.

Schubert, M., P.L. Munday, M.J. Caley, G.P. Jones and L.E. Llewellyn. 2003. The toxicity of skin secretions from coral-dwelling gobies and their potential role as a predator deterrent. *Environmental Biology of Fishes* 67: 359-367.

Schutz, F. 1956. Vergleichende untersuchungen uber die schreckreaktion bei fischen und deren verbreitung. zeitschrift Vergleichende Physiologie 28: 84-135.

Scott, G.R., K.A. Sloman, C. Rouleau and C.M. Wood. 2003. Cadmium disrupts behavioural and physiological responses to alarm substance in juvenile rainbow trout (Oncorhynchus mykiss). Journal of Experimental Biology 206: 1779-1790.

Skelly, D.K. 1994. Activity level and the susceptibility of anuran larvae to predation. Animal Behaviour 47: 465-468.

Smith, R.J.F. 1977. Chemical communication as adaptation: alarm substance of fish. In: Chemical Signals in Vertebrates, D. Müller-Schwarze and M.M. Mozell (eds.). Plenum Press, New York, pp. 303-320.

Smith, R.J.F. 1979. Alarm reaction of Iowa and johnny darters (Etheostoma, Percidae, Pisces) to chemicals from injured conspecifics. Canadian Journal of Zoology 57: 1278-1282.

Smith, R.J.F. 1981. Effect of food deprivation on the reaction of Iowa darters (Etheostoma exile) to skin extract. Canadian Journal of Zoology 59: 558-560.

Smith, R.J.F. 1982. Reaction of Percina nigrofasciata, Ammocrypta beani, and Etheostoma swaini (Percidae, Pisces) to conspecific and intergeneric skin extracts. Canadian Journal of Zoology 60: 1067-1072.

Smith, R.J.F. 1986. The evolution of chemical alarm signals in fishes In: Chemical Signals in Vertebrates, D. Duvall, D. Müller-Schwarze and R.M. Silverstein (eds.). Plenum Press, New York, Vol. 4, pp. 99-115.

Smith, R.J.F. 1989. The response of Asterropteryx semipunctatus and Gnatholepis anjerensis (Pisces, Gobiidae) to chemical stimuli from injured conspecifics, an alarm response in gobies. Ethology 81: 279-290.

Smith, R.J.F. 1992. Alarm signals in fishes. Reviews in Fish Biology and Fisheries 2: 33-63.

Smith, R.J.F. and M.J. Smith. 1989. Predator-recognition behaviour in two species of gobiid fishes, Asterropteryx semipunctatus and Gnatholepis anjerensis. Ethology 83: 19-30.

Smith, R.J.F. and B.J. Lawrence. 1992. The response of a bumblebee goby, Brachybovius sabanus, to chemical stimuli from injured conspecifics. Environmenta Biology of Fishes 34: 103-108.

Toa, D.G., L.B.B. Afonso and G.K. Iwama. 2005. Stress response of juvenile rainbow trout (Oncorhynchus mykiss) to chemical cues released from stressed conspecifics. Fish Physiology and Biochemistry 30: 103-108.

Vokoun, J.C. and D.B. Noltie. 2002. Evidence for inheritance of alarm substance recognition in johnny darter (Etheostoma nigrum). American Midland Naturalist 147: 400-403.

Whitear, M. 1981. Secretions in the epidermis of polypteriform fish. Zeitschrift für Mikroskopische-Anatomische Forschung 95: 531-543.

Whitear, M. and A.K. Mittal. 1986. Structure of the skin of Agonus cataphractus Teleostei. Journal of Zoology (London) A: 551-574.

Wisenden, B.D. 1995. Reproductive behaviour of free-ranging convict cichlids, Cichlasoma nigrofasciatum. Environmental Biology of Fishes 43: 121-134.

Wisenden, B.D. 2003. Chemically mediated strategies to counter predation. In: Sensory Processing in the Aquatic Environment, S.P. Collin and N.J. Marshall (eds.). Springer-Verlag, New York, pp. 236-251.

Wisenden, B.D. and R.C. Sargent. 1997. Antipredator behaviour and suppressed aggression by convict cichlids in response to injury-released chemical cues of conspecifics but not to those of an allopatric heterospecific. *Ethology* 103: 283-291.

Wisenden, B.D., D.P. Chivers and R.J.F. Smith. 1994. Risk-sensitive habitat use by brook stickleback (*Culaea inconstans*) in areas associated with minnow alarm pheromone. *Journal of Chemistry and Ecology* 20: 2975-2983.

Wisenden, B.D., D.P. Chivers, G.E. Brown and R.J.F. Smith. 1995. The role of experience in risk assessment: Avoidance of areas chemically labelled with fathead minnow alarm pheromone by conspecifics and heterospecifics. *Écoscience* 2: 116-122.

Wisenden, B.D., J. Klitzke, R. Nelson, D. Friedl and P.C. Jacobson. 2004. Predator-recognition training of hatchery-reared walleye (*Stizostedion vitreum*) and a field test of a training method using yellow perch (*Perca flavescens*). *Canadian Journal of Fisheries and Aquatic Sciences* 61: 2144-2150.

Wisenden, B.D., M.S. Pollock, R.J. Tremaine, J.M. Webb, M.E. Wismer and D.P. Chivers. 2003. Synergistic interactions between chemical alarm cues and the presence of conspecific and heterospecific fish shoals. *Behavioural Ecology and Sociobiology* 54: 485-490.

Witkowski, A., K. Kaleta and J. Kuryszko. 2004. Histological structure of the skin of Arctic charr, *Salvelinus alpinus* (L.) from Spitsbergen. *Acta Ichthyologica Piscatoria* 34: 241-251.

Zaccone, G. 1981. Studies on the structure and histochemistry of the epidermis in the air-breathing teleost *Mastacembelus armatus* (Mastacembelidae, Pisces). *Zeitschrift für Mikroskopische-Anatomische Forschung* 95: 809-826.

Zaccone, G., B.G. Kapoor, S. Fasulo and L. Ainis. 2001. Structural, Histochemical and Functional aspects of the Epidermis of Fishes. In: *Advances in Marine Biology*, A.J.H. Southward, P.A. Tyler, C.M. Young and L.A. Fuiman (eds.), Academic Press, London, Vol. 40, pp. 255-347.

Zaccone, G., S. Fasulo, P. Lo Cascio and A. Licata. 1986. 5-Hydroxytryptamine immunoreactivity in the epidermal sacciform gland cells of the clingfish *Lepadogaster candollei* Risso. *Cell and Tissue Research* 246: 679-682.

Zaccone, G., L. Ainis, S. Fasulo, A. Mauceri, E.R. Lauriano, A. Licata and P. Lo Cascio. 1999. Paraneurons in the skin epithelium of the fishes. In: *Water/Air Transition in Biology*, A.K. Mittal, F.B. Eddy and J.S. Dataa Munshi (eds.). Oxford & IBH Publishing Co., New Delhi.

Zimmer, R.K. and C.A. Butman. 2000. Chemical signaling processes in the marine environment. *Biological Bulletin* 198: 168-187.

Index

About the Editors

Giacomo Zaccone is Professor of Comparative Anatomy and Endocrinology at the University of Messina. He is a board member of Acta Histochemica, and a reviewer for many authoritative international journals including The Anatomical Record, Journal of Comparative Neurology, Journal of Fish Biology, Acta Zoologica, Aquaculture and Fish Physiology book series. He has authored several invited chapters to books and a special issue and co-edited several books published by Academic Press, Wiley-Liss, Science Publishers and Elsevier covering the autonomic nervous system of the fishes and biology of the fish integument. His studies are now focused on the autonomic innervation of the heart and air-breathing organs in primitive fishes.

Claude Perrière earned his Ph.D. from the University Pierre and Marie Curie (Paris). He teaches Zoology at the University of Paris-Sud (France) where he studies venomous glands and fish venom (Echiichthys vipera and Synanceia verrucosa) by histochemistry, cytochemistry, ultrastructure, biochemistry, electrophysiology, in the Animal Biology Laboratory of the Faculty of Pharmacy.

Alicia Mathis is Professor and Head of the Biology Department at Missouri State University, USA, where she studies antipredator, aggressive, and communication behavior of fishes and amphibians. She teaches courses in behavioral ecology and comparative vertebrate anatomy and has been recognized by the university for excellence in both teaching and research. Dr. Mathis earned a Ph.D. from the University of Louisiana-Lafayette, USA, and was a Post-doctoral Fellow at the University of Saskatchewan, Canada.

B.G. Kapoor was formerly Professor and Head of Zoology in Jodhpur University (India). Dr. Kapoor has co-edited 18 books published by Science Publishers, Enfield, NH, USA. The most recent ones are Fish Reproduction, 2008 (with Maria João Rocha and Augustine Arukwe); Fish Life in Special Environments, 2008 (with Philippe Sébert and D.W. Onyango); Fish Larval Physiology, 2008 (with R.N. Finn); Fish Diseases, 2008 (with J.C. Eiras, Helmut Segner and Thomas Wahli); Fish Behaviour, 2008 (with Carin Magnhagen, Victoria A. Braithwaite and Elisabet Forsgren); Feeding and Digestive Functions of Fishes, 2008 (with J.E.P. Cyrino and D.P. Bureau); Fish Defenses (Volume 1), 2009 (with Giacomo Zaccone, Alfonsa García-Ayala and José Meseguer Peñalver); The Biology of Blennies, 2009 (with Robert A. Patzner, Emanuel J. Gonçalves, Philip A. Hastings); Development of Non-Teleost Fishes, 2009 (with Yvette W. Kunz, Carl A. Luer). Dr. Kapoor has also co-edited The Senses of Fish: Adaptations for the Reception of Natural Stimuli, 2004 (with G. von der Emde and J. Mogdans); and co-authored Ichthyology Handbook, 2004 (with B. Khanna), both from Springer, Heidelberg. He has also been a contributor in books from Academic Press, London (1969, 1975 and 2001). His E-mail is: bhagatgopal.kapoor@rediffmail.com

Milton Keynes UK
Ingram Content Group UK Ltd.
UKHW030900141024
449569UK00025B/1303